T0329579

Phase-Locked Loops

Phase-Locked Loops

System Perspectives and Circuit Design Aspects

Woogeun Rhee and Zhiping Yu
Tsinghua University
Beijing, China

IEEE PRESS
WILEY

Published by John Wiley & Sons, Inc., Hoboken, New Jersey. Published simultaneously in Canada.

For general information on our other products and services or for technical support, please contact our Customer Care Department within the United States at (800) 762-2974, outside the United States at (317) 572-3993 or fax (317) 572-4002.

Wiley also publishes its books in a variety of electronic formats. Some content that appears in print may not be available in electronic formats. For more information about Wiley products, visit our web site at www.wiley.com.

Library of Congress Cataloging-in-Publication Data Applied for:

Hardback: 9781119909040

Cover Design: Wiley
Cover Image: © Tuomas A. Lehtinen/Getty Images

Set in 9.5/12.5pt STIXTwoText by Straive, Chennai, India

To my parents and my wife, Soojung, and
To my acacemic adviser Prof. Bang-Sup Song
Woogeun Rhee

To my academic advisers Prof. Zhijian Li of Tsinghua University, and
Prof. Robert W. Dutton of Stanford University
Zhiping Yu

Contents

Preface

Over 15 years of giving a phase-locked loop (PLL) course to graduate students, the authors felt a strong need for one textbook that covers PLL basics, system perspectives, practical design aspects for integrated circuits, and PLL architecture for both wireless and wireline communication systems. Without such a book, the PLL lecture had to be given based on several textbooks. Even though there are many PLL books available for circuit designers, most of them can be classified into three types. The first one is a theory-oriented book that describes the PLL based on control and communication theories but lacks circuit details. The second type of book deals with more circuits but is mostly based on discrete circuits, not covering practical design issues over on-chip variability or modern PLL architectures such as fractional-N PLLs. The last one is a circuit-oriented book but does not describe a PLL from system basics to circuit design aspects for diverse applications with an integrated step-by-step format.

This book combines bottom-to-top and top-to-bottom approaches to address the system and circuit design aspects of the PLL, covering essential materials for circuit designers, from fundamentals to practical design aspects. Compared with circuit-oriented PLL books, this book has substantial material on system design considerations in addition to circuit design aspects for wireless and wireline applications. Unlike other PLL books from the area of communication systems, this book mainly focuses on the linear behaviors of the PLL and describes them in an intuitive way without deriving mathematical analyses and equations in detail, while touching system analyses tailored for circuit designers. Below are some examples.

- Is the critical damping ratio of loop dynamics ever used for on-chip PLL design?
- Is the natural frequency ω_n from control theory as meaningful as the loop gain to circuit designers?
- Is the type 2 PLL with other phase detectors as well as the phase-frequency detector (PFD) able to provide the infinite range of frequency acquisition if not limited

by circuits? Does the PFD behave like other phase detectors after frequency acquisition?

- Do we implement the second-order type 2 charge-pump PLL in practice? Why should we consider third-order or fourth-order type 2 charge-pump PLLs in most cases?
- How to consider a peak-to-peak jitter budget from random jitter if the random jitter is unbounded in theory?
- How to analyze clock jitter in the frequency domain? How to relate phase noise and sidebands to the time-domain jitter?
- Do we care about frequency-domain sidebands for clock generation if their level is lower than the carrier power by 40 dB?
- Is the digital-intensive phase-locked loop (DPLL) totally a new PLL architecture that requires z-domain analysis?

The first half of the book covers system basics, while the second half deals with hardware implementation. In the first half, PLL basics and system design considerations are discussed. In addition to the linear and transient behaviors of the PLL, analyzing clock jitter in the frequency domain is deeply explained. In addition, the book addresses system design trade-offs for three key applications: frequency synthesis, clock-and-data recovery, and clock generation/synchronization. In the second half, building circuits and PLL architectures for the three applications are discussed by considering system and circuit design aspects. Also, frequency generation and modulation circuits based on analog, digital-intensive, and hybrid PLL architectures are described. Learning system architectures and circuit design trade-offs in wireless and wireline systems, readers will gain the knowledge of where and how to design the PLL for a broad range of applications.

The authors would like to thank Su Han, Xuansheng Ji, Luhua Lin, Longhao Kuang, Qianxian Liao, and Liqun Feng in the School of Integrated Circuits at Tsinghua University for a lot of help drawing figures. Special thanks to Liqun Feng who not only reviewed technical details with valuable comments but also provided many simulation plots.

Beijing, China

Woogeun Rhee
Zhiping Yu

About Authors

Woogeun Rhee, Ph.D., is a Professor at the School of Integrated Circuits, Tsinghua University, Beijing. He has over 25 years of professional career in integrated circuit design with nearly 10 years in industry and 17 years in academia. Dr. Rhee has worked on PLL architectures and circuits not only with different careers (academia and industry) but also over different fields (wireless and wireline systems). He is an IEEE Fellow.

Zhiping Yu, Ph.D., is a Professor at the School of Integrated Circuits, Tsinghua University, Beijing. He is an IEEE Life Fellow with over 400 published papers on subjects related to ICCAD, nanoelectronics, and RF circuit design.

1

Introduction

1.1 Phase-Lock Technique

A basic concept of a phase-locked feedback system for frequency generation was proposed in the early 1930s, but the use of a phase-locked loop (PLL) circuit for mass production began with analog television systems in the 1940s. Since then, the PLL has been one of the most critical building blocks in modern communication IC systems, covering both wireless and wireline applications.

What is the main function of the PLL? From the name and a block diagram shown in Fig. 1.1, it can be deduced that it is the loop that performs a phase lock between a reference clock and an output clock. In the coherent communication systems that use the amplitude and phase information of a signal for modulation and demodulation, interestingly, the phase-lock has not been the primary goal of the PLL in most cases. Let us look at some descriptions of the PLL in other books.

- A circuit synchronizing an output signal (generated by an oscillator) with a reference or input signal in frequency as well as in phase [Best].
- A circuit that synchronizes the signal from an oscillator with a second input signal, so that they operate at the same frequency [Egan].
- When the loop is (phase) locked, the control voltage sets the average frequency of the oscillator exactly equal to the average frequency of the input signal [Gardner].
- Basically, an oscillator whose frequency is locked onto some frequency component of an input signal, which is done with a feedback control loop [Wolaver].

The first description addresses the basic function of the PLL both in phase and frequency domains. In the second or the third description, the goal of the PLL is to achieve the same frequency as the input frequency by using a phase-lock technique. In the last description, the phase lock was not even mentioned, and the PLL was simply defined as an oscillator whose frequency is locked to the input frequency. As implied by those descriptions, we can see that the primary goal of

Phase-Locked Loops: System Perspectives and Circuit Design Aspects, First Edition.
Woogeun Rhee and Zhiping Yu.

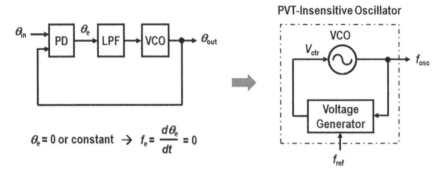

Figure 1.1 Accurate frequency control by a phase-lock technique.

the PLL is not the phase-lock but the frequency-lock. This is because the frequency offset between an input signal and a local oscillator in the coherent receiver system is much more serious than the phase offset problem.

If the main goal of the PLL is to achieve the frequency-lock, we may wonder if a frequency-locked loop (FLL) should be used instead of the PLL. The reason is that the FLL still generates a static frequency error if there exist circuit mismatches, a limited loop gain at DC, or a limited resolution in a frequency detector. To the contrary, the PLL generates a static phase error rather than the frequency error in the presence of a limited loop gain at DC or imperfect matching in a phase detector circuit. Since the frequency error f_e is the derivative of the phase error θ_e, the PLL always achieves a zero-frequency error even with the presence of a static phase error illustrated in Fig. 1.1. In other words, the PLL guarantees that the accuracy of an output frequency is the same as that of a source frequency based on the phase-lock technique. From that point of view, the PLL can be referred as a *phase-locking loop* rather than the *phase-locked loop* since the phase-lock is not the goal but an active method to achieve the frequency-lock. This explains why the PLL has been dominantly employed in the coherent communication system where the frequency offset between a carrier and a local oscillator is critical. When the PLL is used for frequency generation, we may regard the PLL as the oscillator circuit that generates an adaptive DC control voltage V_{ctr} to an internal voltage-controlled oscillator (VCO) so that a stable output frequency is maintained over process-voltage-temperature (PVT) variations as depicted in Fig. 1.1.

1.2 Key Properties and Applications

In addition to the zero-frequency error, there is another important property. The PLL is the only device that performs auto-tracking band-pass filtering with high-quality factor Q and wide tunability. The high-Q band-pass filtering with wide tunability is possible since the bandwidth of a PLL can be independently set without limitation to an output frequency, while the tunability is determined by the tuning range of a VCO regardless of the PLL bandwidth. This feature is well utilized for clock-and-data recovery (CDR) systems to extract a clean clock from a noisy input data. In the CDR system, the phase-lock property is also used to define an optimum edge-of-clock position for data retiming. Besides those two properties, the inherent property of the PLL, phase-lock, makes the PLL play an important role in modern wireline communication systems. As data rate or clock frequency increases, clock de-skewing or phase synchronization has become critical to enhance the data throughput of serial I/O interfaces since the advent of the monolithic PLL implemented with complementary metal-oxide semiconductor (CMOS) technology in the late 1980s. Below is the summary of three fundamental properties of the PLL:

- Zero-frequency error
- High-Q auto-tracking BPF
- Phase synchronization

With those three features, the PLL has been employed for diverse communication systems. We briefly introduce several applications of the PLL, and some key applications will be discussed in detail in later chapters.

1.2.1 Frequency Synthesis

Since the PLL enables the zero-frequency offset between a reference clock and a feedback clock, this feature can be used to generate multiple output frequencies by adding counters in the reference clock path and the feedback clock path of the PLL. As depicted in Fig. 1.2(a), with a fixed reference frequency f_{ref}, the output frequency f_{out} can be set by simply changing the counting values of the digital counters, that is, M and N in the reference-path and the feedback-path counters, respectively. Then, we obtain f_{out} given by $N \times (f_{ref}/M)$ since the feedback frequency $(=f_{out}/N)$ must be equal to the phase-detector frequency f_{PD} $(=f_{ref}/M)$ after the phase-lock. Therefore, the frequency accuracy of the PLL is as good as that of the stable reference source which is typically a crystal oscillator.

1.2.2 Clock-and-Data Recovery

There are three main roles of the PLL for CDR systems. Firstly, a phase detector of the PLL directly extracts a clock information from a non-return-to-zero (NRZ) data without requiring other nonlinear circuits such as a differentiator followed by a squarer as done in traditional CDR systems. Secondly, the PLL acts as a high-Q

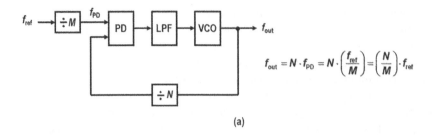

$$f_{out} = N \cdot f_{PD} = N \cdot \left(\frac{f_{ref}}{M}\right) = \left(\frac{N}{M}\right) \cdot f_{ref}$$

(a)

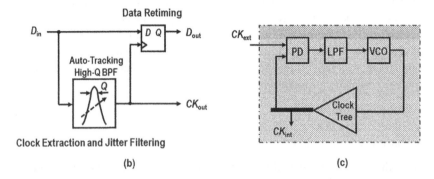

(b)   (c)

Figure 1.2 Three key applications of the PLL: (a) frequency synthesis; (b) CDR; and (c) synchronization.

auto-tracking band-pass filter to recover a clean clock from noisy incoming data by rejecting high-frequency jitter. Thirdly, the PLL recovers the data by re-timing the data with the extracted clean clock. The data retiming is normally performed with a D-type flip-flop (DFF). The phase-lock feature is also utilized for the data re-timing. For example, the falling edge of a recovered clock is used to trigger the DFF when the transition edge of the NRZ data is synchronized to the rising edge of the recovered clock, which gives an optimum clock position for bit slicing, i.e. data retiming as illustrated in Fig. 1.2(b).

1.2.3 Synchronization

Clock jitter has become more important than ever for input/output (I/O) links in recent chip-to-chip communications as clock speed increases. In addition to the clock jitter, a clock skew between an internal clock and an external clock is a concern with high clock frequency. The delay variation due to a big clock tree in a chip significantly increases a worst-case clock skew, making available phase margin much less than expected. By having the clock tree as the part of a PLL, the clock skew variation between the external clock CK_{ext} and the internal clock CK_{int} due to the clock tree can be minimized as illustrated in Fig. 1.2(c). Since the frequency offset is negligible in the chip-to-chip communication, a delay-locked loop (DLL) having a voltage-controlled delay line can also be used to achieve better power supply rejection and more flexible clock control than the PLL.

1.2.4 Modulation and Demodulation

In modern transceiver systems, the PLL plays an important role not only as a local oscillator but also as a frequency/phase modulator. These days, digital frequency/phase modulation based on a fractional-N PLL, as shown in Fig. 1.3(a), greatly simplifies the transmitter architecture. We will discuss how a direct-digital modulation is achieved by the PLL in Chapter 9. Figure 1.3(b) also shows two cases of an FM demodulator and a PM demodulator. For the modulations and the frequency demodulation, the bandwidth of the PLL needs to be wide enough to track the frequency/phase variation. To the contrary, a very narrow bandwidth of the PLL is required to provide an averaged reference phase over a frequency drift for the phase demodulation. The narrow-bandwidth PLL is not desirable for an on-chip design as the phase noise contribution of a VCO becomes too high, which will be learned later.

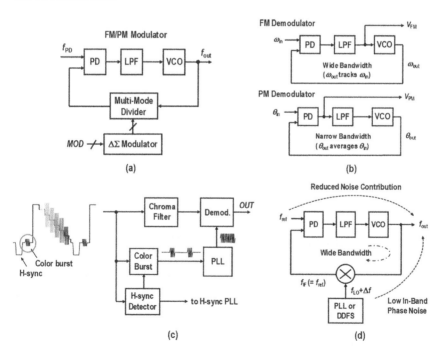

Figure 1.3 Other applications: (a) modulation; (b) demodulation; (c) carrier recovery and (d) frequency translation.

1.2.5 Carrier Recovery

A good example of a PLL-based carrier recovery system can be found in analog television systems. Figure 1.3(c) shows the simplified block diagram of a color-signal demodulator. A color-burst signal is embedded as a sub-carrier and used as a reference phase for the color-signal demodulation. The PLL generates a reference clock whose phase is synchronized to that of the color-burst signal. The phase difference between each color signal and the reference clock is used to provide different color information. Therefore, a narrow-bandwidth PLL using a highly stable VCO such as a voltage-controlled crystal oscillator (VCXO) is designed. In addition to the carrier recovery, additional PLLs are used for horizontal synchronization (*H*-sync) and vertical synchronization (*V*-sync) in the analog television system. Since the PLL was a versatile building block in the television system for mass production, many practical architectures and circuit techniques were developed, including a phase-frequency detector (PFD), a charge-pump PLL, an all-digital PLL with a numerically controlled oscillator (NCO), and so on.

1.2.6 Frequency Translation

A frequency-translation circuit offers a flexible frequency planning for frequency generation systems. As shown in Fig. 1.3(d), instead of having a large-value frequency divider, a mixer is put in a feedback path with another local oscillator (LO), making a phase-detector frequency f_{PD} become an intermediate frequency f_{IF}. Therefore, an LO frequency (f_{LO}) is effectively translated to a desired output frequency. With a high f_{PD} and the absence of a frequency divider, a wide-bandwidth PLL can be designed with low phase noise contribution from the LO and the reference source. Figure 1.3(d) shows an example of how a low-noise frequency synthesizer is implemented with the frequency-translation loop where fine frequency resolution is achieved by having another PLL or a direct-digital frequency synthesizer as the LO.

1.3 Organization and Scope of the Book

This book consists of four major parts, covering basic theories, system and application perspectives, circuit design aspects, and PLL architectures. In the first part, the essential basics of the PLL for circuit designers are described. The linear and transient behaviors of the PLL are discussed in Chapters 2 and 3, respectively. In the second part, Chapter 4 describes system design parameters by discussing the relationship between clock and frequency in the time and frequency domains. Based on the system knowledge gained from previous chapters, Chapter 5 discusses system perspectives for three key applications; frequency synthesis, clock-and-data recovery, and synchronization. The content of Chapter 5 is rather advanced and can be considered a reference for Chapters 9 and 11. Chapters 6–8 of the third part describe building blocks, putting emphasis on basic operation principles and practical design aspects for integrated circuit design. In the last part, various PLL architectures for different applications are discussed. We begin with fractional-N PLL architectures in Chapter 9, move to digital-intensive PLL architectures in Chapter 10, and discuss CDR PLL architectures in Chapter 11. This book puts more weight on the traditional PLL architectures and their analyses for frequency generation and expands the discussion of circuits and design trade-offs for other PLL architectures. This is not the thick PLL book that contains all details of system theories and circuit details but could be one of the PLL books that cover essential materials with balanced system perspectives and circuit design aspects for circuit and system designers. Other valuable resources are listed below.

Bibliography

1 R. E. Best, *Phase-Locked Loops: Design, Simulation, and Applications*, 5th ed., McGraw-Hill, New York, 2003.

2 W. Egan, *Frequency Synthesis by Phase Lock*, 2nd ed., Wiley, New York, 2000.

3 W. Egan, *Phase-Locked Basics*, 2nd ed., Wiley, Hoboken, NJ, 2008.

4 K. Feher, *Telecommunications Measurements, Analysis, and Instrumentation*, Prentice-Hall, Englewood Cliffs, NJ, 1987.

5 F. M. Gardner, *Phaselock Techniques*, 3rd ed., Wiley, Hoboken, NJ, 2005.

6 V. F. Kroupa, *Frequency Synthesis: Theory, Design et Applications*, Wiley, New York, 1973.

7 V. F. Kroupa, *Phase Lock Loops and Frequency Synthesis*, Wiley, Hoboken, NJ, 2007.

8 T. Lee, *The Design of CMOS Radio-Frequency Integrated Circuits*, Cambridge University Press, United Kingdom, 1997.

9 W. C. Lindsey and C. M. Chie (eds), *Phase-Locked Loops*, IEEE Press, New York, 1986.

10 V. Manassewitsch, *Frequency Synthesizers, Theory and Design*, 3rd ed., Wiley, New York, NY, 1987.

11 H. Meyr and G. Ascheid, *Synchronization in Digital Communications, Phase-, Frequency-Locked Loops, and Amplitude Control*, Wiley, New York, 1990.

12 B. Razavi (ed.), *Monolithic Phase-Locked Loops and Clock Recovery Circuits*, IEEE Press, New York, 1996.

13 B. Razavi (ed.), *Phase-Locking in High-Performance Systems*, IEEE Press, New York, and Hoboken, NJ: Wiley, 2003.

14 B. Razavi, *RF Microelectornics*, 2nd ed., Prentice Hall, Upper Saddle River, NJ, 2012.

15 B. Razavi, *Design of Integrated Circuits for Optical Communications*, Wiley, New York, 2012.

16 B. Razavi, *Design of CMOS Phase-Locked Loops: From Circuit Level to Architecture Level*, Cambridge University Press, United Kingdom, 2020.

17 W. Rhee (ed.), *Phase-Locked Frequency Generation and Clocking: Architectures and Circuits for Modern Wireless and Wireline Systems*, The Institution of Engineering and Technology, United Kingdom, 2020.

18 U. L. Rohde, *Microwave and Wireless Frequency Synthesizers: Theory and Design*, Wiley, New York, 1997.

19 K. Shu and E. Sanchez-Sinencio, *CMOS PLL Synthesizers; Analysis and Design*, Springer, New York, 2005.

20 R. B. Staszewski and P. T. Balsara, *All-Digital Frequency Synthesizer in Deep-Submicron CMOS*, Wiley, Hoboken, NJ, 2006.

21 D. H. Wolaver, *Phase-Locked Loop Circuit Design*, Prentice Hall, Englewood Cliffs, NJ, 1991.

Part I

Phase-Lock Basics

2

Linear Model and Loop Dynamics

For circuit designers, it would be more meaningful to consider the practical design aspects of a PLL for various applications than understanding complete mathematical descriptions originated from communication systems (knowing those would be useful though). It is because desired loop parameters for the on-chip PLL design over process, temperature, and voltage (PVT) variations could be different from what we would have obtained based on theoretical analyses. Indeed, designing a robust PLL with optimum system parameters is valuable in integrated-circuit systems rather than designing a best PLL under ideal conditions. Experienced PLL circuit designers seldom design a critically damped loop but consider either an overdamped or underdamped loop in practice.

2.1 Linear Model of the PLL

A feedback system is basically a nonlinear system. Then, why are we interested in the linear model of a PLL? It is because most system performances in which we are interested are determined when the PLL operates within a lock-in range, that is, the PLL maintains a small phase error and does not exceed the linear range of a phase detector. Good examples are phase noise and static phase error performances. In the design of a PLL circuit, nonlinear analyses are used mainly to describe the transient response of the PLL before fully settled. In other words, we are mostly interested in the small-signal behavior of a PLL after a large-signal transient response is fully settled.

Figure 2.1 shows the basic linear model of a PLL. A phase detector (PD) compares the phases of an input signal and a VCO signal, and it generates the voltage that is proportional to the phase difference. The PD gain K_d is measured in units of volts per radian, that is V/rad. Depending on the PD type, a free-running voltage V_{do}, that is, a fixed DC voltage generated with a zero-phase error appears in the PD

Phase-Locked Loops: System Perspectives and Circuit Design Aspects, First Edition.
Woogeun Rhee and Zhiping Yu.
© 2024 The Institute of Electrical and Electronics Engineers, Inc. Published 2024 by John Wiley & Sons, Inc.

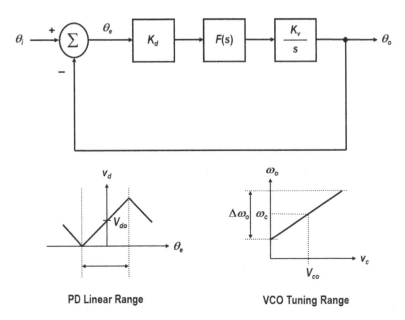

Figure 2.1 Linear model of the PLL.

transfer function as depicted in Fig. 2.1. In the linear model of the PLL, only the slope and the linear range of the PD transfer function are considered. A loop filter is the low-pass filter (LPF) that rejects high-frequency noise in the loop. Ultimately, we need a very narrow bandwidth to tune the output frequency of a VCO with a DC-like voltage, but the 3-dB corner frequency of the loop filter cannot be too low because of other noise considerations. The transfer function of the loop filter $F(s)$ determines the type and order of a PLL. A VCO is the oscillator that modulates the frequency in response to a control voltage. The VCO gain K_v is defined by a frequency change to an input-voltage change, measured in units of radians per second per volt, that is rad/s/V, or hertz per volt, that is Hz/V. A free-running frequency ω_c is the output frequency of the VCO with a floating control voltage but often refers to the center frequency of the VCO tuning range.

In the linear model, we note that there is an integrator, $1/s$, in the VCO model. It is not because the VCO performs integration during voltage-to-frequency conversion but because the phase is used as an error estimate in the loop. Figure 2.2 shows an equivalent liner model of the PLL and explains how the integrator is embedded in the VCO. For example, let us consider a frequency-locked loop (FLL) where a frequency detector generates a frequency error to control the VCO. In the linear model of the FLL shown in Fig. 2.3, the integrator is not put in the VCO even if the same VCO circuit is used. Accordingly, the FLL is more stable than the PLL for a

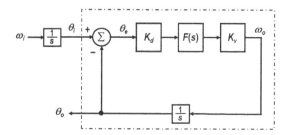

Figure 2.2 Understanding $1/s$ in the linear model of the PLL.

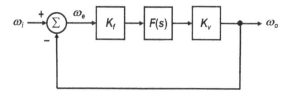

Figure 2.3 Linear model of the FLL.

given loop bandwidth. However, the FLL suffers from the frequency error problem as discussed in Chapter 1.

2.2 Feedback System

Feedback is essential in analog circuit design. One of the reasons is the variation of analog parameters. Therefore, desensitizing the analog parameters is important, which can be done by a feedback topology with a stable reference. In the voltage domain, a bandgap reference voltage is a good reference, while a crystal oscillator is the one in the frequency domain.

2.2.1 Basics of Feedback Loop

An open-loop gain (or open-loop transfer function) is highly useful to analyze the loop dynamics of a feedback system. In Fig. 2.1, the open-loop gain $G(s)$ is

$$G(s) = \frac{K_d K_v F(s)}{s} \qquad (2.1)$$

The order of a loop is defined by the number of poles in $G(s)$. The number of integrators, that is, the number of poles at the origin ($s = 0$) determines the type of the loop. Since the VCO inherently contains $1/s$, the loop type of a PLL is at least one. If the loop filter contains another integrator, we call it a type 2 PLL.

A closed-loop transfer function (also called a system transfer function) $H(s)$ from an input phase to an output phase is given by

$$H(s) = \frac{\theta_o}{\theta_i} = \frac{G(s)}{1 + G(s)} = \frac{K_d K_v F(s)}{s + K_d K_v F(s)} \tag{2.2}$$

For the first-order PLL, the 3-dB corner frequency of $H(s)$ is the same as the unity-gain frequency of $G(s)$. Similarly, an error transfer function $H_e(s)$ from an input phase to a phase error is expressed as

$$H_e(s) = \frac{\theta_e}{\theta_i} = \frac{1}{1 + G(s)} = \frac{s}{s + K_d K_v F(s)} \tag{2.3}$$

Note that $H_e(s)$ exhibits a high-pass filter (HPF) transfer function with the same 3-dB corner frequency of $H(s)$. It means that the PLL tracks the low-frequency components of an input phase, while untracked high-frequency components of the input phase are shown as phase errors.

For the frequency synthesizer that generates multiple frequencies from a low reference frequency, a feedback factor needs to be considered in the linear model of the PLL because a frequency divider is used in the feedback path. For a division ratio N, a feedback factor of N^{-1} is to be used. Figure 2.4 shows the linear model that includes a feedback factor. Since N is constant over frequency, it only plays as a scaling factor in the open-loop gain. Now, we consider a feedforward gain $G_f(s)$ given by (2.1). Then, the system transfer function is expressed as

$$H(s) = \frac{\theta_o}{\theta_i} = \frac{G_f(s)}{1 + G_f(s)/N} \tag{2.4}$$

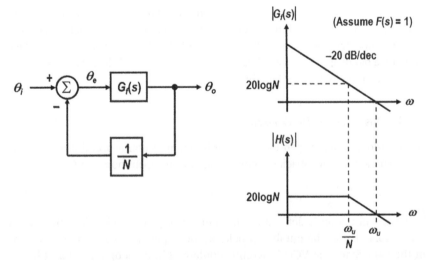

Figure 2.4 Linear model of the PLL including a frequency divider.

The 3-dB frequency is reduced by N, while the DC gain is increased by $20\log N$. Note that the division ratio N not only multiplies the input frequency but also amplifies the input phase variation. For the sake of simplicity, we will exclude the feedback factor to analyze the loop dynamics of the PLL in the rest of this chapter.

2.2.2 Stability

In the negative feedback system, stability is one of the most important things to be considered. For stability analysis, graphical methods are commonly used, including Nyquist diagrams, Evans (root locus) plots, Nicholas charts, and Bode plots. Among them, the Bode plot is well adopted for the stability analysis of a PLL because it gives a straightforward interpretation of the loop dynamics with the distinctive locations of poles and zeros. In addition, circuit designers who know how to design a stable operational amplifier should be familiar with the Bode plot.

For the quantitative analysis of stability, a phase margin or gain margin obtained from the open-loop gain is used. To understand the physical meaning of the phase margin and gain margin, let us consider how a negative feedback system becomes unstable. A negative feedback system becomes a positive feedback system when there is a 180° phase-shift in the negative feedback loop with an open-loop gain equal to or higher than unity as illustrated in Fig. 2.5. Because of the $1/s$ term in the VCO, a phase shift of 90° is already made in the loop. Therefore, the maximum allowable phase margin is 90° with a first-order PLL. If a second-order PLL is designed, additional phase delay occurs due to the pole of a loop filter. The phase margin is defined by an excess phase available from the 180° phase shift when the open-loop gain is unity. For example, a phase margin of 30° means that there is a total phase delay of 150° in the loop at the unity-gain frequency. The less the phase

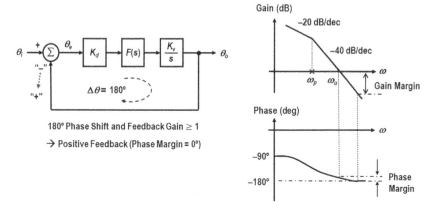

Figure 2.5 Phase margin and gain margin.

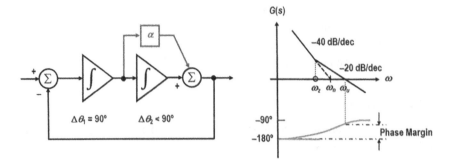

Figure 2.6 Type 2 system with a system zero.

margin is, the less stable is the feedback system. Similarly, the gain margin is an excess gain available from the unity gain when the total phase shift of the loop is 180°. The phase margin is more often used in practice than the gain margin since the gain margin cannot be evaluated if the phase delay in the loop does not reach 180°, for example the first-order PLL. Even though the Nyquist plot or root locus plot shows the detailed information of how a system becomes unstable, the Bode plot is good enough as a starting point for the loop analysis.

Figure 2.6 illustrates why the type 2 feedback system must have a zero in the loop dynamics. Each integrator lags the phase by 90° in the loop, causing a total phase delay of 180°. Accordingly, the type 2 system cannot be stable since the phase margin is 0°, oscillating at a frequency so-called as *natural frequency* ω_n. By providing a fast path in the system, that is, bypassing the integrator, a system zero is formed, and a non-zero phase margin is obtained. Depending on the gain (or strength) of the fast path, the phase margin of the loop is determined.

2.3 Loop Dynamics of the PLL

Depending on the LPF configuration, a PLL has different types and orders. Despite various types and orders, having a good knowledge of the second-order type 2 PLL is enough in the design of PLL circuits for most commercial applications, the grounds of which will be discussed in this chapter.

2.3.1 First-Order Type 1 PLL

The simplest PLL would be a first-order type 1 PLL with a constant LPF gain. Assuming $F(s) = K_f$, the open-loop gain becomes

$$G(s) = \frac{K_d K_f K_v}{s} \tag{2.5}$$

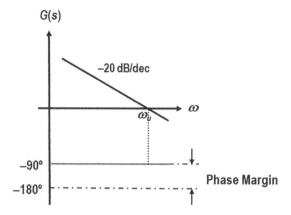

Figure 2.7 Open-loop gain of the first-order PLL with $F(s) = 1$.

In the first-order PLL, a loop gain K is defined by

$$K = K_d K_f K_v \tag{2.6}$$

Then, the closed-loop transfer function is given by

$$H(s) = \frac{K}{s + K} \tag{2.7}$$

In the first-order feedback system, the unity-gain frequency of $G(s)$ is the same as the 3-dB corner frequency of $H(s)$. The first-order type 1 PLL is unconditionally stable with a constant phase margin of 90° as illustrated in Fig. 2.7. The unity-gain frequency ω_u in $G(s)$ is given by

$$\omega_u = K \tag{2.8}$$

We can see that the unity-gain frequency and the loop gain are related with each other. That is, to have a wide bandwidth, a high loop gain is required. With a finite loop gain, transient behaviors such as a lock-in or pull-in range exhibit worse performance than those of higher-order PLLs. More importantly, the type 1 PLL has the problem of a static phase error for the change of an input frequency. There-fore, the use of the first-order PLL by itself is limited in many applications. Those problems will be discussed in the next chapter.

2.3.2 Second-Order Type 1 PLL

An easy way of implementing a second-order PLL is to add an RC LPF in the loop. When the 3-dB corner frequency of the RC filter is put outside the bandwidth of the PLL, the overall behavior of the second-order PLL is similar to that of the

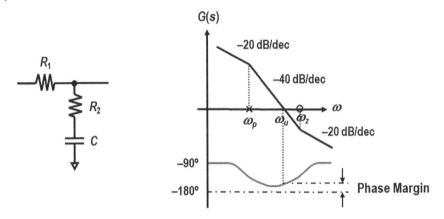

Figure 2.8 Second-order type 1 PLL with a lead-lag filter.

first-order type 1 PLL except second-order filtering outside the bandwidth. In other words, the unity-gain frequency and loop gain are still related with each other.

Now let us consider the case of putting the 3-dB corner frequency of the RC filter within the bandwidth of the PLL. To compensate for the degraded phase margin due to a pole within the bandwidth, a zero should be added. A lead-lag RC filter shown in Fig. 2.8 is commonly used to realize a second-order PLL as a passive filter. The transfer function of the loop filter is

$$F(s) = \frac{1 + s/\omega_z}{1 + s/\omega_p} = \frac{1 + sR_2C}{1 + s(R_1 + R_2)C} \tag{2.9}$$

where

$$\omega_p = \frac{1}{(R_1 + R_2)C}, \quad \omega_z = \frac{1}{R_2C} \tag{2.10}$$

Then, the open-loop gain is given by

$$G(s) = \frac{K_dK_v}{s} \frac{1 + s/\omega_z}{1 + s/\omega_p} = \frac{K_dK_v}{s} \frac{1 + sR_2C}{1 + s(R_1 + R_2)C} \tag{2.11}$$

Since there is only one integrator in $G(s)$, the loop is a second-order type 1 loop. In the second-order PLL, the loop gain K is defined by

$$K = K_dK_{fh}K_v \tag{2.12}$$

where $K_{f,h}$ is the high-frequency loop gain and given by

$$K_{f,h} = \frac{R_2}{R_1 + R_2} = \frac{\omega_p}{\omega_z} \tag{2.13}$$

For the given ω_p and ω_z, the closed-loop transfer function is obtained from (2.2), or

$$H(s) = \frac{K_d K_v \omega_p (1 + s/\omega_z)}{s^2 + \omega_p (1 + K_d K_v/\omega_z)s + K_d K_v \omega_p} = \frac{K\omega_z (1 + s/\omega_z)}{s^2 + (K + \omega_p)s + K\omega_z} \quad (2.14)$$

A Bode plot is shown in Fig. 2.8 with $\omega_p < \omega_u < \omega_z$. As illustrated, ω_z helps increase the phase margin. Compared with the simple case of the second-order PLL with $\omega_u < \omega_p$, the loop dynamics shown in Fig. 2.8 offers a way of controlling ω_u independently for the given loop gain K. Note that the use of the passive filter cannot create a pole at DC to realize a type 2 PLL regardless of the order of the loop.

2.3.3 Second-Order Type 2 PLL

To have a type 2 PLL, an active loop filter should be designed to realize an integrator. An example of an active loop filter is shown in Fig. 2.9. The transfer function of the loop filter is

$$F(s) = \frac{1 + s/\omega_z}{s/\omega_p} = \frac{1 + sR_2C}{sR_1C} \quad (2.15)$$

where

$$\omega_p = \frac{1}{R_1C}, \quad \omega_z = \frac{1}{R_2C} \quad (2.16)$$

Then, the open-loop gain becomes

$$G(s) = \frac{K_d K_v}{s^2} \frac{1 + sR_2C}{R_1C} \quad (2.17)$$

As shown in a Bode plot in Fig. 2.9, the open-loop gain has a slope of -40 dB/dec from DC. If a system zero is not added, the loop will be unstable with a zero phase

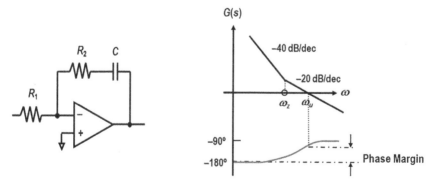

Figure 2.9 Second-order type 2 PLL with an active integrator.

margin and oscillate at a natural frequency ω_n. The loop gain of the second-order type 2 PLL is given by

$$K = K_d K_{f,h} K_v = K_d K_v \frac{R_2}{R_1} = K_d K_v \frac{\omega_p}{\omega_z} \tag{2.18}$$

where

$$K_{f,h} = \frac{R_2}{R_1} \tag{2.19}$$

The closed-loop transfer function is

$$H(s) = \frac{K_d K_v \omega_p (1 + s/\omega_z)}{s^2 + K_d K_v (\omega_p/\omega_z)s + K_d K_v \omega_p} = \frac{K \omega_z (1 + s/\omega_z)}{s^2 + Ks + K\omega_z} \tag{2.20}$$

In the second-order type 2 PLL, the loop filter transfer function can be expressed as the combination of two transfer functions as shown in Fig. 2.10, that is,

$$F(s) = \frac{1 + sR_2 C}{sR_1 C} = \frac{R_2}{R_1} + \frac{1}{sR_1 C} = \alpha + \frac{\beta}{s} \tag{2.21}$$

where

$$\alpha = \frac{R_2}{R_1}, \quad \beta = \frac{1}{R_1 C} \tag{2.22}$$

It implies that two separate control paths in parallel can be defined in the loop filter; a proportional-gain path with a gain of α and an integral path with a gain of β. Note that the loop gain K is determined by α from (2.18) and (2.19). Even though both the single-path and the two-path configurations bring the same loop transfer function, the two-path configuration makes it easy to understand the transient behavior of the second-order type 2 PLL, which will be discussed in the next chapter.

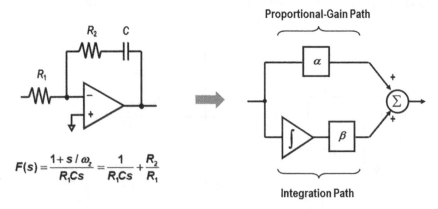

$$F(s) = \frac{1 + s/\omega_z}{R_1 Cs} = \frac{1}{R_1 Cs} + \frac{R_2}{R_1}$$

Figure 2.10 Two-path control of the type 2 PLL.

2.3.4 Natural Frequency and Damping Ratio

In the control theory, the closed-loop transfer function of a second-order feedback system can be fully characterized by two loop parameters, a natural frequency ω_n and a damping ratio ζ with the form of a standard equation given by

$$H(s) = \omega_n^2 \frac{1 + s/\omega_z}{s^2 + 2\zeta\omega_n s + \omega_n^2} \tag{2.23}$$

For a second-order type 2 PLL, ω_n and ζ are obtained from (2.20) and (2.23), or

$$\omega_n = \sqrt{K\omega_z}, \quad \zeta = \frac{1}{2}\left(\frac{\omega_n}{\omega_z}\right) \tag{2.24}$$

Then, we will have the following relation as

$$\omega_n = \frac{K}{2\zeta} \text{ or } K = 2\zeta\omega_n \tag{2.25}$$

In the second-order feedback system, the properties of a pole pair can be described based on ζ. For example, poles are obtained as a complex conjugate pair for $\zeta < 1$ (underdamped), while poles are real and located apart for $\zeta > 1$ (overdamped). When $\zeta = 1$ (critically damped), poles are real and merged.

Figure 2.11 shows different transient behaviors with $\zeta = 0.25, 0.5, 0.707, 1$, and 2. For $\zeta < 0.5$, the loop exhibits clear overshooting. For $\zeta > 1$, the overshooting is not observed, but the loop takes a longer time to be settled. Table 2.1 shows the relationship among the unity-gain frequency ω_u, the loop gain K, and the phase margin. For $\zeta > 0.707$, the phase is margin greater than 65°, and ω_u is close to K.

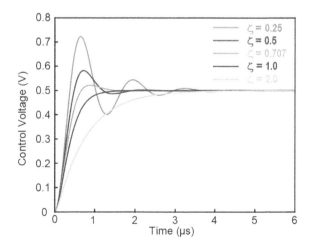

Figure 2.11 Transient behavior with $\zeta = 0.25, 0.5, 0.707, 1$, and 2.

Table 2.1 Relationship among ζ, ω_u, and K.

Damping ζ	ω_u/K	Phase Margin (deg)
0.25	2.13	28.0
0.5	1.27	51.8
0.707	1.10	65.5
1	1.03	76.3
2	1.002	86.4

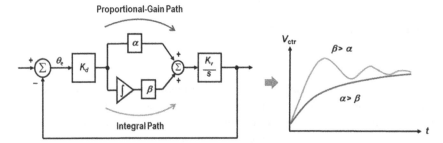

Figure 2.12 Visual insight on loop dynamics of the type 2 PLL.

For an intuitive understanding, the loop dynamics of a second-order type 2 PLL could be considered based on the gain partition of the proportional-gain path and integral path. Figure 2.12 shows the illustrative diagram of the second-order type 2 PLL. If the proportional-gain path is removed, the second-order feedback system contains two integrators without any zero compensation, becoming unstable with 0° phase margin. As a result, the feedback system oscillates with a natural frequency ω_n. If the integral path is removed, there is no other integration block except the VCO, resulting in a type 1 PLL. When the proportional-gain path is dominant over the integral path, the loop behavior is close to that of the type 1 PLL, showing an overdamped loop behavior. If the proportional-gain path is not dominant over the integral path, two separate paths compete for the control of the VCO, exhibiting overshooting and rippling. As the integral path becomes dominant, the overshooting becomes larger, and the rippling period gets close to $1/\omega_n$. Below is the summary.

- When α is zero, the loop becomes an oscillator with an oscillation frequency of ω_n.
- When β is zero, the loop becomes a type 1 PLL.
- When α is bigger than β, the loop becomes an overdamped loop.
- When β is much bigger than or similar to α, the loop becomes an underdamped loop.

2.3.5 High-Order PLLs

Even though ω_n and ζ are valid parameters for the second-order feedback system only, they are extensively used for higher-order PLLs for the sake of convenience. It is because the third- or higher-order PLLs are designed mostly by having additional poles outside the unity-gain frequency ω_u. The high-order poles outside ω_u slightly degrade a phase margin but does not alter the transient behavior and noise transfer function of the PLL much. Therefore, we can roughly analyze the loop dynamics of the high-order PLLs with ζ and ω_n, while considering the impact of those high-order poles. The exact values of system parameters such as the phase margin, the unity-gain frequency, and the closed-loop bandwidth should be obtained with accurate analyses or simulations by including those poles in the transfer function. The open-loop gain of a third-order PLL is shown in Fig. 2.13(a) as an example. A type 3 PLL shown in Fig. 2.13(b) is seldom used in commercial applications due to design complexity for stability. Therefore, a good understanding of the 2nd-order type 2 feedback system is essential for PLL circuit designers and probably enough for most applications.

2.3.6 Bandwidth of PLL

There are several parameters that could be considered to define the bandwidth of a PLL. Those are the natural frequency ω_n, the loop gain K, the unity-gain frequency ω_u of the open-loop gain, the 3-dB corner frequency ω_{3dB} of the closed-loop transfer function, and the noise bandwidth B_n. In the design of the PLL, the noise transfer function of each circuit and a small-signal phase tracking are primarily considered. Therefore, the loop gain K is a good candidate for the bandwidth of the PLL even if it does not contain the integral gain of the second-order type 2 PLL. General definition of the loop gain for high-order PLLs can be found in Gardner's book.

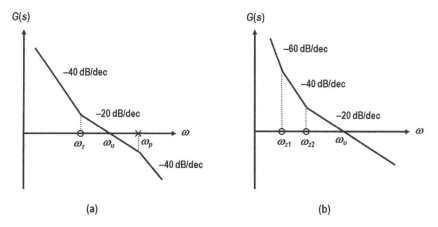

(a)   (b)

Figure 2.13 Third-order PLLs: (a) type 2 and (b) type 3.

Unit of the loop gain K is confusing since we usually presume that the unit of a gain should be unitless. In fact, K has the unit given by

$$K = K_d K_{f,h} K_v = \left[\frac{V}{rad}\right] \cdot \left[\frac{V}{V}\right] \cdot \left[\frac{rad/\sec}{V}\right] = \left[\frac{1}{\sec}\right] \tag{2.26}$$

As seen in (2.26), the unit of K is $\sec^{-1}$ even though K is defined as the gain of a loop. It is because we are mostly interested in how much *frequency change* in units of radians per second occurs at the output even though we begin with *phase change* in units of radian at the input. Strictly speaking, expressing K in dB is not correct since K is not a dimensionless quantity. Note that the numerical value of K is close to the numerical value of ω_u for an overdamped loop even though ω_u is expressed in units of radians per second. Not to make a mistake in calculating a loop gain or a loop bandwidth, it is important to use consistent units for all parameters. For example, if we want to express the loop gain in units of radians per second, the VCO gain should also be expressed in units of radians per second per volt instead of hertz per volt. Otherwise, it is likely to have a wrong scaling factor of 2π in the loop bandwidth, which could cause a serious stability problem in an actual circuit.

Another gain we may consider is a DC loop gain K_{DC} defined by

$$K_{DC} = |\lim_{s \to 0} sG(s)| = K_d |F(0)| K_v \tag{2.27}$$

K_{DC} is the same as the loop gain K for type 1 loops but becomes infinite for type 2 or higher type loops. Having an infinite value for K_{DC} is important for a zero static phase error in the steady state, which will be discussed in the next chapter.

2.3.7 Loop Gain and Natural Frequency

Originated from a control theory, ω_n and ζ have been considered important system parameters in the analysis of the PLL. However, the role of the natural frequency ω_n is not as significant as the loop gain K in the PLL design unless an underdamped loop is to be designed for fast settling time. As a matter of fact, the fast settling time is achieved by the initial tuning of a VCO frequency with digital calibration rather than relying on the underdamped loop in the design of modern integrated PLLs. Moreover, ω_n and ζ are valid parameters for the second-order feedback system only. Therefore, we consider K as a main parameter to characterize the loop dynamics of the PLL and ω_n as an auxiliary parameter for the analysis of a settling behavior. For the second-order type 2 PLL, recall that K and ω_n are related by

$$K = 2\zeta\omega_n \tag{2.28}$$

Apparently, the value of K becomes several times higher than ω_n for an overdamped loop. The role of ω_n for the settling time of the PLL will be discussed in the next chapter.

2.3.8   3-dB Bandwidth

The closed-loop transfer function $H(s)$ represents a system transfer function from an input phase to an output phase, and the 3-dB corner frequency ω_{3dB} indicates the bandwidth of $H(s)$. In a type 2 PLL, the zero frequency causes a gain peaking near ω_{3dB} in $H(s)$ regardless of a damping ratio. With a lower damping ratio, higher gain peaking occurs, resulting in increased ω_{3dB}. When a PLL is designed for a clock-and-data recovery system, the gain peaking problem could be important for jitter transfer function, and the parameters ω_n and ζ become critical, which will be discussed in Chapter 11. The value of ω_{3dB} for a second-order type 2 PLL is formulated as

$$\omega_{3dB} = K'_d K_v R_1 \left(\frac{1}{2} + \frac{1}{4\zeta^2} + \frac{1}{2}\sqrt{1 + \frac{1}{\zeta^2} + \frac{1}{2\zeta^4}} \right)^{\frac{1}{2}} \tag{2.29}$$

The 3-dB corner frequency of the closed-loop transfer function is always higher than the unity-gain frequency of the open-loop gain. For an overdamped loop, ω_{3dB} is close to ω_u like a type 1 PLL.

2.3.9   Noise Bandwidth

Another bandwidth separately defined is the noise bandwidth, the concept of which is well employed in communication systems. As depicted in Fig. 2.14, for the given system transfer function $H(f)$ with a 3-dB bandwidth f_{3dB}, the noise bandwidth B_n is the equivalent bandwidth that gives the same noise power if a flat passband gain G_{nom} is assumed. That is, it is the bandwidth of a rectangular filter to have the same total noise power as the actual filter has. Suppose that a passband gain of 1 is assumed for the system transfer function, that is, $G_{nom} = 1$. Then, the noise bandwidth of a first-order PLL is given by

$$B_n = \int_0^\infty \left| \frac{K/j\omega}{1 + K/j\omega} \right|^2 df_m = \frac{\text{cycle}}{2\pi} \int_0^\infty \frac{K^2}{K^2 + \omega^2} d\omega_m = \frac{K}{2\pi} \tan^{-1} \frac{\omega}{K} \Big|_0^\infty$$

$$\cdot \,\text{cycles} = \frac{K}{4} \,[\text{Hz}] \tag{2.30}$$

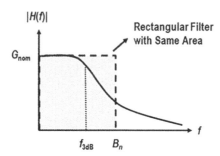

Figure 2.14   Noise bandwidth.

Since K has units of $\sec^{-1}$, we have

$$B_n = \frac{K}{4} \approx \frac{\omega_u}{4} = \frac{\pi}{2} f_u \text{ [Hz]} \tag{2.31}$$

Note that B_n is typically expressed in Hz, not in rad/sec, even though it is expressed with K or ω_u in (2.31). For a second-order type 2 loop, the noise bandwidth is given by

$$B_n = \frac{K}{4}\left(1 + \frac{1}{4\zeta^2}\right) = \frac{\pi K}{2}\left(1 + \frac{1}{4\zeta^2}\right) \text{ [Hz]} \tag{2.32}$$

In the second-order type 2 PLL, the noise bandwidth is wider than the 3-dB bandwidth, so we have the following relation: $B_n > \omega_{3dB} > \omega_u > \omega_n > \omega_z$. The noise bandwidth is useful when we consider the signal-to-noise ratio (SNR) of a PLL or calculate an integrated phase error, which will be discussed in Chapter 4.

2.4 Noise Transfer Function

One important reason of learning the linear model of a PLL is to understand the noise transfer function (NTF) of each noise source. As illustrated in Fig. 2.15, we can easily see that the PLL works as a LPF, a band-pass filter (BPF), and a HPF to an input noise $\theta_{n,i}$, a loop filter noise $\theta_{n,f}$, and a VCO noise $\theta_{n,v}$, respectively in the phase domain. Note that the 3-dB corner frequencies of the LPF and HPF are

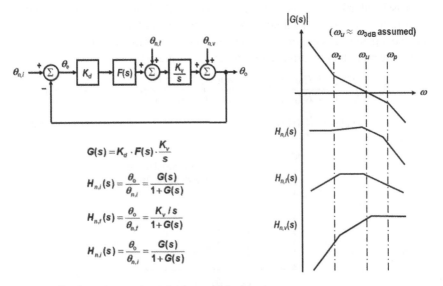

Figure 2.15 Open-loop gain and noise transfer functions.

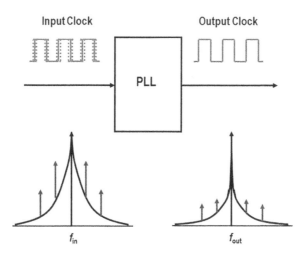

Input Clock   Output Clock

PLL

f_{in} f_{out}

Figure 2.16 Input jitter clean-up PLL with low-pass NTF.

determined by ω_u. A pole ω_p outside the loop bandwidth works as an extra pole outside ω_{3dB} in the LPF transfer function but does not affect the passband gain of the HPF transfer function. With a complementary relationship, the zero ω_z of the LPF transfer function becomes the pole of the HPF transfer function.

The LPF characteristic of a PLL is useful to reject the high-frequency noise of an input signal. Figure 2.16 shows that a PLL is used to suppress a clock jitter with the LPF characteristic to an input jitter. This kind of a clean-up PLL is more effective when a low-noise VCO is available for the design of a narrow-bandwidth PLL. Otherwise, an optimum bandwidth needs to be carefully chosen by considering the noise contribution of both the input clock and the VCO, which will be discussed in Chapter 4. Another good example is the clock-and-data recovery system where a stable clock is recovered from noisy input data and utilized for data retiming.

The fact that the PLL works as the HPF to the VCO phase noise is highly valuable in the design of integrated PLLs. It is because the poor phase noise of a low-Q on-chip VCO can be significantly improved by designing a wide-bandwidth PLL. Figure 2.17 shows how the PLL improves the noise performance of the VCO in the time and frequency domains. The simulated time jitter and phase noise of a closed-loop VCO are shown in comparison with those of a free-running VCO. The unity-gain frequency f_u and the zero frequency f_z in the open-loop gain are set to 2% and 0.5% of the reference frequency, respectively. Phase jitter performance in the time domain is shown in the left figure where X-axis represents the number of clock periods and Y-axis represents the magnitude of the phase jitter. As shown in Fig. 2.17(a), the free-running VCO exhibits a

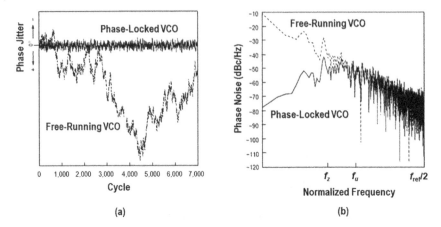

Figure 2.17 Simulated noise performance of open-loop and closed-loop VCOs: (a) long-term jitter in the time domain and (b) phase noise in the frequency domain.

high peak-to-peak jitter due to a long-term phase drift of the free-running VCO. For the phase-locked VCO, the long-term drift of the phase jitter is suppressed by the PLL, and only the high-frequency jitter is shown with much reduced peak-to-peak jitter. Corresponding phase noise performance in the frequency domain is shown in Fig. 2.17(b). The phase noise of the free-running VCO shows a slope of $-20\,\mathrm{dB/dec}$, but it becomes nearly flat within the unity-gain frequency and further goes down with a slope of $+20\,\mathrm{dB/dec}$ below the zero frequency as expected.

Example 2.1 *High-Q BPF characteristics of the PLL*
Let us consider the specific example of the input jitter clean-up PLL shown in Fig. 2.16 where the PLL works as an auto-tracking BPF. Suppose that a 1-GHz divider-less PLL is designed with a second-order type 1 loop having $f_u = 100\,\mathrm{kHz}$ and $f_p = 400\,\mathrm{kHz}$ in the open-loop gain $G(s)$. As depicted in Fig. 2.18, assuming first-order approximation, the NTF to an input phase is characterized as the LPF with $f_{3\mathrm{dB}} = 100\,\mathrm{kHz}$ and $f_p = 400\,\mathrm{kHz}$. The LPF characteristic of the PLL in the phase domain can be viewed as a BPF in the frequency domain with $f_o = 1\,\mathrm{GHz}$, $f_{3\mathrm{dB}} = 100\,\mathrm{kHz}$ and $f_p = 400\,\mathrm{kHz}$ as illustrated in Fig. 2.18. Like the conventional BPF, an effective quality factor Q_{eff} can be defined by

$$Q_{\mathrm{eff}} = \frac{\omega_o}{2\Delta\omega_{3\mathrm{dB}}} \approx \frac{f_o}{2f_u} = \frac{1\,\mathrm{GHz}}{2 \times 100\,\mathrm{kHz}} = 5{,}000$$

The above equation implies that a very high Q_{eff} can be obtained since the PLL bandwidth is set independently of an output frequency. In addition, the tunability

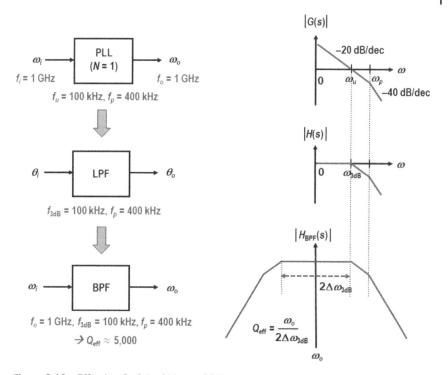

Figure 2.18 Effective Q of the PLL as a BPF.

of the BPF is determined by the PLL bandwidth that exceeds 20% of the output frequency. In other words, we can realize the auto-tracking BPF that has the 20% tunability with a quality factor of 5,000! This unique feature makes the PLL so popular in carrier recovery and clock-and-data recovery applications.

2.5 Charge-Pump PLL

The PLL that employs a phase-frequency detector [1] (PFD) and a charge pump (CP) is the so-called charge-pump PLL (CP-PLL). The CP-PLL has been dominantly used in modern analog PLLs since it realizes an integrator in the loop without implementing an active amplifier. In other words, a type 2 PLL can be formed with a passive RC filter only, which substantially simplifies the PLL design.

Figure 2.19 shows the basic function of the PFD and CP. The PFD generates an up (U) or down (D) pulse whose pulse width is proportional to a phase difference

---

1 We will learn why it is called a phase-frequency detector instead of a phase detector in Chapter 6.

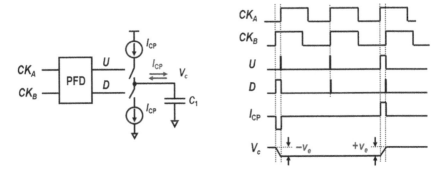

Figure 2.19 Basic function of the PFD and charge pump.

between two input clocks, that is, the rising edge difference of CK_A and CK_B. Then, the charge pump converts the phase error information into a corresponding voltage by enabling the up and down currents that generate an accumulated charge on a capacitor C_1. For a given phase-error time Δt, the amount of an error voltage v_e generated on C_1 at $t = \Delta t$ for each reference clock period is given by

$$|v_e| = \int_0^{\Delta t} \frac{I_{CP}}{C_1} dt = \frac{I_{CP}}{C_1} \Delta t \text{ at } t = \Delta t \tag{2.33}$$

where I_{CP} is a charge pump current. If CK_B is ahead of CK_A, then the down pulse from the PFD turns on the down current only, producing a negative error voltage, or $-v_e$. Equation (2.33) shows that an integrator can be formed in the loop even without having an active amplifier when the charge pump is employed.

As discussed and depicted in Fig. 2.6, a zero must be added to stabilize the type 2 feedback system. In the CP-PLL, the zero can be formed by adding a resistor R_1 in series with C_1 as shown in Fig. 2.20. The resistor provides a way of bypassing the integral path with an instantaneous phase error information, generating the zero of the system for stability. Similar to (2.33), the amount of an error voltage is expressed as

$$|v_e| = I_{CP}R_1 + \int_0^{\Delta t} \frac{I_{CP}}{C_1} dt = I_{CP}R_1 + \frac{I_{CP}}{C_1} \Delta t \text{ at } t = \Delta t \tag{2.34}$$

The capacitor provides the gain of an integral path by accumulating phase errors, while the resistor determines the gain of a proportional-gain path where a pulse-width modulation waveform is generated by the PFD. Unless an under-damped loop is designed with a low phase margin, the loop gain is dominated by the voltage V_{ctr} across R_1 instead of the voltage V_c integrated over C_1 since the voltage increment at the capacitor is much less than the voltage of $I_{CP}R_1$ for

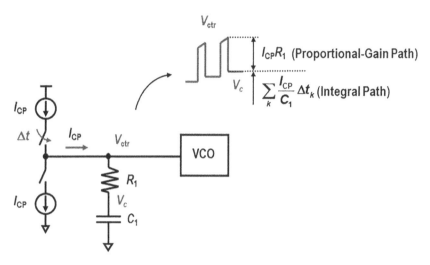

Figure 2.20 Charge pump with an RC loop filter.

each reference clock period. Therefore, in the case of the charge pump with an RC filter, the amplitude of the error voltage waveform can be approximated as

$$|v_e| \approx I_{CP}R_1 \text{ at } t = \Delta t \tag{2.35}$$

With the fixed amplitude of $I_{CP}R_1$, the amount of an effective voltage to control a VCO is determined by the pulse width generated by the PFD. Since the maximum linear range of the PFD is $+2\pi$ or -2π radian, the PFD and CP generate an error voltage of $+I_{CP}R_1$ or $-I_{CP}R_1$, respectively, for the whole reference clock period. Then, we define an effective phase detector gain K'_d of the PFD and charge pump in units of amperes per radian by

$$K'_d = \frac{I_{CP}}{2\pi} \text{ [A/rad]} \tag{2.36}$$

In second-order or higher-order PLLs, we use the high-frequency gain of the loop filter as done in (2.12). For the convenience of the CP-PLL design, we also define the phase detector gain K_d in units of volts per radian given by

$$K_d = \frac{I_{CP}R_1}{2\pi} \text{[V/rad]} \tag{2.37}$$

As to the open loop gain $G(s)$, we still need to consider the impedance of the loop filter $Z(s)$ instead of R_1. That is,

$$G(s) = \frac{K'_d Z(s)K_v}{s} = \frac{I_{CP}}{2\pi} \frac{1 + sR_1C_1}{sC_1} \frac{K_v}{s} = \frac{I_{CP}K_v(1 + sR_1C_1)}{2\pi C_1 s^2} \tag{2.38}$$

Note that $G(s)$ can also be written as

$$G(s) = \frac{K_d' K_v}{s}\left(R_1 + \frac{1}{sC_1}\right) = \frac{K_d' K_v}{s}\left(\alpha + \frac{\beta}{s}\right) \tag{2.39}$$

where

$$\alpha = R_1, \quad \beta = \frac{1}{C_1} \tag{2.40}$$

If the loop dynamics is dominated by the proportional-gain path, that is, $\alpha > \beta$, then the unity gain frequency ω_u is approximated by

$$\omega_u \approx \frac{I_{CP} R_1 K_v}{2\pi} \tag{2.41}$$

The zero frequency ω_z is given by

$$\omega_z = \frac{1}{R_1 C_1} \tag{2.42}$$

With (2.38), the closed-loop transfer function $H(s)$ is given by

$$H(s) = \frac{G(s)}{1 + G(s)} = \frac{K_d' K_v (1 + sR_1 C_1)}{s^2 C_1 + sK_d' K_v R_1 C_1 + K_d' K_v} \tag{2.43}$$

It can also be expressed as

$$H(s) = \omega_n^2 \frac{1 + s/\omega_z}{s^2 + 2\zeta\omega_n s + \omega_n^2} \tag{2.44}$$

where

$$\omega_n = \sqrt{K_d' K_v / C_1} = \sqrt{\frac{I_{CP} K_v}{2\pi C_1}}, \quad \zeta = \frac{\omega_n}{2\omega_z} = \frac{R_1}{2}\sqrt{\frac{I_{CP} C_1 K_v}{2\pi}} \tag{2.45}$$

As seen in (2.41) and (2.45), R_1 is a key parameter to determine ω_u, while C_1 plays an important role for ω_n and ζ. Basic characteristics and detailed operation of the PFD and the charge pump will be learned in Chapter 6.

2.5.1 High-Order CP-PLL

Figure 2.21(a) shows the block diagram of a fourth-order type 2 CP-PLL. For the sake of simplicity, the frequency divider is not considered here. Instead of using equations, let us learn the loop dynamics of the fourth-order PLL in a straightforward way based on what we have learned so far. The open-loop gain $G(s)$ in Fig. 2.21(b) is drawn with corresponding loop parameters from Fig. 2.21(a).

We now easily understand that there are two integrators in the loop; one from the VCO, and the other from the integral path associated with a capacitor C_1.

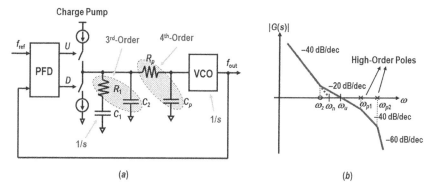

Figure 2.21 Fourth-order type 2 CP-PLL: (a) block diagram and (b) open-loop gain.

Having two poles at DC makes $G(s)$ exhibit a slope of -40 dB/dec from DC in the Bode plot. For the type 2 system, we need a zero frequency ω_z in $G(s)$, which is normally put below the unity-gain frequency ω_u for good phase margin. From (2.41) and (2.42), we learned that the resistor R_1 determines not only ω_z but also ω_u. If an overdamped loop is assumed, the unity-gain frequency is mainly determined by I_{CP}, R_1, and K_v. If the gain of the integral path is comparable to that of the proportional-gain path, that is, C_1 is not large enough to warrant the overdamped loop, the value of ω_u is deviated from (2.41) as the integral path also plays a role in determining ω_u. Note that extending the -40 dB/dec line from DC hits the X axis where $G(s) = 0$ dB and that the frequency is close to the natural frequency ω_n.

Now consider high-order poles that are formed by a shunt capacitor C_2, a series resistor R_p, and another shunt capacitor C_p as depicted in Fig. 2.21(a). The first pole frequency ω_{p1} is mainly determined by R_1 and C_2, while the second pole frequency ω_{p2} is determined by R_p and C_p. In most cases, the main purpose of having high-order poles is to provide additional low-pass filtering outside the loop bandwidth. The high-order poles in the PLL only degrade the phase margin and do not significantly alter the basic properties such as the noise transfer function. For that reason, the third-order or the fourth-order PLL is analyzed in a similar way we do for the second-order PLL as discussed previously. We even describe those high-order PLLs with the natural frequency ω_n and the damping ratio ζ which are originally defined for second-order feedback systems. If necessary, we can put a pole within the loop bandwidth or at DC. For example, a type 3 PLL is sometimes designed for special applications where not only frequency jump but also frequency ramping of an input signal needs to be considered. We will discuss the type 3 PLL in Chapter 3.

2.5.2 Control of Loop Parameters

Design of loop dynamics lies mainly in how to control unity gain, zero, and pole frequencies in the open-loop gain. In the overdamped CP-PLL, those frequencies are approximated as

$$\omega_u \approx \frac{I_{CP}R_1 K_v}{2\pi}, \quad \omega_z \approx \frac{1}{R_1 C_1}, \quad \omega_{p1} \approx \frac{1}{R_1 C_2}, \quad \omega_{p2} \approx \frac{1}{R_p C_p} \tag{2.46}$$

Note that ω_u, ω_z, and ω_{p1} are all affected by R_1. For ω_u, it is difficult to control I_{CP} or K_v since they are much related with other PLL performance (we will discuss later). Accordingly, it is good to control R_1 for optimum ω_u based on fixed I_{CP} and K_v values. Then, ω_z and ω_{p1} need to be controlled by C_1 and C_2 for the given R_1. Before obtaining optimum loop parameters by using a software program, it is important for circuit designers to play with loop parameters in the initial design by drawing a desired open-loop gain with several design aspects. By doing so, the designers can have a visual insight of the loop parameters, which would also be helpful to find a problem or improve the chip performance during hardware testing.

Figure 2.22 illustrates how the loop parameters affect the PLL performance. The unity-gain frequency ω_u plays an important role in optimizing phase noise at the PLL output since it determines the 3-dB corner frequency of both the LPF and the HPF noise transfer functions. As mentioned previously, the zero frequency ω_z is controlled by C_1 rather than by R_1. With lower ω_z, the loop becomes more stable but requires larger C_1, resulting in increased area and longer settling time. Two pole frequencies ω_{p1} and ω_{p2} provide further noise filtering with respect to input phase noise and give good suppression of the spurious tones that are caused by unwanted modulation effects of the PLL. Quantitative analyses of the phase noise and spur performance with loop parameters will be discussed in Chapter 4.

2.5.3 Another Role of Shunt Capacitor

The shunt capacitor C_2 in the LPF is to provide a third pole in the open-loop gain. In practice, the shunt capacitor is needed not only to realize the third pole but also to prevent the charge pump output and the VCO input from being saturated. A specific example is given in Fig. 2.23 to illustrate another role of the shunt capacitor in the PLL. Let us assume that the charge pump is turned on with an up pulse generated by a PFD for each reference clock period of $1/f_{ref}$. We also assume following values for loop parameters; $f_{ref} = 50\,\text{MHz}$, $I_{CP} = 1\,\text{mA}$, $R_1 = 10\,\text{k}\Omega$, $C_1 = 1\,\text{nF}$, and $C_2 = 100\,\text{pF}$. If the shunt capacitor C_2 is not put, a peak voltage of $10\,\text{V}$ set by $I_{CP}R_1$ will be generated at the output of the charge pump, that is, the input of

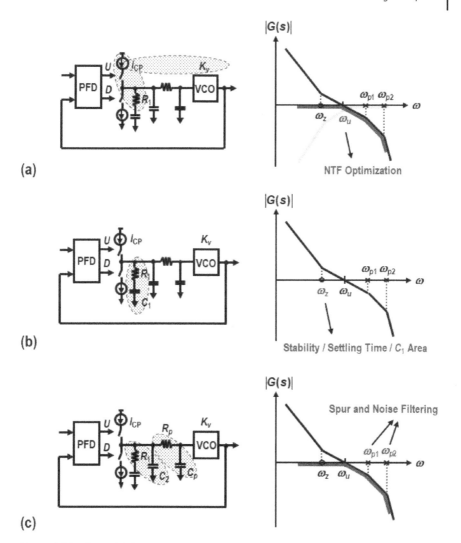

Figure 2.22 Role of *RC* parameters: (a) ω_u for NTF; (b) ω_z for stability, settling time, and area; and (c) spurious tone reduction and noise filtering.

the VCO. At that moment, the output stage of the charge pump will be saturated to a supply voltage, and the up current of the charge pump cannot be turned on. When the shunt capacitor is put, the 50-MHz voltage ripple of the LPF is reduced with an amplitude less than 50 mV as depicted in Fig. 2.23. For this reason, the second-order type 2 CP-PLL is hardly seen for most integrated circuits, and at least a third-order loop is designed for the CP-PLL in practice.

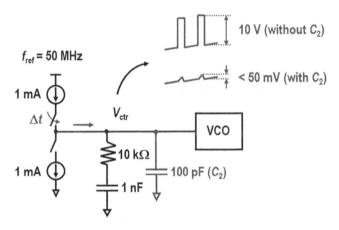

Figure 2.23 Another role of the shunt capacitor.

Example 2.2 *Loop dynamics of the CP-PLL*

A third-order type 2 PLL is shown in Fig. 2.24. A VCO oscillates at 9.9 and 10.1 MHz with control voltages of 1.0 and 1.2 V, respectively. Let $f_{ref} = 10$ MHz, $I_{CP} = 100\,\mu$A, $R_1 = 32$ kΩ, $C_1 = 32$ pF, and $C_2 = 2.5$ pF. From (2.37),

$$K_d = \frac{I_{CP}R_1}{2\pi} = \frac{100\,\mu A \times 32\,k\Omega}{2\pi} = 0.51\,[V/rad]$$

The VCO gain K_v in hertz per volt is obtained by

$$K_v = \frac{10.1\,MHz - 9.9\,MHz}{0.2V} = 1 \times 10^6\,[Hz/V]$$

Then, the loop gain K is

$$K = K_d K_v = 0.51\,\left[\frac{V}{rad}\right] \times 1\,\left[\frac{MHz}{V}\right] = 510\,[kHz/rad]$$

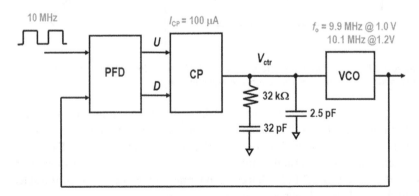

Figure 2.24 Third-order type 2 CP-PLL.

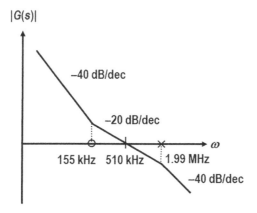

Figure 2.25 Open-loop gain of the third-order type 2 CP-PLL.

Note that the loop gain can also be written in radian, that is,

$$510 \left[\frac{\text{kHz}}{\text{rad}} \right] = 510 \times 10^3 \left[\frac{\text{cycle}}{\text{rad} \cdot \text{sec}} \right] = 510 \times 10^3 \times 2\pi \, [\text{sec}^{-1}]$$

where one cycle is equivalent to 2π radians. Equation (2.42) gives

$$f_z \approx \frac{1}{2\pi R_1 C_1} = \frac{1}{2\pi \times 32 \, \text{k}\Omega \times 32 \, \text{pF}} = 155 \, [\text{kHz}]$$

and

$$f_{p1} \approx \frac{1}{2\pi R_1 C_2} = \frac{1}{2\pi \times 32 \, \text{k}\Omega \times 2.5 \, \text{pF}} = 1.99 \, [\text{MHz}]$$

The Bode plot of the open-loop gain is shown in Fig. 2.25. A phase margin ϕ_M can be estimated by calculating the phase shift at the unity-gain frequency based on tangential approximation. The phase margin is given by

$$\phi_M \approx \tan^{-1} \left(\frac{510 \, \text{kHz}}{155 \, \text{kHz}} \right) - \tan^{-1} \left(\frac{510 \, \text{kHz}}{1.99 \, \text{MHz}} \right) \approx +73.1^\circ - 14.4^\circ = +58.7^\circ$$

As a rule of thumb, a phase margin of about 60° can be achieved for the third-order type 2 PLL when the ratio of f_u to f_z (or ω_u to ω_z) and the ratio of f_{p1} to f_u (or ω_{p1} to ω_u) are close to three and four, respectively.

Example 2.3   *Monitoring on-chip LPF voltage*
For a CP-PLL implemented with an on-chip LPF, monitoring a voltage of the on-chip LPF by having an extra testing pin, if available, is highly useful to evaluate or debug the PLL chip. The testing pin associated with the on-chip LPF can be used for the following purposes: (i) to obtain a voltage-to-frequency transfer characteristic of a VCO by forcing a DC voltage to the testing pin and measuring the output frequency of the VCO; (ii) to check the locked condition of the PLL

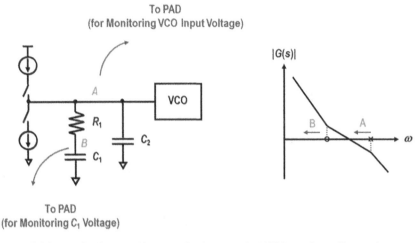

Figure 2.26 Monitoring on-chip control voltage at the VCO input (case A) or at the integral path C_1 (case B).

by measuring the DC voltage of the testing pin; (iii) to evaluate the transient settling performance of the PLL by observing a transient voltage waveform from the testing pin.

As seen in Fig. 2.26, there are two possible nodes in the LPF to be monitored as the testing node that is connected to a testing pad; a node A which is directly connected to the input of the VCO, and a node B which is a top plate of the capacitor C_1. Which node would be better for the monitoring purpose? If we connect the testing pad to the node A, then the parasitic capacitance of the pad will be added to C_2, having the third-pole determined by $1/(R_1 C_2)$ moved to a lower frequency. As a result, the phase margin of the PLL will be reduced. Now, we consider the case that the node B is used as a testing node and connected to the testing pad. The pad connection to the node B increases effective capacitance C_1, shifting the zero of the system given by $1/(R_1 C_1)$ to a lower frequency, resulting in an improved phase margin. In addition, using the node B as the testing node has the following advantages. Firstly, the parasitic capacitance of the testing pad has less effect on the loop dynamics of the PLL since C_1 is much bigger than C_2. Secondly, node connection between the LPF and the pad requires a long routing, possibly suffering from substrate noise coupling. Since the impedance of C_1 is much lower than that of C_2, having the connection to the node B also gives another advantage of achieving good immunity against the noise coupling. Lastly, when the testing pin is connected to the node B, it gives an opportunity of improving stability since an external capacitor can be added to C_1, while adding the external capacitor to the node A degrades the phase margin as discussed.

We may consider adding a MOSFET switch between the LPF and the testing pad to avoid any change of the loop dynamics or the effect of the noise coupling. However, coupling between the MOSFET switch and the on-chip LPF could be a concern since the gate of the MOSFET needs to be connected to either supply or ground. In the case of using the MOSFET switch, we still see the benefit of the node B connection since the bigger capacitor gives a lower impedance.

2.6 Other Design Considerations

2.6.1 Time-Continuous Approximation

There are some other nonideal effects that are considered negligible in most PLL designs but are worth mentioning here. If we look at the operation timing diagram of the PFD and the charge pump in Fig. 2.19, an error information in the feedback system is generated during a very short time for each reference clock period like a sampled system. Hence, the PLL can be analyzed in more accurate way by considering a discrete time model instead of a small-signal linear model. However, it is shown that the small-signal linear model is good enough for the analysis of the PLL when the loop bandwidth is less than one-tenth of the reference clock frequency. In other words, the loop dynamics based on the time-continuous approximation is not much different from that based on the discrete-time analysis when the loop time constant is longer than the reference clock period by at least one decade. If the loop bandwidth is wider than one-tenth of the clock frequency, loop stability is worse than what we expect from the linear model, and the discrete-time model should be employed for accurate analysis.

What about a circuit delay within the feedback loop? In general, the loop delay caused by the circuit delay is negligible, but it still could affect the loop stability for wide bandwidth systems. As a rule of thumb, the circuit delay within the loop can be considered negligible when the delay time is less than one-fiftieth of the loop time constant. In the digital-intensive PLL (DPLL), loop latency caused by additional flip-flop delay may degrade the phase margin, and an accurate analysis need to be considered in a digital domain.

2.6.2 Practical Design Aspects

Design considerations of the PLL for integrated circuit design are somewhat different from those for communication system design. Rigorous mathematical analyses of the SNR performance of the PLL by considering an accurate noise bandwidth and a nonlinear cycle-slipping behavior may not be a must for the design of monolithic PLLs. In fact, modern wireless and wireline transceiver systems well define the system specifications of the PLL based on commercial standards with commonly used (small-signal) parameters such as phase noise, spur, integrated phase error, jitter tolerance, jitter transfer, etc. Therefore, understanding those small-signal parameters with system perspectives is important for circuit designers.

As to the monolithic PLL design, optimum values of the parameters could be chosen not based on how to maximize the system performance in theory but based on how to implement a robust integrated circuit system by considering the variation of circuit parameters over process, voltage and temperature (PVT). For instance, an overdamped loop instead of a critically damped loop is often preferred for a fully integrated PLL design to overcome 20–50% variation of the loop parameters. Therefore, good understanding of how to control loop parameters based on the open-loop gain is important for PLL circuit designers. What we discussed in Section 2.5.2 is depicted in Fig. 2.27, showing that the location of ω_z, ω_u, ω_{p1}, and ω_{p2} plays a critical role in determining the overall performance of the PLL. Among them, high-order poles, ω_{p1} and ω_{p2}, only degrade the phase margin and do not alter the basic property of the PLL much, but ω_{p1} must be incorporated in the design of the type 2 CP-PLL (see Section 2.5.3). When we use a poor quality on-chip VCO, a wide-bandwidth PLL is required to suppress the VCO phase noise at low frequencies. In that case, we may not include ω_{p2} so that enough phase margin can be preserved for good stability. On the other hand, when spurs or out-of-band phase noise needs to be well suppressed with high-order low-pass filtering, ω_{p2} must be put. In general, wireless systems have stringent requirements on spur and out-of-band phase noise, while the wireline

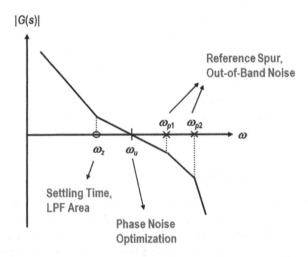

Figure 2.27 PLL design aspect in the open-loop gain.

system often employs a poor-quality on-chip VCO such as a ring VCO. For those reasons, a fourth-order type 2 PLL is adopted for most wireless systems, while a third-order type 2 PLL is well employed for wireline systems.

Below is the summary of some design aspects for monolithic PLL design which have been learned in this chapter.

- The loop gain K is used for the loop bandwidth of the PLL.
- The PLL works as the LPF to an input noise, while it plays as the HPF to a VCO noise.
- A type 2 PLL is commonly used. In the type 2 PLL, the control path is viewed as the combination of a proportional-gain path and an integral path.
- The CP-PLL realizes a type 2 feedback system with a passive filter only.
- In practice, the second-order type 2 CP-PLL is not used. That is, the minimum order of the type 2 CP-PLL should be the third order.
- The basic property of the third- or higher-order PLL is basically similar to that of the second-order PLL if high-order poles are put outside the bandwidth. The fourth-order type 2 PLL and the third-order type 2 PLL are often employed for wireless and wireline applications, respectively.

References

1 F. M. Gardner, *Phaselock Techniques*, 3rd ed., Wiley, Hoboken, NJ, 2005.
2 W. Egan, *Phase-Locked Basics*, 2nd ed., Wiley, Hoboken, NJ, 2008.

3

Transient Response

The transient response of a PLL before settling is quite complicated since a frequency offset between an input and an output causes a time-varying phase error or even cycle-slipping by having a phase error exceed 2π radian. Indeed, the transient behavior of the PLL is a nonlinear process as a whole. To make the transient analysis easier, we decompose the transient settling behavior into two parts; one is frequency acquisition (large signal) and the other is frequency/phase tracking (small signal). A good criterion to distinguish them is to check whether a cycle-slipping occurs or not. Figure 3.1 illustrates the frequency settling behavior of traditional PLLs. The frequency range over which a PLL does not exhibit the cycle-slipping is defined as a lock-in range. When the initial frequency of the PLL is out of the lock-in range, the PLL exhibits a large-signal behavior with the cycle-slipping. The frequency acquisition in the large-signal region is a nonlinear process, making it difficult to analyze the PLL performance. Once the PLL enters the lock-in range, the transient analyses can be simplified with a linear model.

Since the frequency acquisition by the PLL itself takes a very long time, acquisition-aid circuits must be employed in practice. In the design of modern integrated PLLs, an on-chip voltage-controlled oscillator (VCO) is designed mostly with pre-tuning capability, so that the PLL enters the lock-in range immediately after the automatic calibration of the VCO frequency. In frequency synthesizers, the phase-frequency detector (PFD) itself operates not only as a phase detector but also as a frequency acquisition aid circuit. In this chapter, we focus on the self-acquisition property of the PLL. Also, we limit our analyses to the first-order and second-order PLLs. It is shown that the transient behavior of a third-order or higher-order type 2 PLLs is not much deviated from that of a second-order type 2 PLL.

Phase-Locked Loops: System Perspectives and Circuit Design Aspects, First Edition.
Woogeun Rhee and Zhiping Yu.
© 2024 The Institute of Electrical and Electronics Engineers, Inc. Published 2024 by John Wiley & Sons, Inc.

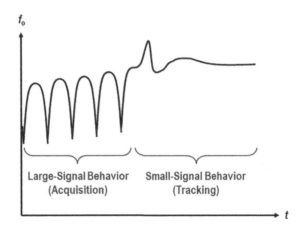

Figure 3.1 Transient settling behavior of traditional PLLs.

3.1 Linear Transient Performance

Within the lock-in range, the transient response of a PLL can be analyzed based on the linear model. One important design metric in the linear transient response is a static phase error. We will learn that the type of a loop is more important than the loop order and discuss why a type 2 PLL is dominantly used in most applications. The answer could be found by examining the phase response of a PLL in the steady state with respect to the phase and frequency changes of an input signal.

3.1.1 Steady-State Phase Response

Let us consider the steady-state phase response of a PLL with three different input conditions; a phase step, a frequency step, and a frequency ramp. We use an error transfer function from (2.3) to express a phase error $\theta_e(s)$ in the frequency domain as

$$\theta_e(s) = \frac{s}{s + K_d K_v F(s)} \theta_i(s) \tag{3.1}$$

To obtain a steady-state phase error in the time domain, we use the final-value theorem of Laplace transform. The final value theorem gives the value of a function at $t = \infty$ based on the value of its frequency component at $f = 0$. That is, for a bounded function $f(t)$ on $(0,\infty)$,

$$\lim_{t \to \infty} f(t) = \lim_{s \to 0} sF(s) \tag{3.2}$$

Then, the steady-state phase error can be calculated as

$$\lim_{t \to \infty} \theta_e(t) = \lim_{s \to 0} s\theta_e(s) = \lim_{s \to 0} \frac{s^2}{s + K_d K_v F(s)} \theta_i(s) \tag{3.3}$$

Now, we assume that there is a phase step $\Delta\theta$ in the input phase. The Laplace transform of the input phase is given by

$$\theta_i(t) = \Delta\theta \quad \leftrightarrow \quad \theta_i(s) = \frac{\Delta\theta}{s} \tag{3.4}$$

From (3.3) and (3.4), we have

$$\lim_{t \to \infty} \theta_e(t) = \lim_{s \to 0} \frac{s \cdot \Delta\theta}{s + K_d K_v F(s)} = 0 \tag{3.5}$$

The result shows that the phase change of an input signal does not cause a phase error in the steady state regardless of the order or type of a PLL.

Let us consider the case that there is a frequency step $\Delta\omega$ in the input frequency. The Laplace transform of the input phase with $\Delta\omega$ is expressed as

$$\theta_i(t) = \Delta\omega t \quad \leftrightarrow \quad \theta_i(s) = \frac{\Delta\omega}{s^2} \tag{3.6}$$

Then, we get

$$\lim_{t \to \infty} \theta_e(t) = \lim_{s \to 0} \frac{\Delta\omega}{s + K_d K_v F(s)} = \frac{\Delta\omega}{K_d K_v F(0)} = \frac{\Delta\omega}{K_{DC}} \tag{3.7}$$

where K_{DC} is a DC loop gain defined in (2.27). Equation (3.7) shows that a steady-state phase error is zero with an infinite DC loop gain, which is possible only when $F(s)$ contains $1/s$, that is, an integrator. Therefore, a first-order or second-order type 1 PLL exhibits a static phase error when there is a frequency change in the input signal. In other words, at least type 2 loop must be considered for the PLL to achieve phase-lock without a static phase error when the frequency of an input signal is not constant over time.

What about an input with a frequency ramp? For the input frequency that changes linearly over time at a rate Λ, the input phase in the time and frequency domains are expressed as $\theta_i(t) = \Lambda t^2/2$ and $\theta_i(s) = \Lambda/s^3$, respectively. The steady-state phase error is obtained by

$$\lim_{t \to \infty} \theta_e(t) = \lim_{s \to 0} \frac{\Lambda/s}{s + K_d K_v F(s)} = \frac{\Lambda}{K_d K_v} \lim_{s \to 0} \frac{1}{sF(s)} \tag{3.8}$$

We clearly see that having the integrator in $F(s)$ is not enough to avoid a static phase error. In that case, we need a type 3 PLL to avoid a static phase error with the frequency ramp variation since the type 2 PLL only mitigates the static phase error but does not completely remove it. The type 3 PLL is seldom employed due to design complexity for stability but could be useful for radar systems to track satellites or missiles whose frequency changes over time due to the Doppler effect.

Table 3.1 Static phase error performance of the type 1 and type 2 PLLs.

	Type 1	Type 2
Phase step	0	0
Frequency step	$\dfrac{\Delta\omega}{K_d K_v F(0)}$	0
Frequency ramp	∞	$\dfrac{\Lambda}{K_d K_v} \lim\limits_{s \to 0} \dfrac{1}{sF(s)}$

Table 3.1 shows the summary of the steady-state phase error performance of the type 1 and type 2 PLLs. Achieving the zero static phase error with the frequency step is important for frequency generation systems. In PLL-based frequency synthesizers, multiple output frequencies need to be generated from a single reference source without causing a large static phase error. In transceiver systems, there exists a frequency offset between a transmitter and a receiver through communication channels. When a PLL-based clock-and-data recovery (CDR) circuit is used, tracking the frequency difference without the static phase error is important for data retiming. On the other hand, dealing with the frequency ramp is not a major concern for those applications. Therefore, the type 2 PLL has been dominantly employed in most commercial applications.

3.1.2 Transient Phase Response

Compared to the steady-state phase error, describing the transient phase response is not straightforward even for a second-order type 1 PLL. In fact, the mathematical derivation of the transient response of a PLL does not give much insight to circuit designers. Relying on transient simulations would be a reliable way to have the visual insight of the PLL settling. Nonetheless, let us take a brief look at the analytic equations of the transient response with different loop types and orders to address how complicated it is to obtain mathematical derivations for high-order PLLs.

The transient phase response with respect to the phase and frequency changes of an input signal can be obtained by using the Laplace transform and converting them to time-domain equations. For example, the transient phase responses with respect to the phase step and the frequency step for a first-order PLL are obtained by

$$\theta_e(s) = \frac{s}{s + K_d K_v F(s)} \frac{\Delta\theta}{s} = \frac{\Delta\theta}{s + K} \leftrightarrow \theta_e(t) = \Delta\theta e^{-Kt} \tag{3.9}$$

and

$$\theta_e(s) = \frac{s}{s + K_d K_v F(s)} \frac{\Delta\omega}{s^2} = \frac{\Delta\omega}{s(s + K)} \leftrightarrow \theta_e(t) = \frac{\Delta\omega}{K}(1 - e^{-Kt}) \tag{3.10}$$

where $F(s) = 1$ is assumed. The transient responses of a second-order or higher-order PLL are typically characterized with natural frequency ω_n and damping ratio ζ. For the second-order type 1 PLL, let $F(s) = 1/(s + \omega_p)$ without ω_z in (2.9) for the sake of simplicity. Then, the Laplace transform of the transient phase responses with respect to the phase step and the frequency step are given by

$$\theta_e(s) = \frac{s(s + 2\zeta\omega_n)}{s^2 + 2\zeta\omega_n s + \omega_n^2} \frac{\Delta\theta}{s} = \frac{(s + 2\zeta\omega_n)\Delta\theta}{s^2 + 2\zeta\omega_n s + \omega_n^2} \tag{3.11}$$

and

$$\theta_e(s) = \frac{s(s + 2\zeta\omega_n)}{s^2 + 2\zeta\omega_n s + \omega_n^2} \frac{\Delta\omega}{s^2} = \frac{(s + 2\zeta\omega_n)\Delta\omega}{s\left(s^2 + 2\zeta\omega_n s + \omega_n^2\right)} \tag{3.12}$$

Transient phase response equations in the time domain are obtained by taking inverse transform of (3.11) and (3.12), respectively. Those equations have different results depending on the value of ζ. Let us take one case of $\zeta < 1$. The transient phase responses with respect to the phase step and the frequency step are given by

$$\theta_e(t) = \Delta\theta e^{-\zeta\omega_n t}\left[\cos\omega_n(1 - \zeta^2)^{1/2}t + \frac{\zeta}{(1 - \zeta^2)^{1/2}}\sin\omega_n(1 - \zeta^2)^{1/2}t\right] \tag{3.13}$$

and

$$\theta_e(t) = 2\zeta\frac{\Delta\omega}{\omega_n} + \frac{\Delta\omega}{\omega_n}e^{-\zeta\omega_n t}\left[\frac{1 - \zeta^2}{(1 - \zeta^2)^{1/2}}\sin\omega_n(1 - \zeta^2)^{1/2}t - 2\zeta\cos\omega_n(1 - \zeta^2)^{1/2}t\right] \tag{3.14}$$

Similarly, for the second-order type 2 PLL, let $F(s) = (1 + s/\omega_z)/sR_1C$ from (2.15). Then, the Laplace transform of the transient phase responses with respect to the phase step and the frequency step are given by

$$\theta_e(s) = \frac{s^2}{s^2 + 2\zeta\omega_n s + \omega_n^2} \frac{\Delta\theta}{s} = \frac{s\Delta\theta}{s^2 + 2\zeta\omega_n s + \omega_n^2} \tag{3.15}$$

and

$$\theta_e(s) = \frac{s^2}{s^2 + 2\zeta\omega_n s + \omega_n^2} \frac{\Delta\omega}{s^2} = \frac{\Delta\omega}{s\left(s^2 + 2\zeta\omega_n s + \omega_n^2\right)} \tag{3.16}$$

Again, transient phase response equations in the time domain are obtained by taking inverse transform of (3.15) and (3.16), respectively. Those equations have different results depending on the value of ζ. For the case of $\zeta < 1$, the transient phase responses with respect to the phase step and the frequency step are given by

$$\theta_e(t) = \Delta\theta e^{-\zeta\omega_n t}\left[\cos\omega_n(1 - \zeta^2)^{1/2}t - \frac{\zeta}{(1 - \zeta^2)^{1/2}}\sin\omega_n(1 - \zeta^2)^{1/2}t\right] \tag{3.17}$$

and

$$\theta_e(t) = \frac{\Delta\omega}{\omega_n}e^{-\zeta\omega_n t}\left[\frac{1}{(1 - \zeta^2)^{1/2}}\sin\omega_n(1 - \zeta^2)^{1/2}t\right] \tag{3.18}$$

It is worth mentioning that the settling behavior of the third-order PLL is similar to that of the second-order type 2 PLL. Hence, understanding the second-order type 2 PLL can be a good basis to understand the transient behavior of high-order PLLs.

3.1.3 Settling Time

Even though the transient analysis is complicated, the analysis of a settling behavior within the lock-in range can be greatly simplified with first-order approximation. For the first-order PLL with a loop gain K, the transient frequency $f(t)$ is given by

$$f(t) = f_1 + \Delta f_{\text{step}}(1 - e^{-Kt}) \tag{3.19}$$

where f_1 is an initial frequency, f_2 is a desired frequency, and Δf_{step} is a step frequency given by $f_2 - f_1$. An illustrative diagram is shown in Fig. 3.3 in which a frequency error f_ε is defined by the difference between the desired frequency and a settling frequency f_s, that is,

$$f_\varepsilon = f_2 - f_s \tag{3.20}$$

From Fig. 3.2, the settling frequency can be expressed as

$$f_s = f_1 + \Delta f_{\text{step}} \left(1 - e^{-Kt_s}\right) \tag{3.21}$$

where t_s is defined as a settling time. From (3.20) and (3.21), the settling time is obtained as

$$t_s = \frac{1}{K} \ln \frac{\Delta f_{\text{step}}}{f_\varepsilon} \tag{3.22}$$

As expected, the loop gain K plays a critical role in determining the settling time. Based on (3.22), the settling time is determined not only by Δf_{step} but also by f_ε. The effect of the frequency accuracy on the settling time is often neglected, but

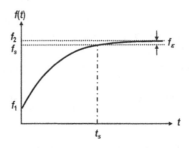

Figure 3.2 Settling time with a frequency error.

some wireless systems put a stringent requirement on the frequency accuracy. Hence, the settling time requirement could be quite different even for the same values of Δf_{step} and K.

Calculating the settling time of a second-order type 2 PLL is not straightforward because it is affected by three parameters, ζ, ω_n, and K. As learned from Chapter 2, the loop gain K is proportional to the natural frequency ω_n, and they are related as

$$K = 2\zeta\omega_n \tag{3.23}$$

Therefore, to see the effect of the damping ratio ζ on the settling time, we need to analyze it for a fixed ω_n or K. Figure 3.3 shows the phase-tracking response of the second-order type 2 PLL with a frequency step. The settling behaviors with a constant ω_n and a constant K are plotted in Fig. 3.3(a) and (b), respectively. In Fig. 3.3(a), the settling time does not vary much over different ζ values even though overshooting amount changes a lot. It is because having a large ζ also increases K based on (3.23). To the contrary, we clearly see the large impact of ζ on the settling time with a constant K as shown in Fig. 3.3(b). With $\zeta > 1$, the settling time dramatically increases as ζ becomes larger. The result clearly shows that the settling time is lagged a lot by the integral path of the second-order type 2 PLL in the case of an overdamped loop. The settling behaviors shown in Fig. 3.3 imply that the role of K is more important than ω_n and that it is important to normalize the loop gain when an optimum ζ is defined. Figure 3.3(b) indicates that designing an underdamped loop is good to achieve a fast settling time. It is because the integral path also actively tunes a VCO in the underdamped loop, which makes the loop agile as discussed in Chapter 2. In practice, a damping ratio of 0.35–0.5 is designed for

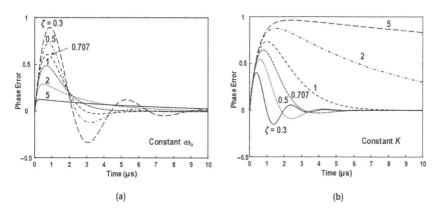

Figure 3.3 Settling time of the second-order type 2 PLL with a frequency step: (a) with fixed ω_n and (b) with fixed K.

fast settling. The settling time of the second-order type 2 PLL with underdamping can be approximated as

$$t_s = \frac{1}{\zeta\omega_n} \ln\left(\frac{\Delta f_{step}}{f_\varepsilon \sqrt{1-\zeta^2}} \right) \tag{3.24}$$

As to the overdamped loop, the settling time is dominated by the slow frequency acquisition of the integral path as indicated in Fig. 3.3(b). If a frequency step is larger than the lock-in range of the PLL, the settling behavior of the integral path needs to be considered separately. Some example is given at the end of this chapter. For higher-order PLLs, analyzing the settling time based on the second-order type 2 PLL can be a good starting point, but an exact analysis should be done based on both numerical and circuit simulations. In modern PLL design, the settling time depends not only on the loop dynamics but also on the digital calibration of the center frequency of a VCO.

Example 3.1 *Settling time of the second-order type 2 CP-PLL*
A frequency synthesizer based on a second-order type 2 PLL is shown in Fig. 3.4. A loop gain is similar to the one in Example 2.2, but an underdamped loop is designed with a smaller value of C_1. To compensate for the reduced loop gain due to a frequency division ratio of 100, the VCO gain is increased by 100 times. For the sake of simplicity, a second-order loop instead of a third-order loop is considered by removing a shunt capacitor in the loop filter. The PLL is initially locked at 1 GHz with a reference frequency of 10 MHz and a division ratio N of 100. Let $I_{CP} = 100\,\mu A$, $R_1 = 32\,k\Omega$, $C_1 = 8\,pF$, and $K_v = 100\,MHz/V$. Suppose that the division ratio is changed from 100 to 101 to tune the output frequency from 1 GHz to 1.01 GHz. Let us calculate an approximated settling time with a frequency accuracy of 100 Hz.

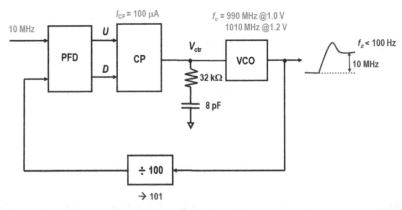

Figure 3.4 Second-order type 2 CP-PLL.

To calculate the settling time, we need to get ω_n and ζ. Note that K_v should be expressed in units of rad/s/V if ω_n (not f_n) is considered. That is,

$$\omega_n = \sqrt{\frac{I_{CP}K_v}{2\pi NC_1}} = \sqrt{\frac{100\,\mu A \times (2\pi \times 100 \times 10^6)\,\text{rad/s/V}}{2\pi \times 100 \times 8\,\text{pF}}} = 3.536 \times 10^6 \quad [\text{rad/s}]$$

and

$$\zeta = \frac{R_1 C_1}{2}\omega_n = \frac{32\,k\Omega \times 8\,\text{pF}}{2} \times 3.536 \times 10^6\,\text{rad/s} = 0.453$$

From (3.24), we get the settling time as

$$t_s = \frac{1}{\zeta\omega_n}\ln\left(\frac{\Delta f_{\text{step}}}{f_\varepsilon\sqrt{1-\zeta^2}}\right) = \frac{1}{0.453 \times 3.536 \times 10^6}$$

$$\times \ln\left(\frac{10^7}{100 \times \sqrt{1-0.453^2}}\right) = 7.259 \times 10^{-6} \quad [\text{sec}]$$

To see the effect of the damping ratio on the settling time, let us use $C_1 = 32\,\text{pF}$ instead of 8 pF. Then,

$$\omega_n = \sqrt{\frac{I_{CP}K_v}{2\pi NC_1}} = \sqrt{\frac{100\,\mu A \times (2\pi \times 100 \times 10^6)\,\text{rad/s/V}}{2\pi \times 100 \times 32\,\text{pF}}} = 1.768 \times 10^6 \quad [\text{rad/s}]$$

and

$$\zeta = \frac{R_1 C_1}{2}\omega_n = \frac{32\,k\Omega \times 32\,\text{pF}}{2} \times 1.768 \times 10^6\,\text{rad/s} = 0.905$$

The settling time becomes

$$t_s = \frac{1}{\zeta\omega_n}\ln\left(\frac{\Delta f_{\text{step}}}{f_\varepsilon\sqrt{1-\zeta^2}}\right) = \frac{1}{0.905 \times 1.768 \times 10^6}$$

$$\times \ln\left(\frac{10^7}{100 \times \sqrt{1-0.905^2}}\right) = 7.721 \times 10^{-6} \quad [\text{sec}]$$

Compared to the case with $C_1 = 8\,\text{pF}$, the settling time is increased by about 6%. As discussed previously, a higher damping ratio means less contribution of the integral path to controlling the VCO, resulting in increased settling time. Now, we consider the case of a relaxed frequency accuracy by having $f_\varepsilon = 10\,\text{kHz}$ instead of 100 Hz. Then, we have

$$t_s = \frac{1}{\zeta\omega_n}\ln\left(\frac{\Delta f_{\text{step}}}{f_\varepsilon\sqrt{1-\zeta^2}}\right) = \frac{1}{0.905 \times 1.768 \times 10^6}$$

$$\times \ln\left(\frac{10^7}{10^4 \times \sqrt{1-0.905^2}}\right) = 4.846 \times 10^{-6} \quad [\text{sec}]$$

The above example shows that the effect of the frequency accuracy on the settling time could be more critical than the damping ratio. As some wireless communication systems have the tight requirement of frequency tolerance, it should be well considered in the settling time.

3.2 Nonlinear Transient Performance

The frequency acquisition and tracking response of a PLL are considered a large-signal nonlinear behavior. The phase and frequency tracking within the lock-in range of the PLL is not categorized as the nonlinear transient behavior unless complicated cyclostationary analyses are involved. Hence, we reserve a term *tracking* for a linear transient region, while using a term *acquisition* for the large-signal or nonlinear transient region where cycle-slipping occurs. The lock-in range ω_L is the frequency range over which the PLL acquires lock without cycle-slipping. In addition to the lock-in range, we define other two regions, the pull-in range[1] ω_P and the hold-in range ω_H as depicted in Fig. 3.5. The pull-in range is the frequency range over which the PLL acquires lock but possibly with cycle-slipping. The hold-in range is the frequency range over with the PLL maintains lock but not necessarily acquires lock. For example, the PLL initially locked with an input frequency can maintain phase lock even if a slowly-varying input frequency exceeds the pull-in range. The hold-in range and the pull-in range of the type 2 PLL are infinite in theory, so they are limited by the tuning range of a VCO in practice. In actual circuit design, the large-signal transient performance is usually determined by an acquisition aid circuit rather than by the self-acquisition property of the PLL. In addition to three ranges, more ranges

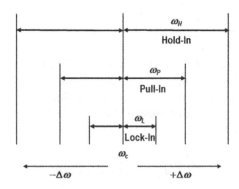

Figure 3.5 Lock-in, pull-in, and hold-in ranges.

---

1 Also called as *capture range* or *acquisition range*.

could be defined as done in some other books, but these three ranges are essential for the design of PLL circuits and will be discussed in this chapter.

3.2.1 Hold-In Range

When a PLL is locked, the maximum frequency range over which the PLL stays in lock is called the hold-in range. An easy way of measuring the hold-in range is to sweep the reference frequency and check the maximum range of the locked frequency at the output. In principle, the hold-in range is determined by the maximum voltage range $V_{d,max}$ of a phase detector (PD) as depicted in Fig. 3.6. Various types of the PD give different linear ranges, which will be discussed in Chapter 6. Accordingly, the hold-in range is fixed for the given PD type. For example, let us assume a triangular PD characteristic as shown in Fig. 3.6. Then, the hold-in range is given by

$$\omega_H = V_{d,max} |F(0)| K_v = \pi K_d |F(0)| K_v = \pi K_{DC} \tag{3.25}$$

where πK_d represents the maximum voltage range $V_{d,max}$ of the PD having a triangular characteristic. For a type 2 PLL, K_{DC} is infinite, giving an infinite hold-in range. In that case, the actual hold-in range of the type 2 PLL is the same as the tuning range of the VCO regardless of the PD types.

3.2.2 Pull-In Range

A frequency pull-in process is an interesting self-acquisition behavior of the PLL. When an initial frequency is out of the lock-in range but within the pull-in range, the PLL can still acquire a desired frequency by itself with a beat tone generated by a PD. That is, any PD conveys frequency information even when cycle-slipping occurs. Figure 3.7 gives a simplified diagram for intuitive understanding of the

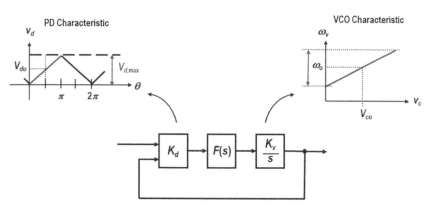

Figure 3.6 Hold-in range.

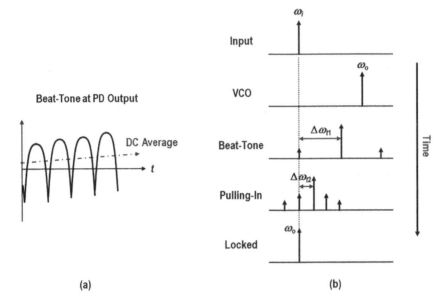

Figure 3.7 Pull-in range: (a) beat-tone generation by the PD and (b) pull-in process in the frequency domain.

pull-in behavior in both time and frequency domains. In the time domain, it is important to observe that the PD generates the asymmetric beat tone that carries DC information when the PLL is out of the lock-in range. The DC voltage tunes the VCO to the right direction, making the PLL enter the lock-in range eventually. This behavior can be interpreted in the frequency domain as well. When an input frequency ω_i and a VCO frequency ω_o are separated by $\Delta\omega$, the PD generates a beat tone with a fundamental frequency of $\Delta\omega$. As the beat-tone modulates the VCO, sidebands located at $\omega_o + \Delta\omega$ and a $\omega_o - \Delta\omega$ are generated. In Fig. 3.7, ω_o is higher than ω_i, and the location of $\omega_o - \Delta\omega$ is the same as ω_i. Therefore, $\omega_o - \Delta\omega$ and ω_i produce DC information at the PD output, making the VCO tuned toward the desired frequency.

The type 1 PLL has a finite pull-in range that depends on the loop order. The pull-in range is the same as the hold-in range for the first-order PLL since there is no attenuation by the loop filter. The pull-in range of the second-order PLLs is approximated as

$$\omega_P \approx \sqrt{2\alpha K_{\mathrm{DC}}K} \approx 2\sqrt{\alpha K_{\mathrm{DC}}\zeta\omega_n} \tag{3.26}$$

where the value of α depends on PD characteristics. Note that the loop gain K of second-order PLLs is given by $K_d K_{f,h} K_v$ where $K_{f,h}$ is a high-frequency gain of the loop filter defined in Chapter 2. The pull-in range of a type 2 PLL is infinite like the hold-in range of the type 2 PLL. Accordingly, the type 2 PLL eventually acquires a desired frequency from any initial frequency regardless of a PD type unless the frequency range is limited by a VCO.

The pull-in range, however, is not considered in the design of monolithic PLLs simply because the pull-in time is too long to make the pull-in range useful in practice. For a second-order type 2 PLL, the pull-in time t_P is approximated as

$$t_P \approx \frac{(\Delta\omega)^2}{2\zeta\omega_n^3} \tag{3.27}$$

For example, the pull-in time of the PLL in Example 3.1 with the case of $C_1 = 32\,\mathrm{pF}$ is calculated as

$$t_P \approx \frac{(\Delta\omega)^2}{2\zeta\omega_n^3} = \frac{(2\pi \times 10^7)^2}{2 \times 0.905 \times (1.768 \times 10^6)^3} = 394.38 \times 10^{-6}\ [\mathrm{sec}]$$

Compared to the settling time of 7.26 μs, the pull-in time is far longer. Such a long pull-in time is not meaningful in analyzing the nonlinear transient behavior of the PLL since the acquisition aid circuitry such as a PFD or a frequency-acquisition loop is mostly employed in the PLL design.

3.2.3 Lock-In Range

A lock-in range is the most meaningful to PLL designers since it gives a good boundary between large-signal and small-signal regions. For a first-order PLL, the lock-in range is the same as the hold-in range or the pull-in range. For a second-order PLL having a PD with a characteristic gain α, the lock-in range is approximated as

$$\omega_L = \alpha K_d |F(\Delta\omega_L)| K_v \approx \alpha K_d K_{fh} K_v = \alpha K \tag{3.28}$$

where $|F(\Delta\omega_L)|$ is the magnitude of a loop filter transfer function at an offset frequency of ω_L from an output frequency. As depicted in Fig. 3.8, the lock-in range of a type 2 PLL is not infinite and is determined mainly by the unity-gain bandwidth ω_u of an open-loop gain. Equation (3.28) also explains why the loop gain K is a good choice for the bandwidth of the PLL rather than the natural frequency ω_n.

3.2.4 Nonlinear Phase Acquisition

Phase acquisition during the nonlinear transient response is considered a part of frequency acquisition as discussed. However, there is another type of nonlinear phase acquisition called *hangup*. The hangup effect occurs when an initial phase error is close to an unstable equilibrium point. For instance, a PD having a sawtooth characteristic as depicted in Fig. 3.9 has a normal equilibrium point at a phase offset of π radian, while exhibiting an unstable equilibrium point at a phase

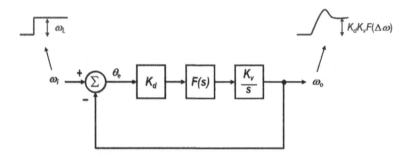

Figure 3.8 Lock-in range.

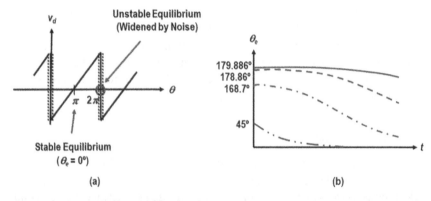

Figure 3.9 Hangup effect: (a) PD with a sawtooth transfer function and (b) hangup effect over different initial phase errors.

offset of 2π radian. If the initial phase is located close to the unstable equilibrium point, the PLL will stay at that point for a while until a certain amount of noise pushes the phase out of that point with a positive feedback. Even though the chance to stay with unstable equilibrium is small, the hangup effect could be substantial statistically when input jitter spreads the unstable equilibrium region. As illustrated in Fig. 3.9, even a PD with a steep transfer function at the unstable equilibrium point can suffer from the hangup effect depending on the initial phase and the amount of input noise. Only the PFD is known as a hangup-free phase detector. We will learn more about the PFD in Chapter 6.

3.3 Practical Design Aspects

In this section, we will address further the difference between the type 1 PLL and the type 2 PLL with a physical insight. Based on that, we will discuss the practical design aspects to understand the transient response of the CP-PLL and also learn the basic structure of a digital loop filter in the digital-intensive PLL (DPLL).

3.3.1 Type 1 and Type 2 PLLs with Frequency-Step Input

Let us revisit the case of the transient response of the PLL with a frequency-step input. Figure 3.10(a) illustrates how a static phase error is generated in a type 1 PLL when there is a frequency change of $\Delta\omega$ at the input. Having the property of frequency-lock, the PLL must produce a DC voltage information of $\Delta\omega/K_v$ since the VCO performs one-to-one mapping between a voltage and a frequency as a voltage-to-frequency converter. In the type 1 PLL that does not have an integrator in the low-pass filter (LPF), there is no way to store the DC voltage information in the LPF. As a result, the only way for the type 1 PLL to provide the DC offset voltage at the input of the VCO is to induce the static phase offset that is equivalent to $\Delta\omega/K$ where K is a loop gain.

On the contrary, a type 2 PLL has the integrator in the LPF, and a desired DC voltage can be stored in the integrator. As discussed in Chapter 2, a control path of the type 2 PLL can be shown as two separate paths, that is, the integral path and the proportional-gain path. Since the integral path stores the DC voltage, the proportional-gain path does not have to generate a static phase error to provide the DC information to the VCO. Therefore, the type 2 PLL does not have the static phase

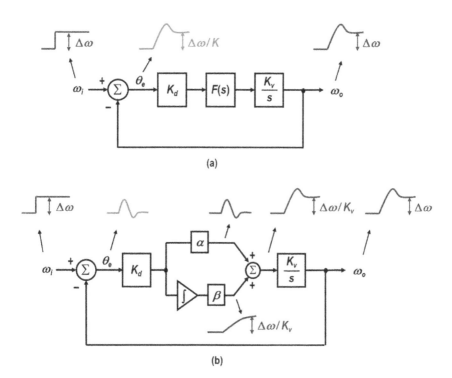

Figure 3.10 Transient performance with a frequency-step input: (a) type 1 and (b) type 2.

error at the PD output even with the frequency step as depicted in Fig. 3.10(b). From Fig. 3.10, we deduce that the proportional-gain path is analogous to a small-signal path to set the gain of an amplifier, while the integral path being analogous to a large-signal path to set the DC operating point of the amplifier.

3.3.2 State-Variable Model

In the previous section, we take it for granted that all the DC information would be stored in the integral path in the case of the type 2 PLL. You may wonder why there is no situation that, for example, 80% of the DC information is stored in the integral path, while the remaining DC is provided with a static phase error by the proportional-gain path. To answer that, we use a state-variable model. Unlike the transfer function model that only considers the relationship between an input and an output, the state-variable model can put additional constraints for the feedback system having integrators. To have the feedback system stable, the input of the integrator must be zero in the steady state. Otherwise, the integrator accumulates the non-zero input over time, and the output will be saturated. In the state-variable model, we define a state variable by the output of the integrator and put the constraint that the state variable should be constant in the steady state. The state-variable description of the PLL helps understand internal system parameter such as the control voltage of the VCO in the steady state, which cannot be described with the input/output (I/O) transfer function.

Figure 3.11 shows a state-variable model of a type 2 second-order CP-PLL with associated state variables. Being outputs of the integrator, a phase error $\theta_e(t)$ and a capacitor voltage $V_c(t)$ in the loop filter can be state variables of the system under the constraint that they must be constant in the steady state. Then, state

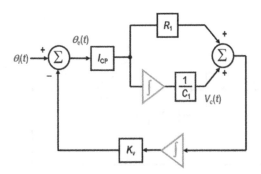

Figure 3.11 State-variable model of the PLL.

variable equations are given by

$$\begin{cases} \theta_e(t) = -K_v \int (V_c(t) + I_{CP}R_1\theta_e(t)) + \theta_i(t) \\ V_c(t) = \frac{I_{CP}}{C_1} \int \theta_e(t) \end{cases} \tag{3.29}$$

If we differentiate the left and right terms in each equation, we get

$$\begin{cases} \dfrac{d\theta_e(t)}{dt} = -K_v V_c(t) - K_v I_{CP}R_1\theta_e(t) + \dfrac{d\theta_i(t)}{dt} \\ \dfrac{dV_c(t)}{dt} = \dfrac{I_{CP}}{C_1}\theta_e(t) \end{cases} \tag{3.30}$$

When the system is in a steady state, the state variables must be constant.

$$\begin{cases} \dfrac{d\theta_{e,s}}{dt} = 0 = -K_v V_{c,s} - K_v I_{CP}R_1\theta_{e,s} + \dfrac{d\theta_i(t)}{dt} \\ \dfrac{dV_{c,s}}{dt} = 0 = \dfrac{I_{CP}}{C_1}\theta_{e,s} \end{cases} \tag{3.31}$$

where $\theta_{e,s}$ and $V_{c,s}$ are the steady-state state variables of $\theta_e(t)$ and $V_c(t)$. In (3.31), the bottom equations show that $\theta_{e,s}$ must be zero. With $\theta_{e,s} = 0$, $V_{c,s}$ is given by

$$V_{c,s} = \frac{1}{K_v}\frac{d\theta_i(t)}{dt} \tag{3.32}$$

The input phase due to a frequency jump is

$$\theta_i(t) = \Delta\omega t \tag{3.33}$$

Then, we get

$$V_{c,s} = \frac{\Delta\omega}{K_v} \tag{3.34}$$

We learned that the output frequency change must be the same as the input frequency change in the PLL system. Therefore, (3.34) verifies that the steady-state voltage stored in the capacitor fully represents an equivalent DC voltage required to tune the VCO frequency by the same amount of the input frequency change.

3.3.3 Two-Path Control in the CP-PLL

Based on what we discussed in the previous section, let us revisit the two-path control of the CP-PLL depicted in Fig. 3.12 that is redrawn from Fig. 2.20. In Chapter 2, we mainly focused on the loop dynamics and the role of the RC filter in stabilizing the type 2 PLL. For every reference clock period, the CP current across a resistor R_1 generates an instantaneous voltage whose pulse width is proportional to a phase error, forming a proportional-gain path. Depending on the polarity of the phase error, either a positive charge or a negative charge is accumulated in a loop capacitor C_1, forming an integral path for frequency

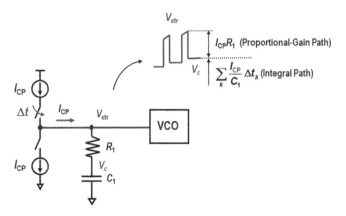

Figure 3.12 Two-path control in the second-order type 2 CP-PLL.

acquisition as proven in (3.34). Having a large C_1 makes the integral path less dominant than the proportional-gain path in the type 2 system since $V_c(t)$ does not change much with the large C_1. In that case, the loop is overdamped, and a loop bandwidth is determined mostly by the proportional-gain path.

Figure 3.13 shows the simulated transient settling behavior of an overdamped third-order type 2 CP-PLL with a damping ratio of 2. Both the VCO input voltage $V_{ctr}(t)$ and the capacitor voltage $V_c(t)$ are plotted. Apparently, no overshooting or rippling is observed, and we clearly see that the settling time of $V_c(t)$ is much longer than that of $V_{ctr}(t)$. As depicted in Fig. 3.13, three regions can be defined for the transient behavior of the overdamped type 2 CP-PLL; region 1 where none of $V_{ctr}(t)$ and $V_c(t)$ is settled, region 2 where only $V_{ctr}(t)$ is settled, and region 3 where both $V_{ctr}(t)$ and $V_c(t)$ are fully settled. The region 2 shows a nearly

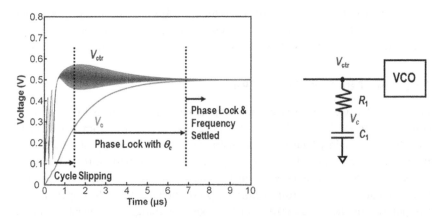

Figure 3.13 Transient settling behavior of an overdamped CP-PLL.

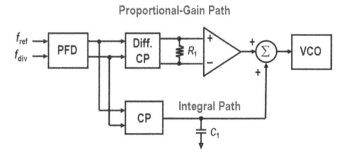

Figure 3.14 Two-path control with the dual charge pumps.

phase-locked state with the static phase offset induced by the proportional-gain path that compensates for the remaining DC voltage required to tune the VCO. Hence, the region 2 is considered an intermediate state between the type 1 loop and the type 2 loop. When both $V_{ctr}(t)$ and $V_c(t)$ are fully settled in the region 3, the phase-tracking performance of the PLL is governed by the proportional-gain path. Therefore, monitoring $V_{ctr}(t)$ only is not a good way to evaluate the settling or the static phase error performance of the type 2 CP-PLL especially for an over-damped loop. Figure 3.13 gives another insight that $V_c(t)$ represents the frequency acquisition of the PLL and also plays like a DC operating point of the PLL where the value of $(V_{ctr}(t) - V_c(t))$ performs small-signal phase tracking like an AC signal. After $V_c(t)$ is fully settled, we can properly evaluate small-signal parameters such as phase noise, reference spur, and static phase error.

Since the VCO control voltage $V_{ctr}(t)$ can be decomposed into $V_c(t)$ and a voltage across R_1, a two-path control with dual charge pumps can also be implemented as shown in Fig. 3.14. To avoid setting a common-mode voltage, a differential topology is employed for the charge pump and the LPF. The use of the dual charge pumps offers additional flexibility in designing loop parameters by controlling the ratio of two charge pump currents. By understanding the two-path configuration shown in Fig. 3.14, we are well positioned to understand the basic structure of a DPLL in the following section.

3.3.4 Two-Path Control in DPLL

Understanding the type 2 PLL with two separate paths is highly useful when we design a DPLL. If a filter transfer function is implemented as a whole in a digital loop filter (DLF), a large number of bits are needed for the DLF to achieve an accurate frequency control. However, such a high-resolution frequency control makes it difficult to perform phase-tracking with low latency. By having two paths that perform the phase tracking and frequency tracking separately, we can put

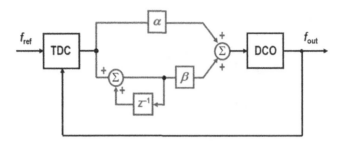

Figure 3.15 Simplified block diagram of the type 2 DPLL.

high number of bits for the integral path to achieve high-resolution frequency control, say > 14-bit resolution, while fast phase tracking being achieved with a small number of bits in the proportional-gain path.

Figure 3.15 shows the simplified block diagram of a type 2 DPLL. Phase error information is generated by a time-to-digital converter (TDC). In the DLF, an accumulator plays like the integration capacitor C_1 of the CP-PLL shown in Fig. 3.12. A digital coefficient β sets the gain of the integral path, which is analogous to $1/C_1$ in the CP-PLL. A digital coefficient of α sets the gain of the proportional-gain path, which is analogous to R_1 in the CP-PLL. We will discuss the DPLL in Chapter 10.

3.3.5 Slew Rate of CP-PLL

The settling time of an overdamped CP-PLL is significantly affected by the integral path of a loop filter, that is, the integration capacitor C_1. If we only consider the frequency acquisition outside the lock-in range of the PLL, the large-signal transient response of the CP-PLL is determined mainly by the charging (or discharging) time of C_1 as the PFD enables the charge pump to generate either up or down pulse depending on the frequency difference.[2] Therefore, the CP-PLL in the frequency acquisition mode can be modeled as a two-stage amplifier whose slew rate is determined by a CP current I_{CP} and C_1. Figure 3.16 shows that the slew rate of the CP-PLL can be defined like an op amplifier. The slew rate of the CP-PLL is based on a frequency-to-time ratio, while the slew rate of the op amplifier being defined by a voltage-to-time ratio. For the sake of simplicity, we assume that the CP current is fully on for the given frequency direction. Then, the maximum slew rate SR_{PLL} of the PLL can be defined by

$$SR_{PLL} = \frac{I_{CP}}{C_1} K_v \quad [Hz/s] \tag{3.35}$$

---

2 This property will be explained in Chapter 6.

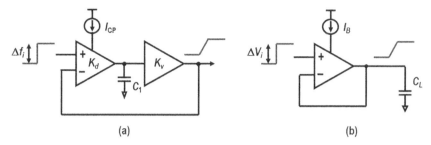

Figure 3.16 Slew-rate comparison: (a) CP-PLL and (b) op amplifier.

Now, let us check how the slew rate affects the settling time of a type 2 CP-PLL. With the first-order approximation, the frequency settling function $f(t)$ over time is given by

$$f(t) = f_o \left(1 - e^{-\frac{t}{\tau}} \right) \tag{3.36}$$

where f_o is a target frequency, and τ is the loop time constant of a PLL, that is, the inverse of a loop gain K. When loop settling is not slew-limited, the slew rate of the PLL with a division ratio of N is higher than the slope of $f(t)$ at $t = 0$, or

$$\frac{I_{\mathrm{CP}}}{C_1} K_v > \left. \frac{df}{dt} \right|_{t=0} = \frac{f_o}{\tau} = f_o \sqrt{\frac{I_{\mathrm{CP}}}{C_1} \frac{K_v}{N}} \tag{3.37}$$

Then,

$$\frac{I_{\mathrm{CP}}}{C_1} > \frac{f_o^2}{N K_v} \tag{3.38}$$

If (3.38) is not satisfied, then the slew rate should be considered in the settling time. Note that the inverse relation of (3.38) also gives the condition of a maximum slope for frequency ramping when the method of ramping a VCO control voltage is used for frequency acquisition aid.

The settling time including the slew rate can be analyzed as illustrated in Fig. 3.17. The frequency over time is described as

$$f(t) = \begin{cases} \frac{I_{\mathrm{CP}} K_v}{C_1} t, & t \leq t_1 \\ f_{t1} + (f_o - f_{t1}) \left(1 - e^{-\frac{t}{\tau}} \right), & t > t_1 \end{cases} \tag{3.39}$$

where f_{t1} is a boundary frequency between the linear region and the nonlinear region and given by

$$f_{t1} = \frac{I_{\mathrm{CP}} K_v}{C_1} t_1 \tag{3.40}$$

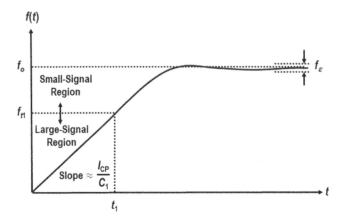

Figure 3.17 Transient settling behavior including the large-signal region.

In the lock-in range, the time for the PLL to be settled to f_o within the frequency error f_ε is given by

$$(f_o - f_{t1})e^{-\frac{t}{\tau}} < f_\varepsilon \tag{3.41}$$

That is,

$$t > \ln\left(\frac{f_o - f_{t1}}{f_\varepsilon}\right)\tau \tag{3.42}$$

Then, the total settling time t_{settle} of the PLL is given by

$$t_{\text{settle}} = t_1 + \ln\left(\frac{f_o - f_{t1}}{f_\varepsilon}\right)\tau \tag{3.43}$$

where t_1 is obtained from Eq. (3.39).

3.3.6 Effect of the PFD Turn-On Time

When a PFD operates as a frequency detector during cycle-slipping, reset delay can cause a gain leakage. The gain leakage problem becomes more significant as comparison frequency increases, and the efficiency of frequency detection is further reduced in high frequencies. Accordingly, the effect of the PFD reset delay needs to be considered as a parameter when nonlinear settling time is considered. Detailed analyses will be given when the PFD circuit is discussed in Chapter 6.

Example 3.2 *Settling time with VCO pre-tuning*
In the case of an overdamped loop, the settling time of a PLL can be significantly reduced if the initial frequency of a VCO is adjusted close to the desired output

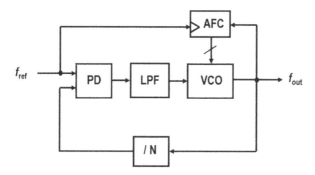

Figure 3.18 Settling time reduction with VCO pre-tuning.

frequency of the PLL. For that reason, an on-chip VCO is usually designed with a digitally tunable center frequency. To tune the center frequency automatically, an automatic frequency calibration (AFC) circuit can be designed as shown in Fig. 3.18. The use of the VCO pre-tuning is helpful to reduce the settling time even for an underdamped loop. Suppose that we have a VCO with 5-bit tunability. With the same loop parameters and conditions given in Example 3.1, we calculate the settling time for comparison. With the 5-bit pre-tuning, the step frequency is reduced to 312.5 kHz ($= 10\,\text{MHz}/2^5$). Then, the settling time is given by

$$t_s = \frac{1}{0.453 \times 3.536 \times 10^6} \times \ln\left(\frac{10^7 \times \frac{1}{32}}{100 \times \sqrt{1 - 0.453^2}}\right) = 5.096 \times 10^{-6} \ [\text{sec}]$$

The settling time is reduced from 7.3 to 5.1 μs with the 5-bit AFC, resulting in nearly 30% reduction without changing the loop dynamics of the PLL. Note that the logic calibration time of the AFC circuit needs to be added for the total settling time and could be substantial if clock frequency is not high.

References

1 F. M. Gardner, *Phaselock Techniques*, 3rd ed., Wiley, New York, 2005.
2 W. Egan, *Phase-Locked Basics*, 2nd ed., Wiley, New York, 2008.
3 V. F. Kroupa, *Phase Lock Loops and Frequency Synthesis*, Wiley, Hoboken, NJ, 2007.
4 K. Shu and E. Sanchez-Sinencio, *CMOS PLL Synthesizers: Analysis and Design*, Springer, New York, 2005.
5 D. Banerjee, *PLL Performance, Simulation, and Design*, Dean Banerjee, 2001. https://www.amazon.com/Performance-Simulation-Design-Dean-Banerjee/dp/

0970820704/ref=sr_1_1?crid=2TD5Q3U6QWI6L&keywords=Banerjee+pll&
qid=1699284032&s=books&sprefix=banerjee+p%2Cstripbooks%2C560&sr=1-1

6 H. Meyr and G. Ascheid, *Synchronization in Digital Communications, Vol. 1, Phase-, Frequency-Locked Loops, and Amplitude Control*, Wiley, New York, 1990.

7 W. Rhee, *Multi-Bit Delta-Sigma Modulation Technique for Fractional-N Frequency Synthesizers*, Ph.D. Thesis, University of Illinois, Urbana-Champaign, IL, Aug. 2000.

8 R. B. Staszewski and P. T. Balsara, *All-Digital Frequency Synthesizer in Deep-Submicron CMOS*, Wiley-Interscience, Hoboken, NJ, 2006.

Part II

System Perspectives

4

Frequency and Spectral Purity

Clock generation and frequency generation are basically the same, but each has different application aspects for integrated circuits and systems. In general, the term *clock generation* is used for synchronous digital systems. On the other hand, the term *frequency generation* contains a broader meaning but has often been reserved for wireless systems that deal with multiple channels. Therefore, spectral purity is important for frequency synthesis, while peak-to-peak jitter is mainly considered for clock generation. For either case, the noise performance of clock and frequency generation circuits is critical to determine the overall performance of communication systems. The phase uncertainty of a periodic signal can be characterized by random jitter (RJ) or deterministic jitter (DJ). The RJ in the time domain corresponds to integrated phase noise in the frequency domain, while the DJ in the time domain corresponding to sidebands in the frequency domain. It is important for PLL designers to gain a good knowledge of those relations, which will be learned in this chapter and the next chapter. We begin with the sideband and DJ since quantifying the phase noise and RJ can be understood from the analysis of the sideband and DJ when narrowband frequency modulation (FM) is assumed.

4.1 Spur Generation and Modulation

4.1.1 Spurious Signal (Spur)

The term *spurious signal,* namely *spur* refers to any undesirable non-harmonic sideband present at the output spectrum. While phase noise represents the randomness of frequency stability, the spur is generated by a deterministic behavior. Spur generation in a PLL comes mostly from the periodic modulation of the VCO that is in essence a voltage-to-frequency converter.

Phase-Locked Loops: System Perspectives and Circuit Design Aspects, First Edition.
Woogeun Rhee and Zhiping Yu.
© 2024 The Institute of Electrical and Electronics Engineers, Inc. Published 2024 by John Wiley & Sons, Inc.

4.1.1.1 Narrowband FM

An instantaneous frequency $f(t)$ at an output frequency f_o with sinusoidal frequency modulation is expressed as

$$f(t) = f_o + \Delta f_{pk} \cos(2\pi f_m t + \theta_c) \tag{4.1}$$

where Δf_{pk} is the peak frequency deviation, f_m is the modulation frequency, and θ_c is the constant phase offset. Then, the modulated phase $\theta(t)$ is

$$\theta(t) = \int 2\pi [f_o + \Delta f_{pk} \cos(2\pi f_m t + \theta_c)] dt = 2\pi f_o t + \Delta \theta_{pk} \sin(2\pi f_m t + \theta_c) \tag{4.2}$$

where $\Delta \theta_{pk}$ is the peak phase deviation and given by

$$\Delta \theta_{pk} = \frac{\Delta f_{pk}}{f_m} \tag{4.3}$$

The ratio of Δf_{pk} to f_m is also called a modulation index denoted by m, that is

$$m = \frac{\Delta f_{pk}}{f_m} = \Delta \theta_{pk} \tag{4.4}$$

Then, a frequency-modulated signal $v(t)$ with an amplitude A and the modulation index m can be expressed as

$$v(t) = A \cos \theta(t) = A \cos[2\pi f_o t + m \sin(2\pi f_m t + \theta_c)] \tag{4.5}$$

When m is small, typically < 0.1, we can classify an FM signal as a narrowband FM signal whose carrier power is well defined. We assume $\theta_c = 0$ for simplicity. Then, the FM signal with sinusoidal modulation is written as

$$
\begin{aligned}
v(t) &= A \cos[2\pi f_o t + m \sin(2\pi f_m t)] \\
&= A\{\cos 2\pi f_o t \cdot \cos(m \sin 2\pi f_m t) - \sin 2\pi f_o t \cdot \sin(m \sin 2\pi f_m t)\}
\end{aligned} \tag{4.6}
$$

For the small value of x, $\sin x \approx x$ and $\cos x \approx 1$. Then, with the narrowband FM assumed ($m \ll 1$), $v(t)$ can be approximated as

$$
\begin{aligned}
v(t) &\approx A[\cos 2\pi f_o t - \sin 2\pi f_o t \cdot m \sin 2\pi f_m t] \\
&= A \left\{ \cos 2\pi f_o t - \frac{m}{2} [\cos 2\pi (f_o - f_m) t - \cos 2\pi (f_o + f_m) t] \right\}
\end{aligned} \tag{4.7}
$$

Equation (4.7) shows that the spectrum of a narrowband FM signal is simply represented by a carrier and symmetric sidebands located at $f_o + f_m$ and $f_o - f_m$ with levels of $+A(m/2)$ and $-A(m/2)$, respectively. The spectrum of the narrowband FM signal is very similar to that of an amplitude modulation (AM) signal but has the phase reversal for the other sideband component as depicted in Fig. 4.1. In fact, test equipment such as the spectrum analyzer that measures power spectral density is not able to tell the difference between the narrowband FM signal and the AM signal.

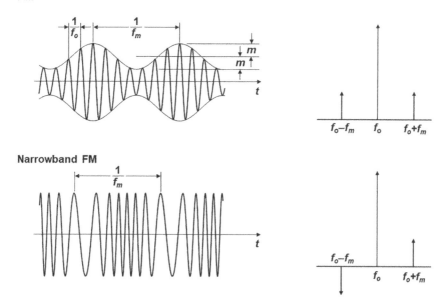

AM

Narrowband FM

Figure 4.1   Spectrum comparison of AM and narrowband FM signals.

4.1.1.2   Generation of a Single Sideband

If a single sideband (SSB) or asymmetric double sideband (DSB) is observed in the output spectrum of a PLL, it is likely that there is a direct coupling at the output of a VCO. As mentioned previously, the spectrum of a narrowband FM signal has a symmetric sideband like an AM signal except the phase reversal in the other sideband. Therefore, the SSB can be decomposed into the AM and narrowband FM signals in the frequency domain as shown in Fig. 4.2. After decomposition, the level of the sidebands is decreased by 6 dB for both AM and FM signals to keep the same signal power. Note that carrier power remains unchanged since neither the AM nor the FM alters the level of the carrier power.

Example 4.1   *Effect of the limiter on the SSB*

Let us consider the case that a 1-GHz signal having a SSB at 1.1 GHz is fed to a limiter. Suppose that the power levels of the carrier and the SSB before the limiter are 0 and −40 dBm, respectively, and that the limiter has a gain loss of 6 dB. What would be the output spectrum of the limiter like? What are the power levels of the carrier and the sidebands after the limiter?

To calculate the power level of the sidebands after the limiter, we first decompose the SSB signal into AM and FM signals as illustrated in Fig. 4.2. The level of the

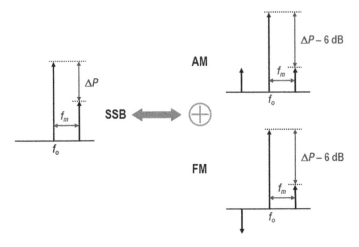

Figure 4.2 SSB decomposition into FM and AM.

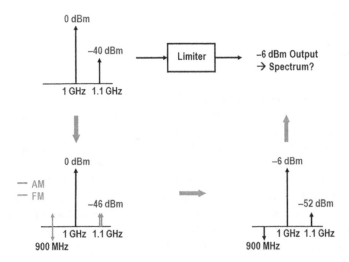

Figure 4.3 Effect of limiter on SSB signal.

DSB for each signal will be −46 dBm, which is 6-dB lower than the SSB. After the limiter, the AM component will be suppressed, and only the FM component will be shown. Since the limiter has a gain loss of 6 dB, the power level of the carrier and the DSB is reduced by 6 dB. As a result, a carrier power of −6 dBm and a DSB power of −52 dBm will be observed after the limiter if the relative strength between the carrier and spur is assumed to be unaltered. Note that the FM property of the input signal is not affected by the limiter except the gain loss. Figure 4.3 shows the comprehensive spectra before and after the limiter.

4.1.1.3 Estimation of Spur Level

For frequency synthesis and clock generation, spurs should be substantially lower than the carrier. It is not meaningful to calculate the spur level for a wideband FM signal whose carrier power is not well defined in the frequency domain. Therefore, it is natural to assume narrowband FM to consider the effect of spurs in the PLL design. When the narrowband FM is assumed, the calculation of the spur estimate is rather straightforward from (4.7) in which the FM signal is represented mainly by a carrier and a single dominant sideband. From (4.7), the spur level P_{spur} with the narrowband FM is obtained by the ratio of the carrier power $P_{carrier,dBm}$ to the dominant spur power $P_{spur,dBm}$ with the following equation:

$$P_{spur} = 10 \log \left(\frac{P_{spur,dBm}}{P_{carrier,dBm}} \right)^2 = 10 \log \left(\frac{A\left(\frac{m}{2}\right)}{A} \right)^2 = 10 \log \left(\frac{m}{2} \right)^2 \qquad (4.8)$$

The equation shows that the level of the dominant spur is simply given by the modulation index m. The spur level is measured in units of dBc, that is, the spur level is referenced to the carrier power. For instance, a spur level of -40 dBc means that the spur power in units of dBm is 40 dB lower than the carrier power in units of dBm.

Example 4.2   Spur calculation

Let us assume that the input of a 10-GHz VCO having a sensitivity K_v of 1 GHz/V is modulated by a 100-MHz sinusoidal input with a peak amplitude ΔV_{pk} of 1 mV as illustrated in Fig. 4.4. As discussed, we just need to get a modulation index to calculate the spur level, that is, the ratio of the peak deviation frequency Δf_{pk} to the modulation frequency f_m. The peak deviation frequency is obtained by

$$\Delta f_{pk} = \Delta V_{pk} \times K_v = 1 \, [\text{mV}] \times 1 \left[\frac{\text{GHz}}{\text{V}} \right] = 1 \, [\text{MHz}]$$

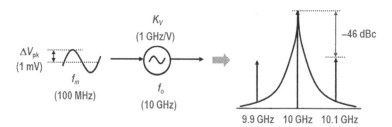

Figure 4.4   Spur generation by VCO modulation.

From (4.8), the spur level is given by

$$P_{\text{spur}} = 10 \log \left(\frac{\Delta f_{\text{pk}}}{2 f_m} \right)^2 = 20 \log \left(\frac{1 \times 10^6}{2 \times 100 \times 10^6} \right) = -46 \, [\text{dBc}]$$

Note that the spur level is not dependent on the carrier frequency f_o as it is fully determined by ΔV_{pk} and K_v. That is, if we use a 1-GHz VCO with the same K_v, we will get the same spur level since the modulation index has nothing to do with the carrier frequency. The example also implies that we can estimate the spur level by directly looking at the input voltage waveform of the VCO without obtaining the VCO spectrum.

4.1.1.4 Deterministic Jitter by Periodic Modulation

Unlike the RJ caused by the random variation of phase, the DJ is the systematic jitter that is bounded and predictable. In general, the DJ includes many kinds of non-random behaviors including data-dependent jitter and duty-cycle jitter. In this chapter, we only consider the DJ caused by periodic modulation. By decomposing the noise-like DJ in the time domain into discrete tones in the frequency domain, we can find a way of reducing the DJ that depends on the noise transfer function of a PLL. Indeed, the frequency-domain analysis is important to understand the effect of the noise coupling that consists of multiple frequency components.

It is good to recall from (4.4) that the modulation index m represents the peak phase deviation due to periodic modulation. Then, the DJ caused by a single sinusoidal jitter is given by

$$DJ = 2 \Delta \theta_{\text{pk}} \frac{T_{\text{CK}}}{2 \pi} = \frac{m}{\pi} T_{\text{CK}} \, [\text{sec}] \tag{4.9}$$

In clock generation, the DJ is often expressed as a portion of a clock period T_{CK}, and one clock period is also noted as a unit interval (UI). Then,

$$DJ = 2 \Delta \theta_{\text{pk}} \frac{1}{2 \pi} = \frac{m}{\pi} \, [\text{UI}] \tag{4.10}$$

Since the spur level is determined by the modulation index, above equations imply that the DJ in the time domain can be estimated by measuring the level of a dominant spur in the frequency domain as depicted in Fig. 4.5. Let us look at a practical example.

Example 4.3 *DJ budget and spur requirement*

Different from wireless communication systems, wireline systems put more emphasis on total jitter in the time domain than on spectral purity in the frequency domain. For instance, in most wireline systems, if the peak-to-peak jitter is less than 1% of a clock period, that is, <0.01 UI, then the DJ is considered to

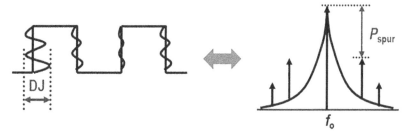

Figure 4.5 DJ in the time domain and spurs in the frequency domain.

have a negligible effect on the total jitter budget. Now let us find the required spur level for the DJ to be less than 0.01 UI. From (4.10),

$$m = \Delta\theta_{\text{pk}} = \pi \cdot \text{DJ} < \pi \times 0.01$$

The required spur level is obtained from (4.8)

$$P_{\text{spur}} = 10\log\left(\frac{m}{2}\right)^2 < 10\log\left(\frac{\pi \times 0.01}{2}\right)^2 = -36.08 \,[\text{dBc}]$$

Therefore, if P_{spur} is lower than -36 dBc regardless of offset frequencies, DJ contribution to the total jitter is less than 0.01 UI. If we add a 3-dB margin to consider the effect of nearby spurs, we safely set -40 dBc as a borderline to determine whether the DJ contribution to the total jitter budget is negligible or not. Note that this condition also holds for the presence of multiple spurs at any frequency. As far as we do not observe a spur beyond the borderline of -40 dBc in the spectrum of an output clock, the DJ contribution to the total jitter is not considered critical.

4.1.1.5 Effect of Frequency Division and Multiplication

To understand the modulation behavior of a PLL, it is important to know the effect of frequency division and multiplication on the spur performance. From (4.5), the phase $\theta(t)$ of the carrier that is modulated with a modulation frequency ω_m at an output frequency ω_o is given by

$$\theta(t) = \omega_o t + \left(\frac{\Delta f}{f_m}\right)\sin\omega_m t \tag{4.11}$$

where the initial phase θ_c is assumed to be zero. Instantaneous frequency at t is defined by the time rate of a phase change

$$\omega(t) = \frac{d\theta(t)}{dt} = \omega_o + \left(\frac{\Delta f}{f_m}\omega_m\right)\cos\omega_m t = \omega_o + 2\pi\Delta f\cos\omega_m t \tag{4.12}$$

After the frequency divider that divides the frequency by N, the output phase $\theta_N(t)$ is given by

$$\theta_N(t) = \int \frac{\omega(t)}{N}dt = \frac{\omega_o t}{N} + \left(\frac{2\pi\Delta f}{N\omega_m}\right)\sin\omega_m t = \frac{\omega_o t}{N} + \frac{m}{N}\sin\omega_m t \tag{4.13}$$

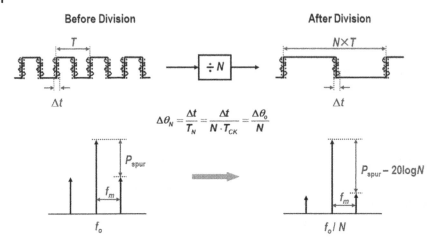

Figure 4.6 Effect of frequency division.

We see that the divider reduces the peak phase deviation $\Delta\theta_N$ by N but does not change the modulation frequency ω_m. Accordingly, the spur occurs at an offset frequency f_m with the level $P_{\text{spur},N}$ given by

$$P_{\text{spur},N} = 10\log\left(\frac{m}{2N}\right)^2 = P_{\text{spur}} - 20\log N \text{ [dBc]} \qquad (4.14)$$

Therefore, the spur level is reduced by 20logN at the output of the divide-by-N circuit, while the offset frequency of the spur from the carrier frequency remains the same, as shown in Fig. 4.6.

One thing we note here is that jitter performance should be normalized to a target clock frequency. If we express an amount of jitter with a time unit, that is seconds, the absolute value of the timing jitter after a divide-by-N circuit $\Delta\tau_N(t)$ is expressed by

$$\Delta\tau_N(t) = \frac{T_N}{2\pi}\theta_N(t) = \frac{NT_{CK}}{2\pi}\left(\frac{\omega_o t}{N} + \frac{m}{N}\sin\omega_m t\right) = \frac{T_{CK}}{2\pi}\theta(t) = \Delta\tau(t) \quad (4.15)$$

where T_{CK} is the clock period before frequency division, T_N is the clock period after frequency division, and $\Delta\tau(t)$ is the timing jitter before frequency division. From (4.15), it is seen that the absolute amount of the timing jitter is not changed after the frequency division, resulting in improved DJ in units of UI at the output of the frequency divider. Therefore, if the amount of jitter is to be expressed in seconds, the clock frequency must be considered together to properly evaluate the jitter performance of a PLL.

As to the spur performance with frequency multiplication, the spur behavior is opposite to the case of the frequency division as illustrated in Fig. 4.7. A peak

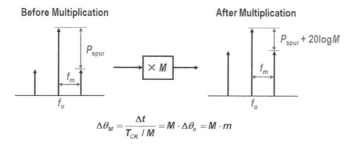

Figure 4.7 Effect of frequency multiplication.

phase deviation $\Delta\theta_M(t)$ after a frequency multiplier with a multiplication factor M is given by

$$\theta_M(t) = \frac{\Delta t}{T_{CK}/M} = M \cdot \Delta\theta_o = M \cdot m \tag{4.16}$$

Accordingly, the spur occurs at offset frequency f_m with the level $P_{spur,M}$ is

$$P_{spur,M} = 10\log\left(\frac{M \cdot m}{2}\right)^2 = P_{spur} + 20\log M \ [dBc] \tag{4.17}$$

Note that the PLL having a frequency divider in the feedback path works as a frequency multiplier. Therefore, any spur in a reference source will be amplified at the output of a VCO, and the increased spur amount depends on the value of the division ratio.

4.1.2 Reference Spur

When a PLL is designed with an integer-N divider, we may observe a spur located at an offset frequency that is the same as the reference frequency. This spur, named *reference spur*, is caused by the characteristic of a PD or by a static phase error associated with an unbalanced large-signal behavior of the PLL. The former case will be discussed in Chapter 6, and we will consider the latter case in this section. Before discussing nonideal effects that cause the unbalanced large-signal behavior, let us understand why the static phase error generates a periodic tone at the rate of reference frequency in the PLL. Interestingly, the fundamental reason is the frequency-locking property of the PLL. If there is an unbalanced voltage at the PD output due to the static phase error, an unwanted DC voltage information will be delivered to the low-pass filter (LPF). To prevent the output frequency of the VCO from being shifted because of the unwanted DC voltage that is originated from the static phase error, the PLL induces a static phase error with an opposite polarity and compensates for the unbalanced DC voltage. As a result, both positive and negative voltages are generated for each reference period, thus causing a

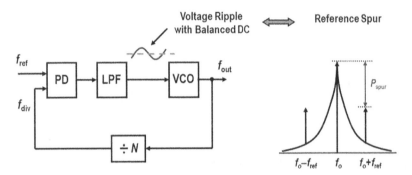

Figure 4.8 Reference spur generation in PLL.

voltage ripple as depicted in Fig. 4.8. Since this action occurs at every reference clock period, the VCO is modulated at the rate of the reference clock. Therefore, the reference spur is considered the systematic behavior of the PLL to compensate for the frequency offset caused by the unbalanced large-signal operation of the PLL. Unless the PLL is designed with an ideal PFD, we expect to observe the reference spur and its harmonics located at offset frequencies of the integer multiples of the reference frequency. If a PLL is properly designed with the conventional frequency divider, we should not see any spur between the carrier and the reference spur. If observed, it means that the PLL has a serious design fault or a coupling problem in hardware.

For the given nonideal effects of circuits, a way of reducing the spur is either to reduce a loop bandwidth or to use a high-order LPF. By having poles outside the loop bandwidth of the PLL, a high-order low-pass filtering can be achieved at the cost of degraded phase margin. Figure 4.9 shows the open-loop gain of a type 2 fourth-order PLL. Two out-of-band poles located at f_{p1} and f_{p2} provide additional spur suppression ΔP_{spur} given by

$$\Delta P_{\text{spur}} = 20 \log \left(\frac{f_{\text{ref}}}{f_{p1}} \right) + 20 \log \left(\frac{f_{\text{ref}}}{f_{p2}} \right) \tag{4.18}$$

Example 4.4 *Reference spur reduction by high-order poles*
Let us consider a 1-GHz fourth-order PLL having a loop bandwidth of 1 MHz and a reference frequency of 32 MHz and assume that the PLL exhibits a reference spur of -30 dBc at the output. If we have f_{p1} and f_{p2} at 4 and 8 MHz, respectively, we achieve additional spur reduction given by

$$\Delta P_{\text{spur}} = 20 \log \left(\frac{32\,\text{MHz}}{4\,\text{MHz}} \right) + 20 \log \left(\frac{32\,\text{MHz}}{8\,\text{MHz}} \right) = 18\,\text{dB} + 12\,\text{dB} = 30\,\text{dB}$$

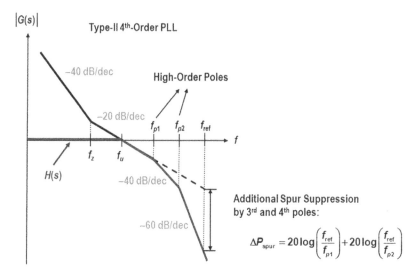

Figure 4.9 Reference spur reduction with the high-order pole.

As a result, the reference spur level is reduced to $-60\,\text{dBc}$. It is also good to check how much phase margin is degraded due to two out-of-band poles. Phase margin reduction $\Delta\phi_M$ is obtained by

$$\Delta\phi_M = \tan^{-1}\left(\frac{1\,\text{MHz}}{4\,\text{MHz}}\right) + \tan^{-1}\left(\frac{1\,\text{MHz}}{8\,\text{MHz}}\right) = 14.04^\circ + 7.13^\circ = 21.17^\circ$$

Accordingly, the spur reduction of 30 dB is obtained at the cost of the phase margin reduction of about 22° by having two out-of-band poles.

Now, we discuss what causes the reference spur in the design of the PLL. There could be two sources for the cause of the static phase error; one is a leakage current in the LPF, and the other is a PD mismatch. For the CP-PLL that is commonly adopted for frequency synthesis applications, a PFD and a charge pump need to be separately considered for the PD mismatch. We will examine those nonideal effects with quantitative analyses in the following sections.

4.1.2.1 Leakage Current

When an on-chip LPF is designed, a leakage current needs to be carefully considered especially when a PLL is implemented with advanced CMOS technology. It is because the MOSFET capacitor that is typically employed in the integral path of the LPF could incur a gate-leakage current. Suppose that there is a leakage current I_{leak} from C_1 as shown in Fig. 4.10. The leakage current lowers the capacitor voltage V_c, which could pull down the VCO frequency. To maintain frequency-lock, the reduced DC component needs to be compensated for by the

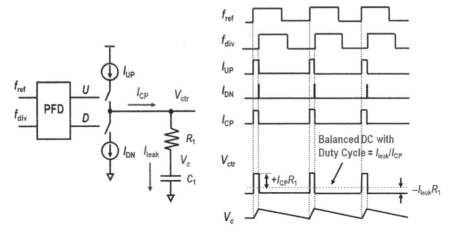

Figure 4.10 Phase offset due to leakage current.

PLL itself. Accordingly, the PLL induces a static phase error to provide the up current that nullifies the negative change generated by I_{leak}. As illustrated in Fig. 4.10, the static phase error θ_e due to the leakage current I_{leak} is given by

$$\theta_e = 2\pi \frac{I_{\text{leak}}}{I_{\text{CP}}} \tag{4.19}$$

We assume a second-order CP-PLL for the sake of simplicity. Then, the waveform of a control voltage V_{ctr} looks like a rectangular pulse with an amplitude of $I_{\text{CP}}R_1$ and a duty ratio of $(I_{\text{leak}}/I_{\text{CP}})$ as illustrated in Fig. 4.10. The magnitude of a fundamental tone can be found from the first coefficient a_1 of Fourier series

$$a_1 = \frac{2I_{\text{CP}}R_1}{\pi} \sin\left(\pi \frac{I_{\text{leak}}}{I_{\text{CP}}}\right) \approx \frac{2I_{\text{CP}}R_1}{\pi} \cdot \pi \frac{I_{\text{leak}}}{I_{\text{CP}}} = 2I_{\text{leak}}R_1 \tag{4.20}$$

where $I_{\text{leak}} \ll I_{\text{CP}}$ is assumed. From (4.20), it is shown that a factor of 2 should be included to calculate the effective amplitude of a fundamental tone for V_{ctr}. Then, the peak frequency deviation Δf caused by θ_e is expressed by

$$\Delta f = \theta_e K_d K_v \approx 2\left(2\pi \frac{I_{\text{leak}}}{I_{\text{CP}}}\right)\left(\frac{I_{\text{CP}}R_1}{2\pi}\right)K_v = 2I_{\text{leak}}R_1 K_v \tag{4.21}$$

Using (4.8) and (4.21), the amount of a reference spur for a second-order CP-PLL can be obtained by

$$P_{\text{spur}} = 20\log\left(\frac{\Delta f_{\text{pk}}}{2f_m}\right) = 20\log\left(\frac{I_{\text{leak}}R_1 K_v}{f_{\text{ref}}}\right) \tag{4.22}$$

where f_{ref} is the reference frequency. When an overdamped loop is assumed, the loop bandwidth f_{BW} with a division ratio N is approximated by

$$f_{\text{BW}} \approx \frac{I_{\text{CP}} R_1 K_v}{2\pi N} \tag{4.23}$$

Then, the amount of the reference spur due to the leakage current for the given ratio of $(f_{\text{BW}}/f_{\text{ref}})$ can be written as

$$P_{\text{spur}} \approx 20 \log \left(2\pi N \frac{f_{\text{BW}}}{f_{\text{ref}}} \frac{I_{\text{leak}}}{I_{\text{CP}}} \right) \tag{4.24}$$

If the reference frequency and the output frequency are fixed by a system, possible ways of reducing the reference spur caused by the leakage current is to decrease the loop bandwidth. Note that the reduction of the loop bandwidth needs to be done by changing either R_1 or K_v. If the charge pump current is reduced, it will increase the static phase error, having no effect on the reference spur as implied in (4.21). Also, note that the spur level increases with a high division ratio or a low reference frequency even if the leakage current remains unchanged as expressed in (4.23).

When a shunt capacitor C_2 is added to form a third-order CP-PLL, the spur level will be further reduced by the third pole of the PLL, and the amount of additional spur reduction is given by (4.18). Applying an approximated pole frequency from (2.46), the reference spur level of the third-order CP-PLL is obtained by

$$P_{\text{spur}} = 20 \log \left(\frac{I_{\text{leak}} R_1 K_v}{f_{\text{ref}}} \right) - 20 \log(2\pi f_{\text{ref}} R_1 C_2) = 20 \log \left(\frac{I_{\text{leak}} K_v}{2\pi C_2 f_{\text{ref}}^2} \right) \tag{4.25}$$

Above results imply that increasing f_{ref} for the given output frequency is the most effective way of reducing the spur, which led to the development of a so-called fractional-N PLL.

4.1.2.2 Charge Pump Mismatch

As the charge pump converts timing information into an analog voltage in the CP-PLL, the current mismatch in the charge pump can significantly contribute to the generation of a reference spur. In this section, we discuss the effect of the current mismatch in the case of a single-ended charge pump. Having up/down current sources and up/down switches implemented with PMOS and NMOS transistors, the single-ended charge pump could have the current mismatch as well as the timing mismatch when both the up and down switches are turned on with the phase-lock condition.

Figure 4.11 illustrates how a static phase offset occurs when there is a mismatch between the up and down currents denoted by I_{UP} and I_{DN}, respectively.

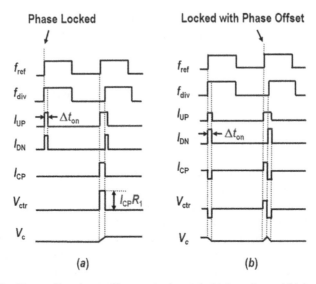

Figure 4.11 Phase offset due to CP current mismatch: (a) $I_{UP} = I_{DN}$ and (b) $I_{UP} < I_{DN}$.

Unlike the case of the leakage current, it is important to consider the turn-on time of the PFD that enables both the up and down switches even when there is no static phase error, which is to mitigate the dead-zone problem of the PFD.[1]

Let the turn-on time of the PFD, the reference clock period, and the current mismatch of the charge pump denoted by Δt_{on}, T_{ref}, and Δi, respectively. As illustrated in Fig. 4.11, the magnitude of the phase error due to the current mismatch is proportional to the ratio of the current mismatch to the charge pump current I_{CP} and given by

$$|\theta_e| = 2\pi \frac{\Delta t_{on}}{T_{ref}} \left(\frac{I_{CP} + \Delta i}{I_{CP}} - 1 \right) = 2\pi \frac{\Delta t_{on}}{T_{ref}} \frac{\Delta i}{I_{CP}} \tag{4.26}$$

Assuming $\Delta i \ll I_{CP}$ and $\Delta t_{on} \ll T_{ref}$, the spur level can be derived in a similar way as done for the leakage current, that is,

$$P_{spur} = 20 \log \left(\frac{2\pi \frac{I_{CP}R_1}{2\pi} |\theta_e| \frac{\Delta t_{on}}{T_{ref}} K_v}{2 f_{ref}} \right) = 20 \log \left(\frac{\pi \Delta i R_1 K_v}{f_{ref}} \left(\frac{\Delta t_{on}}{T_{ref}} \right)^2 \right) \tag{4.27}$$

where a factor of $2\pi(\Delta t_{on}/T_{ref})$ counts for the effective amplitude of a fundamental tone. Compared with the case of the leakage current, the ratio $(\Delta t_{on}/T_{ref})$ plays

---

1 The PFD dead-zone will be discussed in Chapter 6.

an important role for the spur level, which can be implicitly understood from Fig. 4.11. For an overdamped loop, using (4.23), the spur level is given by

$$P_{\text{spur}} \approx 20 \log \left(2\pi^2 N \frac{f_{\text{BW}}}{f_{\text{ref}}} \left(\frac{\Delta t_{\text{on}}}{T_{\text{ref}}} \right)^2 \frac{\Delta i}{I_{\text{CP}}} \right) \tag{4.28}$$

Equation (4.28) shows how important it is to design a charge pump not only with the minimum current mismatch but also with the minimum turn-on time of the PFD. The minimum turn-on time also reduces the noise contribution of the charge pump current, which will be discussed later. Since the minimum turn-on time in the PFD depends on the loading capacitance and the switching time of the charge pump, it is important to design the PFD and the charge pump together as a single phase detector.

When a shunt capacitor is added for a third-order CP-PLL, the analysis of the reference spur can be derived directly from the waveform of a control voltage. Figure 4.12 shows the transient waveforms of the charge pump current I_{CP} and the control voltage V_{ctr} with a shunt capacitor C_2 included in the loop filter. A control voltage V_{ctr} approximates to a sawtooth waveform with a peak-to-peak amplitude of $(\Delta i \Delta t_{\text{on}}/C_2)$ when a phase offset in time t_ε is much smaller than the reference period T_{ref}, equivalently, $\Delta i \ll I_{\text{CP}}$. By considering the first coefficient of the sawtooth function in the Fourier transform, the peak voltage ΔV_{pk} of the fundamental tone is obtained as

$$\Delta V_{\text{pk}} = \frac{\Delta i}{C_2 f_{\text{ref}}} \left(\frac{\Delta t_{\text{on}}}{T_{\text{ref}}} \right)^2 \tag{4.29}$$

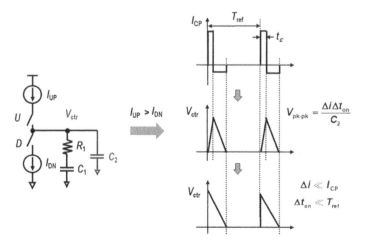

Figure 4.12 Transient waveforms of I_{CP} and V_{CTR} with shunt capacitor.

Then, the spur level is given by

$$P_{\text{spur}} = 20\log\left(\frac{\Delta V_{\text{pk}}K_v}{2f_{\text{ref}}}\right) = 20\log\left(\frac{\Delta i K_v}{2C_2 f_{\text{ref}}^2}\left(\frac{\Delta t_{\text{on}}}{T_{\text{ref}}}\right)^2\right)$$

$$= 20\log\left(\frac{\Delta i \Delta t_{\text{on}}^2 K_v}{2C_2}\right) \tag{4.30}$$

We can also calculate the spur level directly from (4.27) by considering the third pole at $1/R_1 C_2$, that is,

$$P_{\text{spur}} = 20\log\left(\frac{\pi \Delta i R_1 K_v}{f_{\text{ref}}}\left(\frac{\Delta t_{\text{on}}}{T_{\text{ref}}}\right)^2\right) - 20\log(2\pi f_{\text{ref}} R_1 C_2)$$

$$= 20\log\left(\frac{\Delta i \Delta t_{\text{on}}^2 K_v}{2C_2}\right) \tag{4.31}$$

which gives the same result as (4.30). Note that the spur analysis based on the waveform with C_2 can also be applied to the case of the leakage current with the same result as (4.25).

Above results give an idea that a phase offset can be digitally controlled if the charge pump is equipped with a current digital-to-analog converter (DAC). As shown in Fig. 4.13, the fine control of the phase offset can be achieved by controlling the mismatch between the up and the down currents. In frequency synthesis, this technique could give additional flexibility to reduce the reference spur with post-trimming or automatic calibration.

4.1.2.3 PFD Mismatch
Timing mismatch is inherent in the design of a PFD with a single-ended charge pump since the up and the down outputs have to drive PMOS and NMOS switches

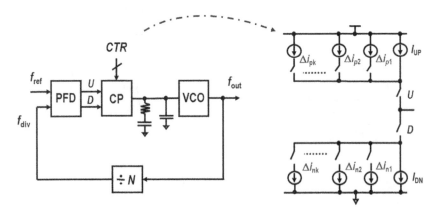

Figure 4.13 Phase offset control by the mismatch of the charge pump currents.

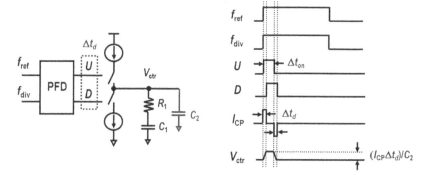

Figure 4.14 Phase offset by the mismatch of the PFD delay.

off the charge pump. Compared with the current mismatch effect of the charge pump in (4.28), the timing mismatch effect of the PFD is considered negligible as the small amount of a gate delay mismatch is further mitigated by the ratio of the turn-on time of the PFD to the reference period.

Let us consider a third-order CP-PLL. With a shunt capacitor C_2, the waveform of the control voltage becomes a trapezoidal waveform with a peak-to-peak amplitude of $(I_{CP}\Delta t_d/C_2)$ as shown in Fig. 4.14. The peak voltage ΔV_{pk} of the fundamental tone can be calculated as

$$\Delta V_{pk} = \frac{2I_{CP}}{C_2 f_{ref}}\left(\frac{\Delta t_d}{T_{ref}}\right)\left(\frac{\Delta t_{on}}{T_{ref}}\right) \tag{4.32}$$

Then, the spur level is given by

$$P_{spur} = 20\log\left(\frac{\Delta V_{pk}K_v}{2f_{ref}}\right) = 20\log\left(\frac{I_{CP}\Delta t_d\Delta t_{on}K_v}{C_2}\right) \tag{4.33}$$

Now, let us calculate back to the spur level of the second-order CP-PLL by subtracting the effect of a third pole. That is,

$$\begin{aligned}
P_{spur} &= 20\log\left(\frac{I_{CP}\Delta t_d\Delta t_{on}K_v}{C_2}\right) + 20\log(2\pi f_{ref}R_1C_2) \\
&= 20\log\left(\frac{2\pi I_{CP}R_1K_v}{f_{ref}}\cdot\frac{\Delta t_d}{T_{ref}}\cdot\frac{\Delta t_{on}}{T_{ref}}\right)
\end{aligned} \tag{4.34}$$

which implies that the magnitude of a static phase error due to Δt_d is given by

$$|\theta_e| = 2\pi\frac{\Delta t_d}{T_{ref}}\frac{\Delta t_{on}}{T_{ref}} \tag{4.35}$$

For an overdamped second-order loop, we have

$$P_{spur} \approx 20\log\left(4\pi^2 N\frac{f_{BW}}{f_{ref}}\frac{\Delta t_d\Delta t_{on}}{T_{ref}^2}\right) \tag{4.36}$$

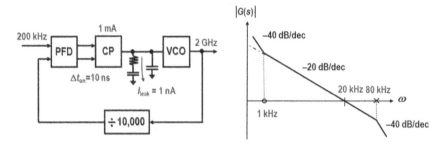

Figure 4.15 Reference spur generation in the third-order CP-PLL.

In practice, the PFD delay mismatch is much less than the PFD turn-on time, that is $\Delta t_d \ll \Delta t_{on}$. Therefore, the timing mismatch effect of the PFD is not as critical as the mismatch effect of the charge pump or the leakage current.

Example 4.5 *Reference spur calculation*
Let us consider a type 2 third-order CP-PLL with the following parameters: $f_{ref} = 200\,\text{kHz}, f_{out} = 2\,\text{GHz}, N = 10,000, I_{CP} = 1\,\text{mA}, \Delta t_{on} = 10\,\text{ns}, f_{BW} = 20\,\text{kHz},$ $f_z = 1\,\text{kHz},$ and $f_{pl} = 80\,\text{kHz}$. The block diagram of the PLL and the Bode plot of the open-loop gain are shown in Fig. 4.15. Estimate the reference spur level by considering: (a) a leakage current of 1 nA; (b) a charge pump current mismatch of 10%; and (c) a PFD timing mismatch of 20 ps.

(a) Leakage current

From (4.24), we get

$$P_{spur} \approx 20\log\left(2\pi \times 10^4 \times \frac{20\,\text{kHz}}{200\,\text{kHz}} \times \frac{1\,\text{nA}}{1\,\text{mA}}\right) = -44\,[\text{dBc}]$$

Including the third pole, we get

$$P_{spur} \approx -44 - 20\log\left(\frac{200\,\text{kHz}}{80\,\text{kHz}}\right) = -52\,[\text{dBc}]$$

(b) CP current mismatch

From (4.28), we get

$$P_{spur} \approx 20\log\left(2\pi^2 \times 10^4 \times \frac{20\,\text{kHz}}{200\,\text{kHz}} \times \left(\frac{10\,\text{ns}}{5\,\mu\text{s}}\right)^2 \times 0.1\right) = -42\,[\text{dBc}]$$

Including the third pole, we get

$$P_{spur} \approx -42 - 20\log\left(\frac{200\,\text{kHz}}{80\,\text{kHz}}\right) = -50\,[\text{dBc}]$$

(c) PFD timing mismatch

From (4.36), we get

$$P_{\text{spur}} \approx 20 \log \left(4\pi^2 \times 10^4 \times \frac{20 \text{ kHz}}{200 \text{ kHz}} \times \frac{20 \text{ ps}}{5 \text{ μs}} \times \frac{10 \text{ ns}}{5 \text{ μs}} \right) = -70 \, [\text{dBc}]$$

Including the third pole, we get

$$P_{\text{spur}} \approx -70 - 20 \log \left(\frac{200 \text{ kHz}}{80 \text{ kHz}} \right) = -78 \, [\text{dBc}]$$

Another possible cause of spur generation is noise coupling. Depending on the location of coupling within a PLL, different spur behaviors occur, and the spur level may not depend on the loop bandwidth. When the coupling occurs at the input of a VCO, spurs are located symmetrically around a carrier, while a single-sideband spur or asymmetric spurs are likely to appear when the VCO output is directly coupled by noise. Spur generation due to coupling is more critical in the design of a fractional-N PLL, which will be discussed in Chapter 9.

4.2 Phase Noise and Random Jitter

The period of an oscillator output cannot be constant all the time in real circuits. Short-term frequency variation is considered phase fluctuation, and random phase fluctuation is defined as phase noise. The phase noise representing the randomness of short-term frequency instability is one of the most important system parameters in PLL-based clock generation and frequency synthesis.

4.2.1 Phase Noise Generation and Measurement

4.2.1.1 Spectral Analysis of Real Signal

By performing the Fourier transform of the autocorrelation of a random signal, we get a power spectral density (PSD) in the frequency domain. The Fourier transform, however, is based on complex computation and not suitable to obtain the PSD information of a real-time signal. A practical approach is to use a band-pass filter (BPF) with a very narrow bandwidth. Suppose that $y(t)$ is an output of an ideal BPF with an input signal $x(t)$ and a transfer function $H(f)$ as illustrated in Fig. 4.16. Then, an output transform $Y(f)$ is given by

$$Y(f) = \begin{cases} X(f), & f_L < |f| < f_H \\ 0, & \text{otherwise} \end{cases} \tag{4.37}$$

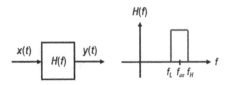

Figure 4.16 Spectral information of a real-time signal with a narrow BPF.

where $X(f)$ is the PSD of the input signal $x(t)$, and f_L and f_H are the low and high cut-off frequencies of an ideal BPF, respectively. The output signal $y(t)$ is obtained by the inverse transform of $Y(f)$, that is,

$$y(t) = \int_{f_L}^{f_H} X(f)e^{j2\pi ft}df + \int_{-f_H}^{-f_L} X(f)e^{j2\pi ft}df \tag{4.38}$$

If f_H is very close to f_L, the integrand is approximately constant over the integration range. Then, $y(t)$ can be approximated as

$$y(t) \approx (f_H - f_L)\left[X(f_{av})e^{j2\pi f_{av}t} + X(-f_{av})e^{-j2\pi f_{av}t}\right] \tag{4.39}$$

where f_{av} is the center frequency of the BPF. Since $X(-f_{av}) = X^*(f_{av})$ for a real signal $x(t)$, we have

$$y(t) \approx (f_H - f_L)|X(f_{av})|\cos[2\pi f_{av}t + \angle X(f_{av})] \tag{4.40}$$

Hence, the magnitude of $y(t)$ is proportional to the magnitude of $X(f_{av})$ with the phase shifted by the phase of $X(f_{av})$ for the given t, implying that the PSD of a real signal can be obtained by directly measuring an output power of $y(t)$ from the narrowband BPF. Therefore, by sweeping the center frequency of the BPF, the PSD values over the entire frequency band could be measured.

4.2.1.2 Phase Noise Analysis

Based on the spectral analysis discussed in the previous section, noise power $P_n(f_m, B_n)$ at a certain offset frequency f_m with a noise bandwidth B_n is to be measured at the output of a narrowband BPF centered at f_m as illustrated in Fig. 4.17. In fact, if we consider the narrowband noise around f_m as a signal whose power is equivalent to noise power at f_m with an infinitesimal bandwidth, phase noise calculation at f_m is somewhat similar to what we did for the spur calculation at f_m with the narrowband FM assumption. Then, by sweeping offset frequencies and regarding the phase noise as the continuum of sidebands, we can obtain the entire phase noise values.

Let us consider an oscillator output $v(t)$ given by

$$v(t) = A(t)\cos[2\pi f_0 t + \theta(t)] \tag{4.41}$$

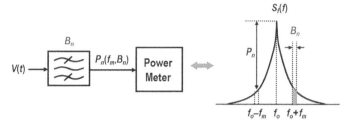

Figure 4.17 Measurement of phase noise.

where $A(t)$ describes the amplitude variation as a function of time, and $\theta(t)$ is the phase variation. For the sake of simplicity, let $A(t)$ be constant by assuming that the amplitude of the oscillator output is stable. Similar to (4.5) and (4.6), $v(t)$ with the constant amplitude A is expressed as

$$v(t) = A\cos(2\pi f_0 t + m\sin 2\pi f_m t) \tag{4.42}$$

and

$$v(t) \approx A\left\{\cos 2\pi f_0 t - \frac{m}{2}[\cos 2\pi(f_0 + f_m)t - \cos 2\pi(f_0 - f_m)t]\right\} \tag{4.43}$$

Then, noise power A_n^2 at $f_0 \pm f_m$ is given by

$$A_n^2 = \left(\frac{m}{2}\right)^2 A^2 \tag{4.44}$$

Phase noise in a 1-Hz bandwidth has the noise power-to-power ratio $L(f_m)$ given by

$$L(f_m) = \frac{A_n^2}{A^2} = \frac{m^2}{4} = \frac{\Delta\theta_{pk}^2}{4} = \frac{\Delta\theta_{rms}^2}{2} \ [\text{rad}^2/\text{Hz}] \tag{4.45}$$

The total noise is the double-sideband noise denoted by $S_\theta(f_m)$ and given by

$$S_\theta(f_m) = 2L(f_m) = \Delta\theta_{rms}^2 \ [\text{rad}^2/\text{Hz}] \tag{4.46}$$

$S_\theta(f_m)$ can be expressed in units of dBr/Hz, that is,

$$S_\theta(f_m) = 10\log(\Delta\theta_{rms})^2 \ [\text{dBr/Hz}] \tag{4.47}$$

For the measurement of the phase noise, a single-sideband noise power ratio in units of dBc/Hz is normally used. That is,

$$L(f_m) = 10\log\left(\frac{m}{2}\right)^2 = 10\log\left(\frac{\Delta\theta_{pk}}{2}\right)^2 = 10\log\left(\frac{\Delta\theta_{rms}}{\sqrt{2}}\right)^2 \ [\text{dBc/Hz}] \tag{4.48}$$

It is interesting to know that the modulation index contains the meaning of the signal-to-noise ratio (SNR) as implied in (4.45). It is because m or $\Delta\theta_{pk}$ is normalized to 2π radian, that is, the ratio of the peak jitter to the clock period, which can be regarded as the time-domain SNR.

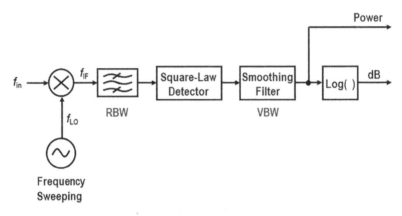

Figure 4.18 Simplified block diagram of a spectrum analyzer.

4.2.1.3 Spectrum Analyzer

A spectrum analyzer is the testing equipment that displays the PSD of a real-time signal by measuring the noise power of an input signal over a narrow bandwidth as discussed in Section 4.2.1.1. Figure 4.18 shows the simplified block diagram of the spectrum analyzer based on a swept-tuned receiver architecture that avoids multiple BPFs. When an input signal is received, it is down converted to an intermediate frequency (IF) signal by a local oscillator (LO) and a mixer. By sweeping the LO frequency, noise power at different offset frequencies is measured. The bandwidth of the BPF sets an effective noise bandwidth. Once all the noise powers are collected with frequency sweeping, the PSD over interesting frequencies is displayed.

In the spectrum analyzer, there are two knobs that control two kinds of bandwidths; one is a resolution bandwidth (RBW), and the other is a video bandwidth (VBW). The RBW sets the noise bandwidth, while the VBW is used for smoothing the PSD output. When the phase noise is measured with the spectrum analyzer, we assume that the AM noise or AM-to-PM conversion is negligible. For an accurate measurement of the phase noise, a phase noise analyzer should be used instead of the spectrum analyzer. The phase noise analyzer measures the phase noise based on delay-line-based FM demodulation, so that the effect of the AM-to-PM conversion is minimized for the phase noise measurement.

Example 4.6 *Phase noise calculation from the spectrum analyzer*
In this example, we learn how we calculate the phase noise at a certain offset frequency from the PSD plot of a spectrum analyzer. In Fig. 4.19, a 1-GHz input signal with a carrier power of -10 dBm is shown. In the spectrum analyzer, both X-axis and Y-axis have 10 grids shown in the display. "SPAN 100 MHz" on the bottom

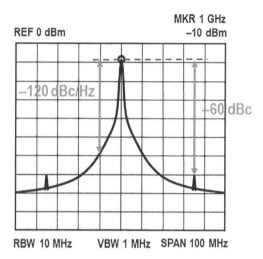

Figure 4.19 Phase noise calculation from a spectrum analyzer plot.

right means a frequency span of 100 MHz over 10 grids in the X-axis. Therefore, each grid in the X-axis represents a frequency step of 10 MHz. Figure 4.19 shows a center frequency of 1 GHz, having the frequency range from 950 to 1050 MHz with the frequency span of 100 MHz. In the Y-axis, each grid represents a power level of 10 dB. "REF 0 dBm" means that the first grid in the Y-axis indicates a power of 0 dBm. Since the carrier power is located on the second grid in Fig. 4.19, the carrier power should be -10 dBm. "VBW 1 MHz" means that a video bandwidth of 1 MHz is used to reduce high-frequency noise in the display. "RBW 10 MHz" shows that the PSD shown in the spectrum analyzer is measured with a noise bandwidth of 10 MHz.

Now let us calculate the phase noise at 10-MHz offset as an example. First, we measure the noise power at a 10-MHz offset. In the plot, the noise power at 10-MHz offset frequency from the carrier is -60 dBm. Since the carrier power is -10 dBm, we have a noise power level of -50 dBc at a 10-MHz offset. We normalize the noise power in units of dBc/Hz by considering the noise bandwidth of 10 MHz. The noise power of -50 dBc is measured with a noise bandwidth of 10 MHz, which is set by the RBW. To have a normalized phase noise in units of dBc/Hz, we divide the noise power by the noise bandwidth, or subtract 10log(RBW). That is,

$$\text{Phase noise at 10 MHz offset} = -60\,\text{dBm} - (-10\,\text{dBm}) - 10\log(10\text{MHz})$$

$$= -50\,\text{dBc} - 10\log(10\text{MHz}) = -120\,\text{dBc/Hz}$$

What happen if we change the value of RBW from 10 MHz to 1 MHz? Then, we will see that the overall phase noise floor will decrease by 10 dB. That is, we get the noise power of -70 dBm instead of -60 dBm at a 10-MHz offset frequency. Equivalently, we get the noise power of -60 dBc instead of -50 dBc. Does it mean that phase noise performance is improved? If the phase noise is normalized with a 1-Hz bandwidth, we will get the same phase noise value in units of dBc/Hz regardless of the value of RBW. In this case, we will get -60 dBc $- 10\log10^6 = -120$ dBc/Hz, which is the same as the measured value with the RBW of 10 MHz.

In the plot, we also see a spur at 35-MHz offset frequency. To measure the spur level, we directly calculate the difference between the carrier power and spur power, which is -60 dBc in this case. Note that the spur level does not depend on the RBW setting since the spur is not random noise but deterministic noise.

4.2.1.4 Effect of Frequency Division and Multiplication

As the phase noise calculation is based on the narrowband FM, the effect of frequency division and multiplication on the phase noise is similar to what we had on the spur performance. Let us investigate the effect of frequency multiplication this time since the case of the frequency division was discussed previously. From (4.12), the output phase $\theta_M(t)$ with a frequency multiplication factor M is given by

$$\theta_M(t) = \int M\omega(t)dt = M\omega_0 t + \left(\frac{2\pi M\Delta f}{\omega_m}\right) sin\omega_m t = M\omega_0 t + Mm\sin\omega_m t$$

(4.49)

We see that the multiplier increases the carrier frequency by M but does not change the modulation frequency ω_m. Then, the phase noise after frequency multiplication $L_M(f_m)$ is given by

$$L_M(f_m) = 10\log\left(\frac{Mm}{2}\right)^2 = L(f_m) + 20\log M \text{ [dBc/Hz]}$$

(4.50)

As depicted in Fig. 4.20, the effective peak deviation of the phase or modulation index is increased by M as the period of an output clock is M times shorter than that of an input clock, which results in M times as much phase noise.

Even though the phase noise degradation due to frequency multiplication can easily be understood for the case of open-loop frequency multiplication, it is not straightforward to understand how input phase noise is amplified within the PLL that directly controls the phase of a VCO by feedback. Let us consider a PLL with a divide-by-M circuit and an output frequency f_0. Since a phase detector operates at f_0/M, it generates an output voltage at every Mth period of the VCO clock. In other words, the VCO phase is corrected by the PLL at every Mth period, while the VCO is in a free-running mode during the remaining $(M-1)$ periods. The reference noise or charge pump noise will be transferred to the VCO whenever the phase detector generates an output voltage at every Mth period of the VCO clock. The

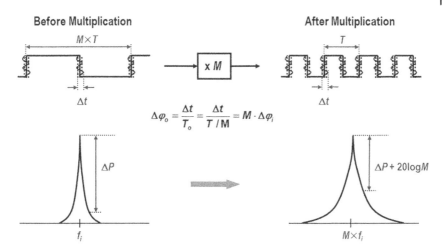

Figure 4.20 Effect of frequency multiplication on phase noise.

injected input noise will be accumulated during the remaining $(M - 1)$ periods when the VCO behaves like a free-running oscillator. In that sense, the behavior of the PLL is somewhat like an injection-locked oscillator. As a matter of fact, the noise transfer function and locking bandwidth of the injection-locked oscillator are analogous to those of a first-order PLL.

4.2.2 Integrated Phase Noise

Integrated phase noise in root-mean-square (RMS) degrees is an important parameter to determine the bit-error-rate (BER) performance in digital communication systems as illustrated in Fig. 4.21. Since integrated noise over the bandwidth of interest can be obtained from spectral density functions, the integrated phase noise $\Delta\theta_n$ or RMS phase jitter is obtained by

$$\Delta\theta_n = \sqrt{\int_a^b 2L(f_m)df_m} \ [\text{rad}] = \frac{180}{\pi}\sqrt{\int_a^b 2L(f_m)df_m} \ [\text{deg}] \quad (4.51)$$

where a and b are lower and upper limit frequencies of the integration bandwidth. Since $L(f_m)$ is a single-sideband noise, $2L(f_m)$ must be considered to get a double-sideband noise power.

Example 4.7 *Integrated phase noise of a first-order PLL*
Figure 4.22 shows a simplified phase noise plot of a first-order PLL with an in-band phase noise of -90 dBc/Hz and a 3-dB bandwidth of 400 kHz. Note that the 3-dB frequency of a closed-loop transfer function $H(s)$ is the same as the unity-gain frequency f_u of an open-loop gain $G(s)$ for the first-order PLL. Instead of integrating

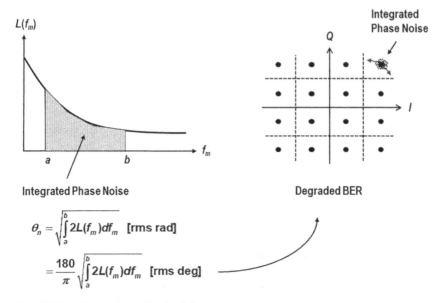

Figure 4.21 Phase noise effect in digital communication systems.

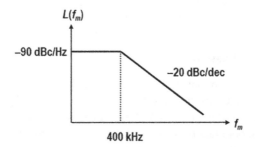

Figure 4.22 Phase noise plot of the first-order PLL.

the phase noise, we consider a noise bandwidth for easy calculation of the integrated phase error. The noise bandwidth B_n of the first-order PLL is obtained from (2.31), that is,

$$B_n = \frac{\pi}{2}f_u = 2\pi \times 10^5 \text{ [Hz]}$$

By the definition of the noise bandwidth, the integrated phase noise can be simply obtained by the product of the in-band phase noise N_o and the noise bandwidth. Then, the approximated value of RMS phase jitter is given by

$$\Delta\theta_n = \frac{180}{\pi}\sqrt{\int_a^b 2L(f_m)df_m} \approx \frac{180}{\pi}\sqrt{2N_oB_n} = \frac{180}{\pi}\sqrt{2 \times 10^{-9} \times 2\pi \times 10^5} = 2.03°$$

Therefore, having a narrow bandwidth is important to reduce the integrated phase noise for the given in-band phase noise. Now let us take a practical example.

Example 4.8 *Integrated phase noise and RJ calculation from the spectrum analyzer*

Figure 4.23 shows the spectrum of a 1-GHz signal in a spectrum analyzer. We will learn how to estimate RJ performance from the single plot of the spectrum analyzer. Here, we clearly see a flat in-band noise without a noise peaking near the bandwidth, which shows that the loop dynamics of a PLL is heavily overdamped and close to that of a first-order loop. In the spectrum, the 3-dB bandwidth is about 5 MHz. Assuming the first-order loop, the noise bandwidth is approximated as

$$B_n = \frac{\pi}{2} \times 5 \times 10^6 = 7.85 \times 10^6 \text{ [Hz]}$$

Within the loop bandwidth, the flat in-band phase noise is 40 dB lower than the carrier power. With the resolution bandwidth of 10 MHz, the in-band phase nose is given by

$$N_o = -40 \text{ dBc} - 10 \log 10^7 = -110 \text{ [dBc/Hz]}$$

As done in the previous example, we get the RMS integrated phase noise given by

$$\Delta\theta_n \approx \frac{180}{\pi} \sqrt{2N_o B_n} = \frac{180}{\pi} \sqrt{2 \times 10^{-11} \times 7.85 \times 10^6} = 0.72°$$

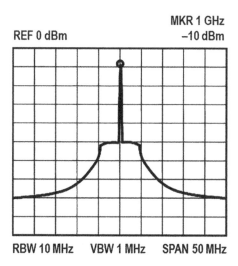

Figure 4.23 RJ calculation from the spectrum.

It can also be expressed in units of seconds. Knowing that carrier frequency is 1 GHz, we have the value of the RMS RJ in second given by

$$RJ \approx 0.72 \times \frac{1\,ns}{360°} = 2.0\,[ps]$$

4.2.3 Optimum Loop Bandwidth for Phase Noise

To have an optimum phase noise performance, it is important to identify the phase noise contribution of each building block within a PLL. In the design of frequency synthesizers, overall phase noise is determined by four noise sources; a reference source, a PD, a frequency divider, and a VCO. Noise from the reference source, the PD, and the frequency divider will be low-pass filtered by the noise transfer function of the PLL, while the VCO noise will be high-pass filtered. Depending on the PLL bandwidth, the VCO noise can be dominant on both the in-band phase noise and out-of-band phase noise or only on the out-of-band phase noise. If a wide bandwidth is designed, it is likely that the in-band phase noise is determined by one of the reference sources, the PD, and the frequency divider. If a narrow band-width is designed, the VCO noise contributes more to the in-band phase noise. Therefore, the optimum PLL bandwidth for phase noise would be determined by the cross section of the VCO phase noise and other low-pass filtered noises.

Now let us take a good example of how to choose a loop bandwidth for the optimum phase noise. Here, we consider an overdamped third-order type 2 PLL with a division ratio of 1,000 as shown in Fig. 4.24. Zero and pole frequencies are set to 1 kHz and 1 MHz, respectively. Suppose that a PD has a noise floor of $-140\,dBc/Hz$ and the divider has a noise floor of $-150\,dBc/Hz$. Also, the phase noise plots of a reference source and an open-loop VCO are shown. First of all, we need to draw each phase noise referred to the PLL output. With the division ratio of 1,000, we add 60 dB to the noises from the reference source, the PD, and the divider. Note that the VCO noise is not affected by the division ratio. Then, we plot all noises together as shown in Fig. 4.25(a). Now, we see that the cross section of the low-pass filtered noise and the VCO noise occurs at 100 kHz.

Once we set the optimum bandwidth to 100 kHz, we draw the phase noise plots of all sources based on the loop dynamics of the PLL as shown in Fig. 4.25(b). An overall phase noise at the PLL output is denoted by a thick line. The VCO noise is high-pass filtered in offset frequencies below 100 kHz and further sup-pressed from 1 kHz, while other noises are low-pass filtered in offset frequencies above 100 kHz and further filtered from 1 MHz. Below 10 kHz, the noise from the reference source is dominant. From 10 kHz to 1 MHz, both the PD noise and the VCO noise have the same contribution at the PLL output with the optimum bandwidth of 100 kHz. The phase noise is increased by 3 dB as two noises are merged. Beyond 1 MHz, the phase noise is fully determined by the VCO. If a loop bandwidth of 10 kHz is chosen, the in-band noise in 1–10 kHz will be governed

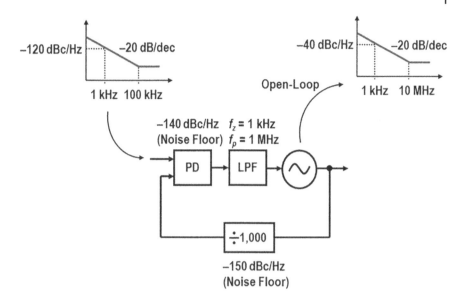

Figure 4.24 PLL with various noise sources.

by the VCO since the VCO noise at 10 kHz will be −60 dBc/Hz with the bandwidth of 10 kHz. On the contrary, if a loop bandwidth of 1 MHz is used, the in-band noise from 10 kHz to 1 MHz will be the same as the PD noise of −80 dBc/Hz as the VCO noise becomes −100 dBc/Hz below 1 MHz. Therefore, the loop bandwidth of 100 kHz is optimal to minimize the integrated phase noise.

This fact gives a useful way of evaluating the in-band phase noise based on the closed-loop response of a PLL. Suppose that the in-band phase noise performance of a PLL chip does not meet a given target during testing. A PLL designer wants to know whether a loop bandwidth is optimally set or not. If the in-band noise is dominated by a VCO, increasing the loop bandwidth further improves the in-band noise. If the in-band noise performance is limited by other sources, increasing the loop bandwidth does not alter the in-band noise. In this case, only the noise bandwidth is widened, resulting in degraded integrated phase noise.

Example 4.9 *Identifying the in-band noise contribution of building blocks*
Let us consider two examples of how to identify the in-band noise contribution of a PD or VCO. Figure 4.26(a) shows the block diagram of a PLL with a reference divider M and a feedback divider N. Suppose that a loop filter can be adjusted or programmable. For the given reference frequency, output frequency, and loop bandwidth, we double the values of the reference divider and the feedback divider as illustrated in Fig. 4.26(b). Then, the PD frequency is reduced by half, but the output frequency remains the same with the division ratio of $2N$. We also adjust the loop filter values so that the same bandwidth is maintained. Then, the

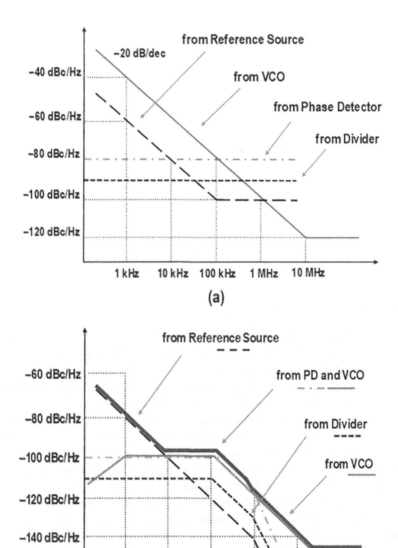

Figure 4.25 Phase noise contribution of each source at the output of the third-order type 2 PLL: (a) open-loop and (b) closed-loop with the optimum bandwidth of 100 kHz.

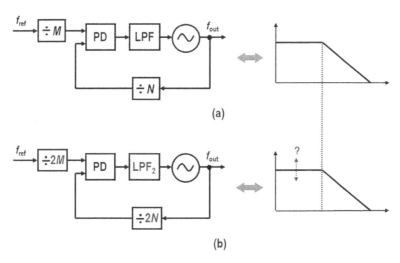

Figure 4.26 Finding the source of in-band phase noise: (a) phase noise with dividers M and N and (b) phase noise with dividers $2M$ and $2N$ for the same loop bandwidth.

VCO will give the same noise contribution to the PLL output. Also, the noise contribution from a reference source is the same as the PLL output. What about the PD noise? As the PD frequency is reduced by half, the multiplication factor of the PD noise is doubled at the PLL output. Accordingly, the PD has 6-dB more noise contribution to the PLL output. As to the in-band noise performance, we reach the following conclusion. If the in-band noise is unchanged by doubling the values of both the reference and the feedback dividers, the in-band noise is not dominated by the PD. On the contrary, if the in-band noise is increased with the doubled values of both dividers even for the same loop bandwidth, we conclude that the in-band noise is dominated by the PD.

Figure 4.27 shows another example of identifying the in-band phase noise by observing the out-of-band phase noise in the case of a third-order PLL. If we observe a noise slope of $-40\,\mathrm{dB/dec}$ after the PLL bandwidth as depicted in Fig. 4.27, then the in-band noise is not limited by the VCO noise but by other low-pass filtered noise sources. It is because the VCO noise is not affected by the high-order poles of the system transfer function $H(s)$, while the PD or the reference noise is further filtered by the out-of-band pole, resulting in the noise slope of $-40\,\mathrm{dB/dec}$. In that case, we need to reduce the loop bandwidth. If we do not observe the noise slope of $-40\,\mathrm{dB/dec}$ after the PLL bandwidth, it means that the in-band noise is determined by the VCO noise or equally contributed by the VCO and other noise sources. In that case, the loop bandwidth should be increased to check whether the in-band phase noise can be further improved or not.

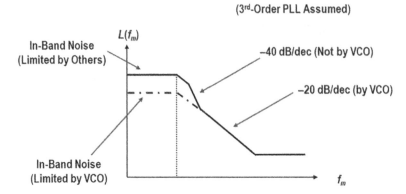

Figure 4.27 Finding the VCO noise contribution to in-band phase noise.

References

1 V. Manassewitsch, *Frequency Synthesizers, Theory and Design*, Wiley, New York, 1987.

2 U. L. Rohde, *Microwave and Wireless Frequency Synthesizers: Theory and Design*, Wiley, New York, 1997.

3 W. Egan, *Frequency Synthesis by Phase Lock*, 2nd ed., Wiley, New York, 2000.

4 H. Meyr and G. Ascheid, *Synchronization in Digital Communications, Phase-, Frequency-Locked Loops, and Amplitude Control*, Wiley, New York, 1990.

5 F. M. Gardner, "Charge-pump phase-locked loops," *IEEE Transactions on Communications*, vol. 28, pp. 1849–1858, 1980.

6 W. Rhee, "Design of high-performance CMOS charge pumps in phase-locked loops," in *Proc. IEEE International Symposium of Circuits and Systems*, June 1999, pp. 545–548.

7 K. Shu and E. Sanchez-Sinencio, *CMOS PLL Synthesizers; Analysis and Design*, Springer, 2005.

8 M. Azarian and W. Ezell, "A simple method to accurately predict PLL reference spur levels due to leakage current," *Linear Technology, Application Notes*, Dec. 2013.

9 W. Rhee, *"Multi-Bit Delta-Sigma Modulation Technique for Fractional-N Frequency Synthesizers,"* Ph.D. Thesis, University of Illinois, Urbana-Champaign, IL, Aug. 2000.

10 K. Feher, *Telecommunications Measurements, Analysis, and Instrumentation*, Prentice-Hall, Englewood Cliffs, NJ, 1987.

11 M. S. Roden, *Analog and Digital Communication Systems*, 4th ed., Prentice Hall, Upper Saddle River, NJ, 1996.

5

Application Aspects

Designing a functional phase-locked loop (PLL) at the circuit level is not so difficult if good circuit schematics are available for reference. However, designing a robust PLL for various applications requires diversified design aspects. If circuit designers do not know how to deal with system parameters, they may choose the wrong PLL topology or overdesign a system with unnecessary power consumption or area cost. In modern integrated circuits and systems, three areas are primarily considered as the main applications of a PLL. Those are frequency synthesis, clock generation, and clock-and-data recovery (CDR).

A frequency synthesizer is one of the key building blocks in wireless transceiver systems to synthesize frequencies for multiple channels. In the frequency synthesizer, high spectral purity is important not only to achieve a low integrated phase error with low in-band phase noise but also to mitigate the effect of interferers with low out-of-band phase noise and spurs. There are three critical performance metrics we need to consider in the design of frequency synthesizers; phase noise, reference spur, and settling time. Since the PLL is a feedback system, those parameters are highly sensitive to a loop bandwidth. Accordingly, defining an optimum bandwidth of the PLL is an important task for PLL designers.

A clock generator in digital systems plays a critical role as a clock multiplier unit (CMU). For clock generation in the digital system, minimizing peak-to-peak jitter over a clock period is a primary goal to provide a maximum phase margin[1] for synchronous digital circuits. Designing a robust clock generation circuit is challenging since it is embedded in a large digital system where billions of transistors generate substantial switching noise. Achieving both high power supply rejection (PSR) and good immunity against substrate noise coupling is important. In addition, the clock generator designed with standard CMOS technology suffers more from process, voltage, and temperature (PVT) variations than the frequency synthesizer implemented with full-featured RF CMOS technology.

---

1 Here, the term *phase margin* refers to a timing margin over a clock period.

Phase-Locked Loops: System Perspectives and Circuit Design Aspects, First Edition.
Woogeun Rhee and Zhiping Yu.

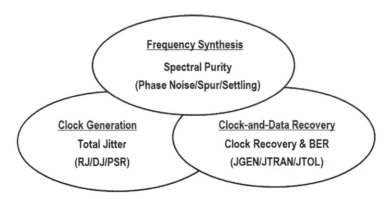

Figure 5.1 Key applications of the PLL with design aspects.

Compared with the frequency and clock generation circuits, a CDR system takes a non-return-to-zero (NRZ) data as an input and performs data retiming by using a recovered clock from the NRZ data itself. The ultimate goal of the CDR is to extract a clean clock from a noisy input data and reproduce the data with the minimum bit error rate (BER). In the design of the PLL-based CDR circuit, there are three important system parameters; jitter generation (JGEN), jitter transfer (JTRAN), and jitter tolerance (JTOL). Three applications with key design aspects are summarized in Fig. 5.1.

5.1 Frequency Synthesis

5.1.1 Direct Frequency Synthesis

A frequency synthesizer is a device that generates one or many frequencies from a single or several reference sources. First-generation frequency synthesizers used the incoherent method that requires multiple reference sources. That is, frequencies were synthesized by mechanically switching several crystal oscillators and filters as depicted in Fig. 5.2(a). Later, the coherent method that generates various output frequencies from a single reference source was developed by employing frequency multipliers, dividers, and mixers as shown in Fig. 5.2(b). In the coherent synthesis, the stability and accuracy of the output frequency are the same as those of the reference source. There are direct and indirect methods in the coherent synthesis. A modern approach for the direct frequency synthesis is based on the all-digital architecture, namely a direct-digital frequency synthesizer (DDFS). A functional block diagram is shown in Fig. 5.2(c). A periodic digital word is generated by a counter, and a read-only memory (ROM) maps the digital word to the digital values that represent a sinusoidal-like waveform. The digital waveform

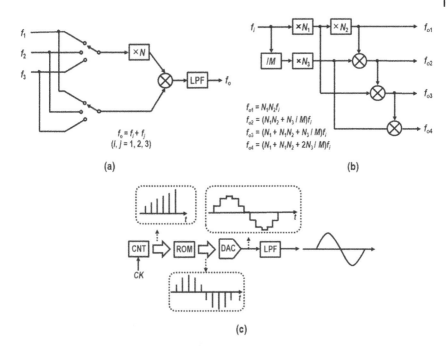

Figure 5.2 Direct frequency synthesis: (a) incoherent; (b) coherent; and (c) DDFS.

generated in the form of a series of digital numbers is converted into an analog waveform by a digital-to-analog converter (DAC) followed by a low-pass filter (LPF). For the DDFS to produce a complete cycle of a sinewave for the lowest frequency with an N-bit counter, it requires 2^N clock cycles. The DDFS features a fine frequency resolution and fast settling time since the output frequency can be synthesized in a single clock period. Unlike the PLL-based frequency synthesizer, the output frequency of the DDFS should be lower than the half of an input frequency to satisfy the Nyquist sampling criterion. Therefore, the DDFS is analogous to a programmable frequency divider, while the PLL-based frequency synthesizer is analogous to a programmable frequency multiplier. Accordingly, most frequency synthesizers employed in wireless transceivers are based on the PLL-based frequency synthesis to generate high frequencies with low power and compact area.

5.1.2 Indirect Frequency Synthesis by Phase Lock

The phase-lock technique guarantees that the accuracy of the output frequency of a PLL is as good as that of a reference source, making an indirect coherent frequency synthesis possible by having digital counters in the reference and

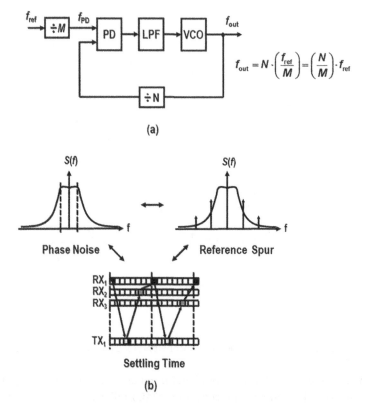

Figure 5.3 PLL-based frequency synthesis: (a) block diagram and (b) key design parameters.

feedback paths as shown in Fig. 5.3(a). Since two input frequencies of a phase detector (PD) should be the same, the output frequency f_{out} and the reference frequency f_{ref} are related by

$$f_{out} = N\left(\frac{f_{ref}}{M}\right) = Nf_{PD} \tag{5.1}$$

where M is the forward divider value, N is the feedback divider value, and f_{PD} is the PD frequency. Equation (5.1) implies that we can control not only the output frequency but also the frequency resolution by simply changing the values of M and N counters. For that reason, the PLL-based frequency synthesizer was considered a digital frequency synthesizer in the early days.

Like other feedback systems, the loop bandwidth is important in the design of the PLL-based frequency synthesizer as it comes from a fundamental trade-off among three system parameters; phase noise, reference spur, and settling time, as illustrated in Fig. 5.3(b). In Chapter 4, we discussed how to choose an optimum

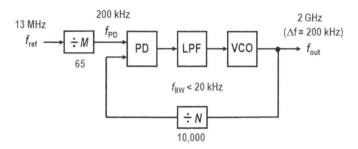

Figure 5.4 Example of a PLL having a large N to meet the frequency resolution requirement for the given reference frequency.

bandwidth for phase noise after identifying the noise contribution of each source and the noise transfer function. As to the reference spur, narrowing the bandwidth is helpful since it comes from a phase detector or the leakage current of a loop filter. However, the narrow bandwidth makes it difficult for the PLL to suppress the VCO noise. In addition, we need a wide bandwidth for fast settling.

We also learned that the frequency division ratio has a significant impact on the overall performance of the PLL. For the given output frequency, the in-band phase noise contribution by a charge pump or reference source would be substantial with a high division ratio, which results in a high integrated phase error. In addition, the mismatch effect of the charge pump on the reference spur becomes more significant with a higher division ratio. Figure 5.4 shows a heuristic example to address the design difficulties of the PLL having a very high division ratio. Suppose that the PLL needs to generate a center frequency of 2 GHz with a frequency resolution of 200 kHz[2]. If a reference frequency of 13 MHz is used, a division ratio of 10,000 has to be used since the PD frequency should be the same as the frequency resolution. Note that the reference frequency from a crystal oscillator is usually fixed by the system and not a negotiable design parameter in most commercial products. Design impact with such a high division ratio is substantial. For example, to achieve an in-band phase noise of -80 dBc/Hz, the phase noise contribution from a PD should be less than -160 dBc/Hz by considering the noise multiplication factor of 20 log(10,000). What about the reference spur? From (4.28), we can easily deduce that the charge pump mismatch should be less than 1% over the entire range of an output voltage. Moreover, the maximum loop bandwidth should be less than 15% of the reference frequency for stability, making it difficult to have agile frequency synthesis with a bandwidth of 30 kHz. If there is a way of synthesizing frequencies directly from the reference frequency of 13 MHz or with

---

2 This requirement is based on the Universal Mobile Telecommunications System (UMTS) 3G wireless standard in which a frequency range of 1.9–2.1 GHz with a 2G-compatible resolution of 200 kHz needs to be covered.

a higher PD frequency for the same frequency resolution of 200 kHz, the design complexity of the PLL-based frequency synthesizer can be significantly relaxed.

5.1.3 Frequency Synthesizer Architectures for Fine Resolution

To overcome the fundamental trade-off between the channel spacing and the PD frequency in the conventional PLL-based frequency synthesizer, several architectures have been proposed. Among them, the following three architectures are noteworthy for integrated circuit design:

- Hybrid architecture based on an integer-N PLL and a DDFS
- Multi-loop PLL architecture
- Fractional-N PLL architecture

In the hybrid architecture shown in Fig. 5.5(a), the DDFS is employed to provide fine frequency resolution and fast settling time, while an integer-N PLL generates

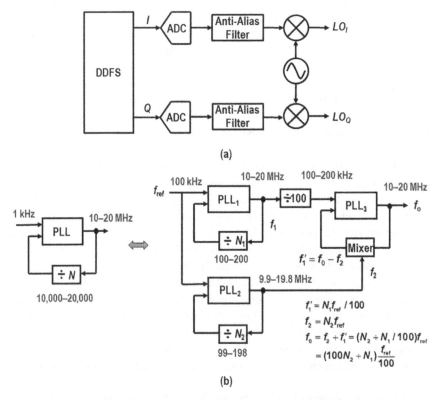

(a)

(b)

Figure 5.5 Frequency synthesis for fine resolution: (a) using a hybrid frequency synthesizer based on the DDFS and the PLL and (b) using a multi-loop PLL.

a carrier frequency for upconversion. Since the integer-N PLL does not need to provide the fine frequency resolution, a wideband PLL can be designed. However, the architecture suffers from high spur generation due to an on-chip mixer and hardware complexity of the DDFS.

Figure 5.5(b) shows an example of the frequency synthesizer based on multiple PLLs. If a single PLL is used to generate the frequency range of 10–20 MHz with the frequency resolution of 1 kHz, the division ratio range of 10,000–20,000 is required. When a multi-loop topology is employed as shown in Fig. 5.5(b), the first-stage PLL (PLL_1) generates a frequency range of 10–20 MHz with a frequency resolution of 100 kHz, and the second-stage PLL (PLL_2) generates a frequency range of 9.9–19.8 MHz with the same frequency resolution of 100 kHz. When the output frequencies of PLL_1 and PLL_2 are combined in the third-stage PLL (PLL_3) with a forward frequency divider and a mixer, the same frequency range of 10–20 MHz with the same frequency resolution of 1 kHz could be obtained with the minimum PD frequency of 100 kHz for all PLLs. The output frequency of the multi-loop PLL is given by

$$f_3 = \left(N_2 + \frac{N_1}{100}\right)f_{\text{ref}} = (100N_2 + N_1)\frac{f_{\text{ref}}}{100} \tag{5.2}$$

where f_1, f_2, and f_3 are the output frequencies of PLL_1, PLL_2, and PLL_3 respectively, and N_1 and N_2 are the values of the frequency division ratios of PLL_1 and PLL_2, respectively. Like the hybrid architecture, hardware complexity and spur generation due to mixer linearity would be a concern in the design of a fully integrated multi-loop PLL.

The fractional-N PLL achieves finer frequency resolution than the PD frequency by modulating the frequency division ratio. Figure 5.6 shows the traditional architecture of a fractional-N frequency synthesizer. By modulating

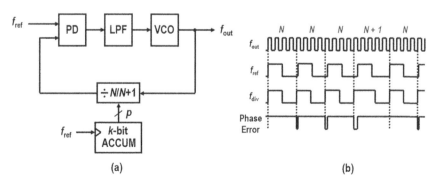

Figure 5.6 Traditional fractional-N PLL: (a) block diagram and (b) timing diagram example for N + 1/4.

a dual-modulus $(N/N+1)$ divider based on the output of an accumulator, an interpolative frequency division is achieved. However, the periodic modulation of the dual-modulus divider generates spurs near the carrier frequency. Accordingly, various spur reduction methods were proposed. Later, the development of a delta-sigma ($\Delta\Sigma$) fractional-N frequency synthesizer made a big leap for wireless transceiver systems. By generating output frequencies based on pseudo-random digital modulation, spur-free frequency generation with arbitrarily fine frequency resolution could be obtained. Nowadays, the $\Delta\Sigma$ fractional-N frequency synthesizer plays an important role in modern transceivers not only as a local oscillator but also as a frequency modulator with direct digital modulation. We will discuss the fractional-N PLL in detail in Chapter 9.

5.1.4 System Design Aspects for Frequency Synthesis

In the wireless transceiver system, a frequency synthesizer is a key building block to generate multiple frequencies for channel selectivity. Figure 5.7 illustrates the effect of the in-band phase noise and out-of-band phase noise of the frequency synthesizer on the transceiver performance. The in-band phase noise is important to achieve a low integrated phase error as discussed in Chapter 4. A large integrated phase error increases the probability of crossing the decision boundary in

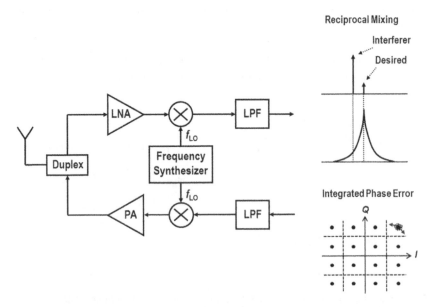

Figure 5.7 Phase noise effect on transceiver performance.

Table 5.1 PLL specifications for GSM standard.

PLL Parameters	Specification
TX Frequency	890–915 MHz
RX Frequency	935–960 MHz
Channel Spacing	200 kHz
Settling Time	$<$250 µs
RMS Phase Error	$<1°$
Out-of-band Phase Noise	<-138 dBc/Hz @ 3 MHz
	<-165 dBc/Hz @ 20 MHz
Spur	<-60 dBc @ 400 kHz

a bit constellation, which degrades the BER performance. In Fig. 5.7, the bit constellation diagram of a 16 quadrature amplitude modulation (16-QAM) is shown as an example. The out-of-band phase noise should also be considered for the transceiver. In the transmitter, it is important to comply with a spectrum mask set by wireless standards. In the receiver, the out-of-band phase noise can mix with strong interferers as illustrated in Fig. 5.7. The interferers downconverted by the out-of-band phase noise can corrupt a baseband signal. This is called a reciprocal mixing effect. The following example shows how to calculate the system parameters of a frequency synthesizer for a specific transceiver system.

Example 5.1 *PLL design considerations for GSM transceivers*
Let us take an example of the system design considerations of a frequency synthesizer for the traditional GSM system (GSM stands for Global System for Mobile communications). Key design specifications are shown in Table 5.1. For a channel spacing of 200 kHz, a frequency synthesizer needs to generate output frequencies with a frequency resolution of 200 kHz. Therefore, the PD frequency should be 200 kHz, requiring a maximum division ratio of 4,800 for an output frequency of 960 MHz. The RMS integrated phase error should be less than 5° for the transceiver system, and typically 1° or less is required for the frequency synthesizer. A settling time should be less than 250 µs. In addition to those parameters, the out-of-band phase noise and reference spur must comply with the spectrum mask set by the GSM standard. In this example, we will consider the PLL parameters for the settling time, loop bandwidth, in-band phase noise, and out-of-band phase noise for in-band blocking. The requirements of the spur and the out-of-band phase noise at 20-MHz offset are mainly dependent on charge pump and VCO circuits.

Settling Time and Bandwidth

For agile frequency synthesis, an underdamped loop is preferred. Let us begin with a damping ratio ζ of 0.6. To achieve a settling time t_s of 250 μs with a frequency step Δf_{step} of 100 MHz and a frequency error f_e of 100 Hz, a required natural frequency f_n is obtained from (3.24)

$$
\begin{aligned}
f_n &= \frac{-1}{2\pi t_s \zeta} \ln\left(\frac{f_\varepsilon \sqrt{1 - \zeta^2}}{\Delta f_{step}}\right) \\
&= \frac{-1}{2\pi \times 250\,\mu s \times 0.6} \ln\left(\frac{0.8 \times 100\,Hz}{100\,MHz}\right) = 14.9\ [\text{kHz}]
\end{aligned}
$$

For comparison, let us obtain the value of a time constant τ_n, that is,

$$
\tau_n = \frac{1}{2\pi f_n} = \frac{1}{2\pi \times 14.9\,kHz} = 10.7\ [\mu s]
$$

Above results show that the settling time is nearly 24 times the time constant related with the natural frequency. This is mainly because the required frequency error is only 100 Hz, which is about 0.1 ppm ($= 10^{-7}$) of the carrier frequency. Therefore, it is important to check the requirement of the frequency error when a settling time is calculated for wireless systems. Assuming a second-order type 2 PLL, a loop bandwidth f_{BW} is calculated as

$$
f_{BW} = 2\zeta f_n = 2 \times 0.6 \times 14.9\,kHz = 17.9\ [\text{kHz}]
$$

This is to get the basic idea of how to obtain critical system parameters by hand calculation. The final values must be obtained by using a PLL-dedicated software. If a programmable VCO or digitally controlled oscillator (DCO) is designed, the center frequency of the VCO or DCO can be pre-tuned to a desired output frequency. As a result, the PLL achieves a faster settling time with reduced f_{step}, but additional digital-calibration time must be included in the total settling time.

In-Band Phase Noise

For a given bandwidth, in-band phase noise is determined by the requirement of the RMS phase error. We first obtain a noise bandwidth from (2.32)

$$
B_n = \frac{\pi f_{BW}}{2}\left(1 + \frac{1}{4\zeta^2}\right) = \frac{\pi \times 17.9\,kHz}{2} \times 1.69 = 47.5\ [\text{kHz}]
$$

For in-band phase noise N_i to have the RMS phase jitter θ_n less than 1°, we have

$$
\theta_n = \frac{180}{\pi}\sqrt{2 B_n N_i} = \frac{180}{\pi}\sqrt{2 \times 47.5\,kHz \times N_i} < 1^\circ
$$

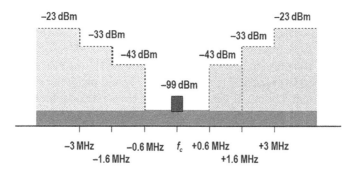

Figure 5.8 In-band blocking for GSM system.

From the above equation, we get the in-band phase noise of $-85\,\text{dBc/Hz}$. If the in-band phase noise better than $-85\,\text{dBc/Hz}$ is achieved, a loop bandwidth can be increased to further improve the settling time. However, increasing the loop bandwidth degrades the spur or out-of-band phase noise performance, so choosing the right bandwidth is critical for the overall performance of a frequency synthesizer.

Spur and Out-of-Band Phase Noise

In the wireless system, mitigating interferences from other channels is important. In the transmitter, a spectrum mask is defined to limit the power emission of a transmitter output. In the receiver, adjacent channel interference, in-band blocking, and out-of-band blocking signals should be considered to minimize the reciprocal mixing problem. In the case of the GSM standard, the required spur level is mainly determined by the transmitter spectrum mask, which requires a typical spur level of $-65\,\text{dBc}$ at 400-kHz offset with an additional system margin of 5 dB, which is equivalent to a reference spur of $-59\,\text{dBc}$ at 200-kHz offset frequency.

In the GSM receiver, the phase noise requirement for the in-band blocking is quite challenging. Figure 5.8 shows the requirement of the out-of-band phase noise for the in-band blocking. In Fig. 5.8, one of the most difficult conditions is to meet the requirement of a blocking signal at a 3-MHz offset frequency where the blocking signal is as high as $-23\,\text{dBm}$, whereas the desired signal power is only $-99\,\text{dBm}$. The required signal-to-noise ratio (SNR) for Gaussian minimum shift keying (GMSK) modulation to satisfy the BER of 10^{-3} is 9 dB. With a system margin of 3 dB, the noise power of the frequency synthesizer at 3-MHz offset should be 88-dB lower than the carrier power. Considering a signal bandwidth of 100 kHz, we calculate the required phase noise as

$$L(f)|_{f=3\,\text{MHz}} = -99\,\text{dBm} - (-23\,\text{dBm}) - 12\,\text{dB} - 20\log(100\,\text{kHz})$$
$$= -138\,[\text{dBc/Hz}]$$

In addition, a very far out-of-band phase noise is also critical for a transmitter since the far-out phase noise of the transmitter can be spread to a receiver band. In the transmitter, a phase noise of $-165\,\text{dBc/Hz}$ is required at 20-MHz offset. Since the phase noise at a far offset frequency is totally determined by the VCO noise, designing a low-noise VCO is critical to meet the out-of-band phase noise requirement of the GSM system. The out-of-band phase noise is a good example of showing the major difference of the PLL design between wireless and wireline systems.

5.2 Clock-and-Data Recovery

5.2.1 Wireline Transceiver with Serial Link

To understand the meaning of recovering clock and data, let us take an example of an optical transceiver system shown in Fig. 5.9. A transmitter combines several low-speed data in parallel to form a high-speed serial data that modulates a laser. The conversion of the parallel data into a serial data is done by a multiplexer (MUX) with a clock whose frequency is the same as the rate of the serial data. The modulated laser emits a light through an optical cable to represent data "1" or "0" with on-off keying (OOK) modulation. A received light is transformed to a current by a photodiode in a receiver. With a transimpedance amplifier (TIA) and limiting amplifier, a healthy serial data is obtained and fed to a data slicer, typically a D-type flip-flop (DFF). A demultiplexer (DMUX) converts the serial output of the DFF into low-speed parallel data for multiple users. Since we convert the low-speed parallel data into the serial data and then reconvert it into the parallel data, it is called as a *SerDes* (serializer and deserializer) system. Overall operation is straightforward if we assume that we have a clean clock with well-controlled frequency and phase for the MUX, the DFF, and the DMUX. However, low-jitter

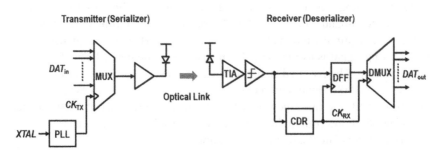

Figure 5.9 Optical transceiver (SerDes) system.

clock generation and good synchronization are quite challenging as the data rate increases.

Since the NRZ data has twice the minimum pulse width of the RZ data, the bandwidth of the NRZ data is half of the RZ data. For that reason, the NRZ data is dominantly employed for most communication systems. The NRZ data, however, does not contain clock information. In optical communication, we only send the data without the clock through an optical cable, clock information must be extracted from a received NRZ data to synchronize the phase and frequency for the DFF and the DMUX in the receiver. Therefore, the CDR system is required to extract the clock information from a noisy NRZ data and retime the data with the extracted clock.

Compared with the optical link, an electrical link can easily provide many wires to send low-speed data in parallel. Nonetheless, we prefer using a single serial link at the cost of high-frequency clock generation. When high-speed data are sent in parallel, the cross-talk among wires becomes substantial. Moreover, synchronizing the multiple data to perform proper demultiplexing is difficult at high speed. Accordingly, the serial-link communication is considered reliable even with the tough requirement of low-jitter high-speed clock generation. In the electrical link, there are two ways of sending the NRZ data as illustrated in Fig. 5.10. One is a clock-forwarded data transmission shown in Fig. 5.10(a), and the other is a clock-embedded data transmission in Fig. 5.10(b). In the clock-forwarded transmission, we send both the NRZ data and the clock. Since a receiver does not need to recover the clock from the NRZ data, data retiming can be done directly with the received clock and a variable delay circuit. The disadvantage of this method is the need of an extra link for the clock whose bandwidth is twice that of the NRZ data. In the case of the clock-embedded system, only the NRZ data is sent without having a dedicated link for the clock. Like the optical link, the clock needs to be recovered from the NRZ data. In the following section, we will discuss the role of the CDR system.

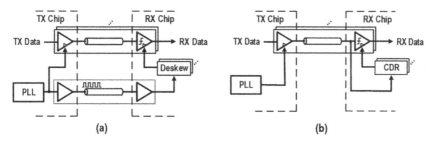

Figure 5.10 Electrical link: (a) clock-forwarding and (b) clock-embedded methods.

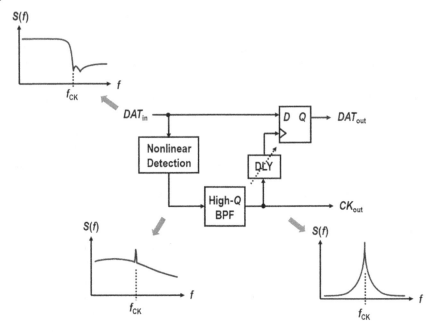

Figure 5.11 Traditional CDR circuit.

5.2.2 Clock Recovery and Data Retiming by PLL

Figure 5.11 shows the functional block diagram of a traditional CDR system. As the PLL-based CDR takes a high-speed non-periodic NRZ data as an input, designing a high-speed high-performance phase detector is important. Since the NRZ data does not contain clock information, the clock information is extracted from the NRZ data by using a nonlinear detection circuit. However, the recovered clock through the nonlinear circuit could be very weak, and a high-Q band-pass filter is used to boost a clock power by rejecting undesired broadband noise as illustrated in Fig. 5.11. This kind of traditional CDR systems have the following difficulties. Firstly, we need an external high-Q band-pass filter such as a surface acoustic wave (SAW) filter. Secondly, such a high-Q band-pass filter suffers from a limited frequency tunability as there is a fundamental trade-off between the quality factor and the tunability. Thirdly, we need a phase shifter for data retiming since a phase offset between the input data and the recovered clock could not be well controlled. Equipped with the external filter and the separate phase shifter, the CDR circuit has difficulty in achieving low-power or high-speed performance.

Now, let us look at the PLL-based CDR system shown in Fig. 5.12. As discussed in Chapter 2, the PLL plays as a band-pass filter to an input signal in the frequency domain (or a low-pass filter to an input phase in the phase domain). Since the

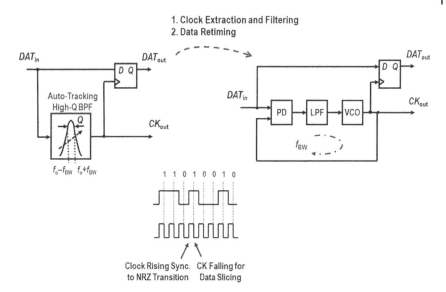

Figure 5.12 PLL-based CDR circuit.

PLL bandwidth can be set independent of the output frequency, a very high-Q band-pass filter can be realized. For instance, we consider a PLL with an output frequency of 10 GHz and a bandwidth of 1 MHz. In the phase domain, the PLL performs as a low-pass filter with a 3-dB corner frequency of about 1 MHz, while working as a band-pass filter with the same 3-dB corner frequency. Then, the effective quality factor for the band-pass filtering function is approximated as (1 GHz/2 MHz) = 500. If the loop bandwidth is reduced to 100 kHz, then the quality factor becomes 5,000! More importantly, the tunability of such a high-Q band-pass filter is mainly determined by the tuning range of a VCO, which is typically more than 20%.

In addition to the auto-tracking high-Q filtering function, the PLL provides phase synchronization between the transition of the NRZ data and the rising or falling edge of the clock. In Fig. 5.12, the rising edge of the clock is synchronized with the data transition so that the mid- point of the NRZ data is sampled for bit decision by the falling edge of the clock. Those two reasons, that is, the high-Q band-pass filtering with wide tunability and the phase synchronization, make the PLL unique for a compact low-power CDR system. In summary, the key features of the PLL-based CDR are listed as follows:

- Auto-tracking high-Q band-pass filter
- Clock recovery from NRZ data
- Data retiming with automatic phase tracking

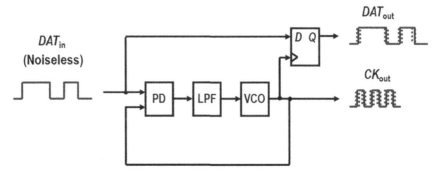

Figure 5.13 Jitter generation.

Being a feedback system, the CDR PLL requires an optimum bandwidth for the best performance. In the CDR PLL design, three key parameters are mainly considered. Those are jitter generation, jitter transfer, and jitter tolerance. It is important to understand those system parameters and design trade-offs among them.

5.2.2.1 Jitter Generation

Jitter generation indicates how much jitter is generated at the output of the PLL when noiseless input data is injected as illustrated in Fig. 5.13. Similar to the case of frequency synthesizer, having an optimum bandwidth is important by considering different noise sources in the PLL as discussed in Chapter 4. Since the CDR PLL does not have a frequency divider, the phase noise degradation due to frequency multiplication does not exist. Therefore, the in-band noise contribution from a high-frequency phase detector is less substantial than that from a VCO, making a wide bandwidth desirable to suppress the phase noise of a VCO in most cases. With a noisy input data, choosing an optimum bandwidth also depends on other two parameters; jitter transfer and jitter tolerance.

5.2.2.2 Jitter Transfer

Jitter transfer indicates how much jitter of an input data is filtered by a CDR circuit, which is basically determined by the system transfer function of the PLL. Figure 5.14 shows the conceptual diagram of an ideal CDR PLL that completely filters the input jitter with an extremely narrow bandwidth. When an output data is retimed with a clean clock, the input jitter is fully suppressed at the output data. The CDR system based on a type 2 PLL always exhibits a gain peaking near the 3-dB bandwidth in the jitter transfer function because of the system zero that is required for stability. In some applications that employ cascaded CDR circuits, the gain peaking in the jitter transfer function could be a concern. It amplifies jitter contribution around the PLL bandwidth, resulting in a jitter accumulation problem for CDR repeaters as illustrated in Fig. 5.15. For instance,

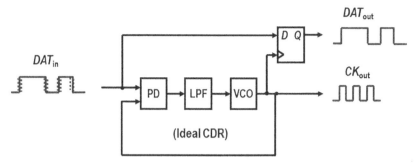

Figure 5.14 Jitter transfer property with perfect jitter filtering.

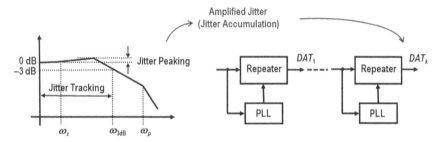

Figure 5.15 Jitter peaking and jitter accumulation.

for the synchronous optical network (SONET) where a single clock handles data retiming across the entire network with repeaters, this kind of jitter peaking should be tightly controlled to be less than 0.1 dB. For that reason, a heavily overdamped PLL must be considered. Some electrical links also require a tight jitter peaking control in the jitter transfer function to filter the input jitter when cascaded CDR PLLs are used.

5.2.2.3 Jitter Tolerance

Jitter tolerance shows the phase tracking capability of the CDR PLL, indicating the quality of data retiming to minimize the BER. As illustrated in Fig. 5.16, if the phase edge of an input data can be fully tracked by the PLL, the phase edge of a clock for a DFF is always synchronized to that of a data, making no error for the DFF to retime the input data with the recovered clock. If the frequency component of the data jitter is within the loop bandwidth, the phase variation of the input data can be tracked by the recovered clock so that the input jitter is tolerable for the data retiming. In contrast to the jitter transfer, the frequency response of the jitter tolerance shows the high-pass filter characteristic of the PLL. Therefore, a wide bandwidth is needed for the PLL to track the fast variation of the phase edge.

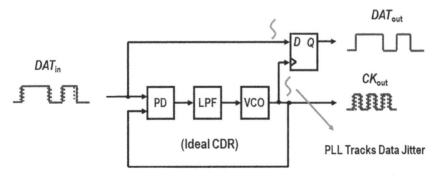

Figure 5.16 Jitter tolerance.

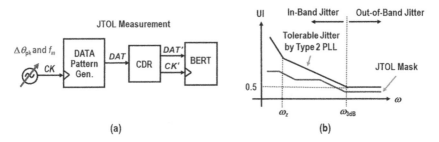

Figure 5.17 Jitter tolerance: (a) measurement setup and (b) JTOL mask.

In other words, the jitter tolerance follows the error transfer function of the PLL with the same 3-dB corner frequency of the system transfer function.

In most transceiver systems, the minimum requirement of data tracking is specified with a frequency response plot, namely jitter tolerance mask. To measure the jitter tolerance, a phase-modulated data is used at the input of the CDR circuit. Figure 5.17(a) shows an example of the jitter tolerance measurement. As learned in Chapter 4, the peak phase deviation $\Delta\theta_{pk}$ is given by the modulation index m that is the ratio of peak frequency deviation Δf_{pk} to modulation frequency f_m. We also learned that $\Delta\theta_{pk}$ for the given f_m can be obtained by measuring the spur level of the clock in the frequency domain. A phase-modulated clock with a variable $\Delta\theta_{pk}$ for a given f_m is used for a data pattern generator to generate a phase-modulated data. The maximum amount of $\Delta\theta_{pk}$ that satisfies the BER requirement is recorded for the given f_m. The measurement of the BER performance can be done with a BER tester (BERT) where the output data and clock of the CDR circuit are used. Figure 5.17(b) shows the typical jitter tolerance performance of a type-2 PLL along with the jitter tolerance mask. Since the jitter tolerance is governed by the open-loop gain of the PLL, it gives the same plot as the open-loop gain curve except that the phase tolerance is expressed by a unit

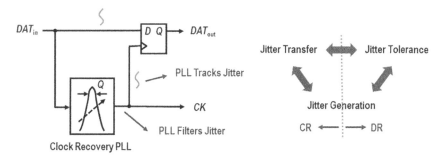

Figure 5.18 Design trade-offs of the CDR PLL.

interval (UI), that is, one clock period. Accordingly, the CDR PLL has good jitter tolerance within the CDR bandwidth, while showing no jitter tracking outside the bandwidth. In practice, 0.5 UI is set to the minimum value for the jitter tolerance by assuming that all other jitter components contribute to less than 50% of the clock period. The jitter tolerance mask implies that the long-term jitter of an input data is tolerable because of the high-pass filter characteristic of a type 2 PLL, while the short-term jitter is critical.

5.2.2.4 Role of Bandwidth and Comparison

Figure 5.18 shows a design trade-off of the CDR PLL with respect to jitter transfer and jitter tolerance performance. As discussed previously, we need a narrow bandwidth for the jitter transfer to recover a clock from a noisy data. On the other hand, a wide bandwidth is desired for the jitter tolerance to perform the data retiming with the minimum BER. As to the jitter generation, an optimum bandwidth by considering the noise contribution of each circuit is needed. Interestingly, the bandwidth requirement for three system parameters of the CDR PLL is comparable to that for three system parameters of the frequency synthesizer. As shown in Fig. 5.19, the bandwidth requirements for the jitter generation, the

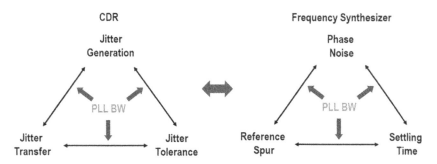

Figure 5.19 Comparison of CDR and frequency synthesis.

jitter transfer, and the jitter tolerance in the CDR are analogous to the bandwidth requirements for the phase noise, reference spur, and the settling time in the frequency synthesis, respectively. Therefore, the PLL bandwidth plays a key role for both frequency synthesizer and CDR circuits.

5.3 Clock Generation

5.3.1 System Design Aspects

Even though the functional roles of the frequency synthesizer and the clock multiplier circuit are the same, there are different system design aspects between wireless and wireline transceiver systems. Figure 5.20 shows a pictorial view of typical design trade-offs of the PLL for wireline systems in comparison with wireless systems. Firstly, the PLL used for microprocessors or I/O links does not have to generate multiple frequencies during normal system operation. Accordingly, fast frequency settling is not needed since the initial lock-in time requirement for wireline systems is much longer than the settling time requirement of wireless systems. Secondly, the in-band phase noise of the PLL in most wireline transceiver systems is not as critical as that in wireless transceiver systems. It is because long-term jitter caused by low-frequency phase noise is tolerable in the wireline system since the effect of the low-frequency jitter can be compensated for by the CDR circuit that provides jitter tolerance within the tracking bandwidth. As discussed previously, the short-term jitter is more important than the long-term jitter to determine BER performance in most wireline transceivers. Thirdly, the spur requirement of the PLL in wireline applications is not as tight as that

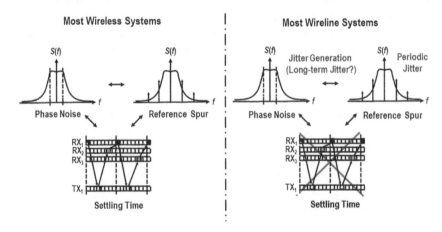

Figure 5.20 System design aspects in wireless and wireline systems.

in wireless applications where channel interference or blocking signals need to be considered. The spur level in the frequency domain is related with the deterministic jitter (DJ) performance in the time domain. For example, a spur level of -40 dBc contributes to <1% of a VCO clock period as discussed in Chapter 4, which is sufficient for most wireline systems. For a similar reason, the required out-of-band phase noise for the wireline system is not as low as that for the wireless systems as long as a PLL meets the short-term jitter requirement.

The relaxed system parameters make it possible to employ a ring VCO with a wide bandwidth PLL in many wireline applications. However, it does not necessarily mean that the PLL design for the wireline system is easier than that for the wireless system. Unlike the wireless system, the clock generation circuit in a microprocessor must operate with billions of digital logic circuits, thus suffering from supply and substrate noise coupling. In addition, the clock generation circuit for the microprocessor is mostly implemented with standard CMOS technology that does not offer good model-to-hardware correlation in the early stage and also lacks linear or high-quality passive devices. Therefore, robust clock generation with those conditions could be quite challenging for some wireline systems.

Example 5.2 *Jitter consideration for transmitter PLL*
As clock speed exceeds GHz range, clock jitter becomes one of the most critical factors to determine overall BER performance. For a full understanding of a clock behavior, it is important to analyze the clock jitter in the frequency domain. As discussed in previous chapters, the RJ is determined by integrating phase noise, while the DJ is given by spurs due to modulation or coupling. In a PLL-based clock generator, both RJ and DJ depend on the bandwidth of the PLL. By considering the frequency response of them, the loop bandwidth of the PLL must be determined.

Figure 5.21 gives a heuristic example of how important it is to understand a clock behavior in the frequency domain. Figure 5.21(a) shows a case that the RJ performance is not good enough to meet the jitter requirement when a long-term jitter is included. However, the same PLL turns out to be satisfactory when only a short-term jitter is considered. In most wireline transceivers, the phase noise within the tracking bandwidth of a CDR circuit does not need to be considered in the RJ budget of the PLL in a transmitter. Therefore, the RJ can be calculated by integrating the phase noise from the tracking bandwidth of the CDR to the maximum frequency specified by the system as illustrated in Fig. 5.21(b).

As discussed, the primary goal of the clock generation in digital systems is to achieve a small peak-to-peak jitter in the time domain. Even though a PLL exhibits an excellent RJ performance, it still has a chance of exceeding the jitter budget when the DJ performance is included. Since a total peak-to-peak jitter can also be

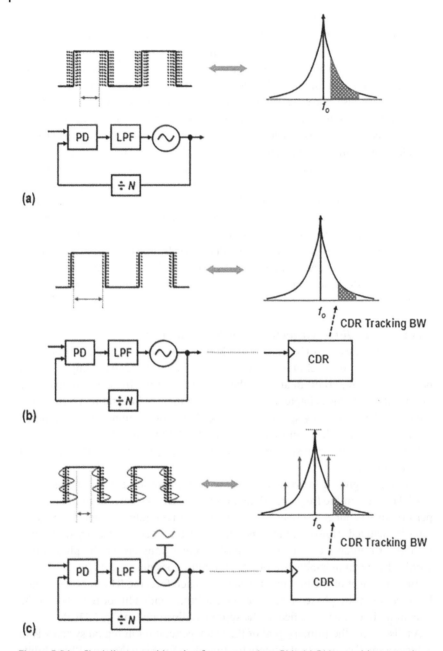

Figure 5.21 Clock jitter consideration for a transmitter PLL: (a) RJ by total integrated phase noise; (b) RJ by considering CDR tracking bandwidth; and (c) dominant contribution of DJ due to supply noise coupling.

dominated by the DJ, the PLL with low RJ still suffers from large peak-to-peak jitter if the PLL is not robust enough to provide good immunity against noise coupling. As illustrated in Fig. 5.21(c), a large spur causes substantial periodic jitter in the time domain, resulting in a poor peak-to-peak jitter performance. This example shows that not only designing good circuits but also understanding the system requirements is important for circuit designers to design PLLs for various applications.

5.3.2 Clock Jitter for Wireline Systems

5.3.2.1 RJ and BER

Integrating phase noise over a certain noise bandwidth gives an RJ value in root-mean-square. Assuming Gaussian distribution in a jitter histogram, the RJ value equals the standard deviation of the normal distribution. Then, the peak-to-peak jitter caused by the RJ depends on how much deviation of the jitter distribution is tolerable for a given BER. As illustrated in Fig. 5.22, the peak-to-peak RJ (RJ_{pk-pk}) is given by

$$RJ_{pk-pk} = n \times RJ \tag{5.3}$$

where n is a multiplication factor given by the BER requirement of a system. The RJ_{pk-pk} becomes at least 14 times the RJ when the BER of 10^{-12} is required, while n of 6 is used for the BER of about 10^{-3}. For the BER less than 10^{-12}, the multiplication factor n does not increase much. For that reason, a typical value of 14 is used for n in most wireline communication systems. For instance, if a 10-GHz clock is used for 10-Gb/s serial I/O links, the required RJ is less than 0.7 ps to have the RJ_{pk-pk} below 0.1 UI, that is 10 ps.

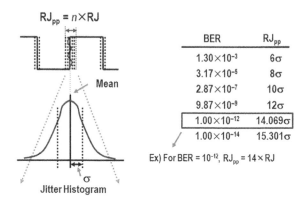

BER	RJ_{pp}
1.30×10^{-3}	6σ
3.17×10^{-5}	8σ
2.87×10^{-7}	10σ
9.87×10^{-9}	12σ
1.00×10^{-12}	14.069σ
1.00×10^{-14}	15.301σ

Ex) For BER = 10^{-12}, $RJ_{pp} = 14 \times RJ$

Figure 5.22 Peak-to-peak jitter due to RJ.

5.3.2.2 Total Jitter

Based on the previous discussion, we can define a total peak-to-peak jitter (TJ_{pk-pk}) by

$$TJ_{pk-pk} = RJ_{pk-pk} + DJ = n \times RJ + DJ \qquad (5.4)$$

Note that the value of the DJ already contains the meaning of the peak-to-peak jitter. Depending on the phase and spur performance of the PLL, TJ_{pk-pk} is dominated either by the RJ or by the DJ as illustrated in Fig. 5.23. If the DJ comes mainly from a reference spur, the loop bandwidth of a PLL must be reduced to mitigate the DJ contribution to the total jitter. If the RJ is more dominant than the DJ, an optimum loop bandwidth should be determined by considering the noise contribution of each block in the PLL.

Example 5.3 *RJ and DJ performance with bandwidth control*

Let us consider the case that a total jitter performance can be improved with the control of a loop bandwidth by analyzing the RJ and DJ performance in the frequency domain. Figure 5.24 shows the example of a third-order type 2 CP-PLL where only two noise sources are considered; a reference clock and a VCO. In the loop filter shown in Fig. 5.24, the value of a capacitor in an integral path is much larger than that of a shunt capacitor, showing that the loop is heavily overdamped. Then, the unity-gain frequency f_u can be obtained by

$$f_u = \frac{I_{CP}R_1K_v}{2\pi N} = \frac{1\,\text{mA} \times 5\,\text{k}\Omega \times 400\pi\,\text{MHz/V}}{2\pi \times 1000} = 1\,[\text{MHz}]$$

Figure 5.25 shows the combined phase noise of the reference source and the VCO at the PLL output. Since the reference frequency is divided by 100 and multiplied by 1,000, an effective multiplication factor of 10 results in the phase noise increment of 20 dB at the 1-GHz output. With f_u of 1 MHz, the in-band noise of -100 dBc/Hz is determined by the phase noise of the VCO. The noise bandwidth B_n is given by

$$B_n = \frac{\pi}{2}f_u = \frac{\pi}{2}\,[\text{MHz}]$$

Knowing the in-band noise and the noise bandwidth, we can calculate the RMS integrated phase error. That is,

$$\Delta\theta_n \approx \frac{180}{\pi}\sqrt{2N_0B_n} = \frac{180}{\pi}\sqrt{2 \times 10^{-10} \times \frac{\pi}{2} \times 10^6} = 1.02° = 2.83\,[\text{ps}]$$

If the BER of 10^{-12} is assumed, the peak-to-peak jitter becomes

$$RJ_{pk-pk} = 14 \times 2.83 = 39.6\,[\text{ps}]$$

Now let us get the value of the DJ caused by spurs. In this example, we see only one spur from the reference source that exhibits -40 dBc at 8-MHz offset frequency.

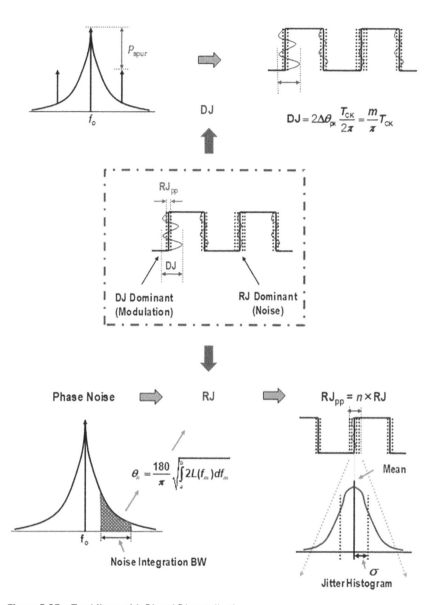

Figure 5.23 Total jitter with RJ and DJ contribution.

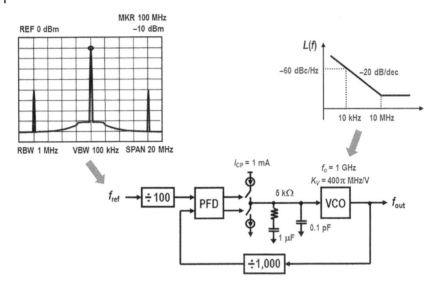

Figure 5.24 CP-PLL with two noise sources.

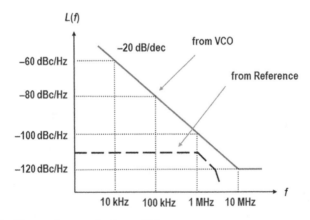

Figure 5.25 Phase noise contribution at PLL output.

Since the 3-dB bandwidth of the system transfer function is close to f_u for the heavily overdamped loop, the PLL works as an LPF with a 3-dB bandwidth of 1 MHz. Then, the spur level due to the reference spur at 8-MHz offset is given by

$$P_{spur} = -40\,\text{dBc} + 20\log 10 - 20\log\frac{8\,\text{MHz}}{1\,\text{MHz}} = -38.1\,[\text{dBc}]$$

Note that the effect of a third-order pole is considered negligible. The DJ is obtained by

$$DJ = \frac{m}{\pi} = \frac{2}{\pi}\left(\frac{\Delta f}{2f_m}\right) = \frac{2}{\pi}10^{-\frac{38.1}{20}} = 0.008\,UI = 8\,[ps]$$

From the values of the RJ and DJ, it is shown that the total jitter is dominated by the RJ and mainly comes from the VCO. Therefore, we reach the conclusion that the loop bandwidth of the PLL should be increased to reduce the peak-to-peak jitter.

5.4 Synchronization

5.4.1 PLL for Clock De-skewing

As clock frequency increases, a clock skew between an internal clock and an external clock becomes critical in chip-to-chip communications, for example data transfer between a microprocessor and a memory chip. The clock skew could be caused by the clock tree that buffers an external clock to drive a large number of internal nodes. Figure 5.26 shows the example of how the clock skew is generated by the clock tree between two chips. Even though both chips, chip1 and chip2, share the same external clock, CK_{ext}, an actual clock edge with which internal logic circuits operate could be different since the size and number of a clock tree in each chip cannot be the same (and two chips could be implemented with different CMOS technologies). In addition, the amount of the clock skew varies a lot over process, temperature, and supply voltage variations. Therefore, the clock skew between chips could substantially degrade the data throughput for high-speed communication.

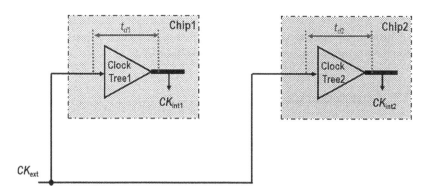

Figure 5.26 Clock skew problem in chip-to-chip communication.

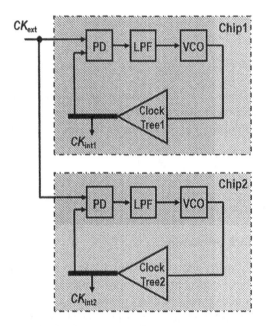

Figure 5.27 Clock de-skewing by an on-chip PLL.

Figure 5.27 shows how the clock skew can be mitigated by employing a PLL. The PLL takes the output of a single gate at the last stage of the clock tree as the feedback input of a phase detector and compares the phase of the feedback clock with that of an external clock. Since both phases at the input of the phase detector are synchronized by phase lock, we can guarantee that the actual clock edge with which leaf logic cells operate is aligned to the phase of the external clock. In other words, regardless of the size of the delay variation of the clock tree, the internal logic circuits have digital transitions at the same phase of the external clock. In that way, both chips work without the clock skew problem, thus maximizing the data throughput. Since the PLL was adopted for on-chip synchronization, the data speed for I/O interfaces has been dramatically improved.

5.4.2 Delay-Locked Loop

One major problem of the PLL employed for clock synchronization in a large digital system is high sensitivity to a supply noise. It is mainly because of the poor supply rejection of a VCO. Since the VCO is essentially a voltage-to-frequency converter and contains an integrator in the linear model of the PLL, jitter propagation in the phase domain due to a supply voltage jump is inevitable. Accordingly, jitter accumulation caused by the supply voltage jump is a big concern when a

PLL with a ring VCO is placed in microprocessors or other digital systems where a large number of transistors generate switching noise. When the PLL is used for clock synchronization, there is no need of frequency acquisition since the clock frequency is well defined by a system clock. In that case, a PLL with a voltage-controlled delay line (VCDL) instead of the VCO could be considered. The PLL that directly controls the output phase with the VCDL is named as a delay-locked loop (DLL).

The simplified block diagram of the DLL along with a linear model is shown in Fig. 5.28. Unlike the PLL-based clock generator, the DLL-based clock synchronization circuit operates at a fixed frequency, thus tracking the phase only. Accordingly, other types of a phase detectors instead of the phase-frequency detector (PFD) can also be designed with a charge pump to improve operation speed or to avoid a dead-zone problem (we will discuss in Chapter 6). Note that there is no $1/s$ in the VCDL in the linear model of the DLL since the output phase of the VCDL is directly controlled by an input voltage. Therefore, the DLL has better stability than the PLL. For a type 1 feedback system, no zero is needed in a loop filter for stability, and only a capacitor C_1 can be used as shown in Fig. 5.28(a). In the linear model of the DLL shown in Fig. 5.28(b), a phase detector gain K_{PD} and a VCDL gain K_{VCDL} are expressed in units of volts per radian and in units of radians per volt, respectively. Then, the open-loop gain of the first-order charge-pump DLL is simply approximated as

$$G(s) \approx \frac{I_{CP}K_{VCDL}}{2\pi sC_1} \qquad (5.5)$$

As the output clock is a delayed input clock, the system transfer function from an input phase to an output phase is close to an all-pass transfer function. An exact analysis can be done by using the z-domain analysis, showing that a small amount of gain peaking is observed due to a system delay. Because of the all-pass transfer characteristic, the jitter filtering of the DLL is not as good as that of the PLL. Since the output of the VCDL is the delayed signal of a reference clock, the duty cycle of the DLL output is directly affected by that of the reference clock. For that reason, a

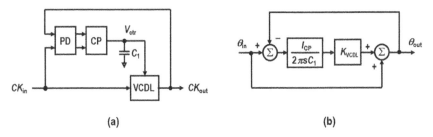

(a)   **(b)**

Figure 5.28 Delay-locked loop: (a) block diagram and (b) linear model.

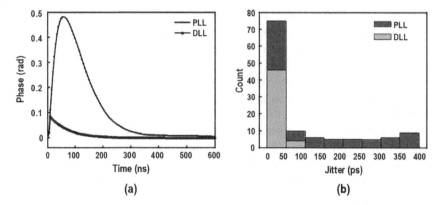

(a) (b)

Figure 5.29 Phase error propagation of the PLL and DLL with supply jump: (a) transient phase error propagation and (b) jitter histogram.

duty-cycle corrector circuit is often employed in the DLL if clock generation with 50% duty cycle is required. On the other hand, the PLL is not sensitive to the duty cycle of an input clock since the duty cycle of an output clock is determined by a VCO.

Figure 5.29 shows the jitter accumulation comparison of the PLL and the DLL. Without having the integration factor of the VCDL, the DLL suffers less from phase error propagation with a supply voltage jump than a type 2 PLL, exhibiting better immunity against the supply voltage variation. In addition, the crosstalk among multiple VCDLs is much less significant than that among multiple VCOs, making the DLL highly useful for multiple I/O links. The DLL is also useful for multiphase generation since the total delay of the VCDL that consists of multiple delay cells can be accurately controlled by feedback.

Unlike the PLL, the DLL is exposed to a potential problem of false-locking. Figure 5.30 shows the example of the false-lock condition that depends on the

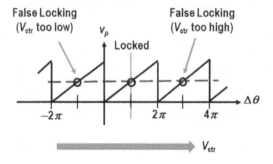

Figure 5.30 False-lock problem of the DLL.

type of a phase detector. In Fig. 5.30, an excess phase error of 360° could be considered another equilibrium point when an initial excess phase error is larger than 180°. Since the VCDL has a limited phase range, the control voltage of the VCDL could be saturated when the DLL is locked with a phase error of 360°. To avoid the false-locking problem, a startup circuit to initialize the VCDL with a proper control voltage is needed.

Compared with the PLL, the DLL cannot achieve frequency acquisition or frequency multiplication. Accordingly, its applications were limited to clock or data synchronization. The lack of frequency acquisition could degrade the synchronization performance even with a slight frequency offset between a transmitter and a receiver. To achieve the frequency acquisition or tracking, a semi-digital DLL with a phase interpolator was proposed. Its operation is based on the phase rotation of digitized phases, which will be discussed in Chapter 10. The frequency-tracking DLL is highly useful for multiple I/O links since multiple CDRs based on the DLL architecture can be employed, thus avoiding multiple VCOs.

As to the frequency multiplication, two architectures are mostly considered. The first approach employs the edge combiner circuit that combines the multiple phases of the DLL to generate high frequency as illustrated in Fig. 5.31(a). The

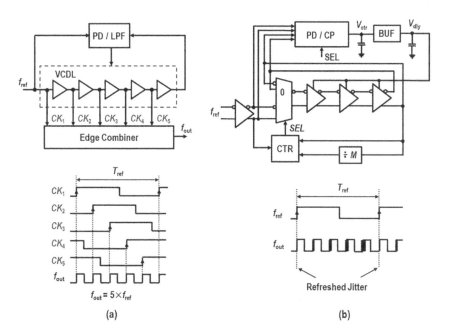

Figure 5.31 DLLs for frequency multiplication: (a) using an edge-combining DLL and (b) using an MDLL.

DLL generates multiphase clocks, and the rising and falling edges of the multiple clocks are combined by an edge combiner circuit to synthesize a high-frequency clock. Since the DLL generates a high-frequency output from a low-frequency crystal oscillator, excellent phase noise performance could be achieved. The main drawback of this architecture lies in the difficulty of phase matching and glitch-less edge combining for high-frequency operation. As a second approach, a multiplying delay-locked loop (MDLL) shown in Fig. 5.31(b) is considered for frequency multiplication. The MDLL employs a ring VCO whose phase is periodically replaced with the phase of a reference clock so that the accumulated jitter in the ring VCO is refreshed by a clean reference clock with periodic synchronization. In some sense, the MDLL can be categorized to one of the PLL architectures employing a gated ring VCO. The jitter performance of the MDLL could be worse than the conventional DLL, but the MDLL can work as a clock multiplier unit. The mismatch in the charge pump or imperfect phase selection of the multiplexed ring VCO generates a reference spur, requiring careful matching and timing like the edge-combining DLL.

References

1 V. Manassewitsch, *Frequency Synthesizers*, Theory and Design, Wiley, New York, 1987.

2 U. L. Rohde, *Microwave and Wireless Frequency Synthesizers: Theory and Design*, Wiley, New York, 1997.

3 W. Egan, *Frequency Synthesis by Phase Lock*, Wiley, New York, 2000.

4 K. Shu and E. Sanchez-Sinencio, *CMOS PLL Synthesizers; Analysis and Design*, Springer, 2005.

5 D. H. Wolaver, *Phase-Locked Loop Circuit Design*, Prentice Hall, Englewood Cliffs, NJ, 1991.

6 H. Meyr and G. Ascheid, *Synchronization in Digital Communications, Phase-, Frequency-Locked Loops, and Amplitude Control*, Wiley, New York, 1990.

7 K. Feher, *Telecommunications Measurements, Analysis, and Instrumentation*, Prentice-Hall, Englewood Cliffs, NJ, 1987.

8 B. Razavi (ed.), *RF Microelectornics*, 2nd ed., Prentice Hall, Upper Saddle River, NJ, 2012.

9 B. Razavi, *Design of CMOS Phase-Locked Loops: From Circuit Level to Architecture Level*, Cambridge University Press, United Kingdom, 2020.

10 B. Razavi, *Design of Integrated Circuits for Optical Communications*, Wiley, New York, 2012.

11 W. Rhee (ed.), *Phase-Locked Frequency Generation and Clocking: Architectures and Circuits for Modern Wireless and Wireline Systems*, The Institution of Engineering and Technology, United Kingdom, 2020.

12 M. Johnson and E. Hudson, "A variable delay line PLL for CPU-coprocessor synchronization," *IEEE Journal of Solid-State Circuits*, vol. 23, pp. 1218-1223, Oct. 1988.

13 M. Horowitz *et al.*, "PLL design for a 500 MB/s interface," in *Dig. Tech. Papers Int. Solid State Circuits Conf.*, Feb. 1993, pp. 160-161.

14 T. Lee, K. Donnelly, J. Ho *et al.*, "A 2.5 V CMOS delay-locked loop for an 18 Mbit, 500 MB/s DRAM," *IEEE Journal of Solid-State Circuits*, vol. 29, pp. 1491-1496, Dec. 1994.

15 G. Chien and P. R. Gray, "A 900-MHz local oscillator using a DLL-based frequency multiplier technique for PCS applications," *IEEE Journal of Solid-State Circuits*, vol. 35, pp. 1996-1999, Dec. 2000.

16 R. Farjad-Rad, W. Dally, H.-T. Ng *et al.*, "A low-power multiplying DLL for low-jitter multigigahertz clock generation in highly integrated digital chips," *IEEE Journal of Solid-State Circuits*, vol. 37, pp. 1804-1812, Dec. 2002.

17 M.-J. E. Lee, W. J. Dally, T. Greer *et al.*, "Jitter transfer characteristics of delay-locked loops - theories and design techniques," *IEEE Journal of Solid-State Circuits*, vol. 38, pp. 614-621, April 2003.

Part III

Building Circuits

6

Phase Detector

A phase detector (PD) is a circuit that compares the phase of an input signal with that of a voltage-controlled oscillator (VCO) and generates an error voltage whose average value is proportional to the phase difference. In the frequency synthesizer, the input signal comes from a stable source, typically a low-frequency crystal oscillator. In the clock-and-data recovery (CDR) circuit, the input signal is a high-rate non-return-to-zero (NRZ) data in which clock information is not contained. For that reason, the design aspects of the PD for the CDR circuit are far different from those of frequency and clock generation circuits. Various PDs for the CDR circuit will be separately covered in Chapter 11.

6.1 Non-Memory Phase Detectors

Nowadays, a PFD along with a charge pump is the most popular choice for phase detection in the design of frequency synthesizers and clock generators. However, other PDs have some features such as high-speed operation or low-spur generation. Since those PDs do not have the memory of previous states and deliver the instantaneous phase error information for every reference period, they can be classified as non-memory PDs. We will discuss several kinds of non-memory PDs before moving to the PFD.

6.1.1 Multiplier PD

A balanced mixer or a multiplier was one of the traditional phase detectors in the early days. When two sinusoidal signals are multiplied, a DC term at the output contains phase error information. Let $A\sin(\omega t + \theta)$ and $B\cos\omega t$ be the inputs of

Phase-Locked Loops: System Perspectives and Circuit Design Aspects, First Edition.
Woogeun Rhee and Zhiping Yu.
© 2024 The Institute of Electrical and Electronics Engineers, Inc. Published 2024 by John Wiley & Sons, Inc.

the multiplier where θ is the phase offset between two signals. Then, the output voltage $v_p(t)$ of the multiplier is described as

$$v_p(t) = A\sin(\omega t + \theta) \cdot B\cos \omega t = \frac{AB}{2}\sin\theta + \frac{AB}{2}\sin(2\omega t + \theta) \tag{6.1}$$

After a low-pass filter, a high-frequency term will be suppressed, and the DC voltage V_p of the multiplier for a small θ becomes

$$v_p \approx \frac{AB}{2}\theta = K_d\theta \tag{6.2}$$

where K_d is an effective conversion gain from an input phase error to an output voltage. From (6.2), we see that the multiplier can work as a PD with a gain of $AB/2$.

Figure 6.1 shows the PD characteristic of the multiplier PD. Phase locking occurs with a static offset of $90°$ between two sinusoidal inputs if the multiplier PD is used. Note that it is a systematic offset and should not be considered a phase error. From Fig. 6.1, the linear range of phase detection is less than π radian, and a negative-slope region is not steep enough to prevent the phase-locked loop (PLL) from getting out of an unstable region immediately. For example, when an initial phase error is close to π radian and clock jitter is substantial, a PLL momentarily has an unstable equilibrium state and exhibits a longer settling time, which results in the hang-up effect discussed in Chapter 3. In addition to the narrow linear range, the PD gain depends on the amplitude of an input signal. As a result, the PLL with this kind of an analog PD suffers from the high variation of a loop gain. Another problem of the multiplier PD is harmonic locking. For example, if a second input signal has a square waveform whose frequency is one-third of the frequency of a first input signal, there is a chance of phase locking with the third harmonic of the second input signal. Despite those issues, the multiplier PD features the fastest operation as it takes analog signals.

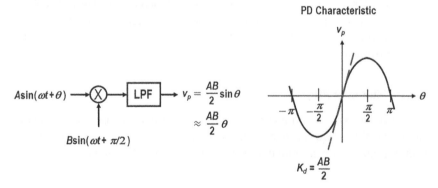

Figure 6.1 Multiplier PD.

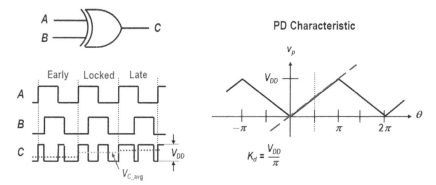

Figure 6.2 XOR PD.

6.1.2 Exclusive-OR PD

An Exclusive-OR (XOR) gate is considered a digital-domain multiplier since the output waveform of the XOR gate is equal to the product of two input digital signals. Therefore, the XOR gate can also work as a PD if the inputs are clock signals. As illustrated in Fig. 6.2, the output of the XOR gate has a balanced output with a 50% duty cycle when there is a 90° phase offset between two inputs, which is the phase-locked status in a PLL. If a feedback clock B comes earlier than a reference clock A, that is, a phase offset is less than 90°, the duty ratio of the XOR PD output becomes lower than 50%, making an effective DC level less than $0.5V_{DD}$. If the clock B comes later than a reference clock A, the duty ratio is higher than 50%, thus increasing the DC level over $0.5V_{DD}$. Since the XOR PD has a monotonic DC control over $\pm90°$, the linear range of the XOR PD is $\pm\pi/2$ radian as depicted in Fig. 6.2. Unlike the analog multiplier PD, the gain of the XOR PD does not depend on the amplitude of an input signal because of the digital operation with a rail-to-rail swing. Hence, the PD gain is given by

$$K_d = \frac{V_{DD}}{\pi} \tag{6.3}$$

Even though the XOR PD is considered the simplest PD, the use of the XOR PD is limited in practice since it has worse PD characteristic than other logic-based PDs and does not provide frequency acquisition like a PFD, which will be discussed later.

6.1.3 Flip-Flop PD

Figure 6.3 shows that a set-reset (RS) flip-flop gate operates as a PD with a balanced duty cycle when there is a 180° phase offset between two inputs. If a feedback clock B comes earlier than a reference clock A, a phase offset is less than 180°, or vice

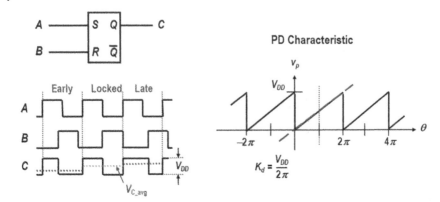

Figure 6.3 Flip-flop PD.

versa. If an inverter is inserted before one of two inputs, then the flip-flop PD has the phase lock with a 0° phase offset. Compared with the PFD, the flip-flop PD does not memorize the previous state and generates an instantaneous phase error based on the phase offset of two inputs for each reference period. The effective DC level centered at $0.5V_{DD}$ is controlled by the duty cycle of an output waveform like the XOR PD, but the linear range of the flip-flop PD is expanded to $\pm\pi$ radian, which is two times the linear range of the XOR PD. For the same supply voltage, the PD gain is reduced by half due to the extended range as shown in Fig. 6.3 and given by

$$K_d = \frac{V_{DD}}{2\pi} \tag{6.4}$$

In addition to the widened linear range, the flip-flop PD features a steep negative slope at $\pm 2\pi$ radian, exhibiting a sawtooth PD characteristic curve as shown in Fig. 6.3. Accordingly, the flip-flop PD is much less likely to have an unstable equilibrium state near $\pm 2\pi$ radian in the presence of noise, thus minimizing the hangup effect. As the non-memory PD having a wide linear range, the flip-flop PD is well employed for the delay-locked loop that does not require frequency acquisition and can also deal with clock-stretching or missing-edge clocks in some digital systems.

6.1.4 Sample-and-Hold PD

A sample-and-hold (S/H) circuit with an integrator can work as a PD if integration time is proportional to a phase error. A basic operation principle is illustrated in Fig. 6.4. A rising-edge detector generates short pulses at the rising edges of reference and divider clocks, f_{ref} and f_{div}. Once f_{ref} is triggered, a capacitor C_{int} is charged and generates an integrated voltage until the arrival of f_{div}. The integrator output is sampled and delivered to a hold capacitor C_{hold} with a S/H clock ϕ_{SH}.

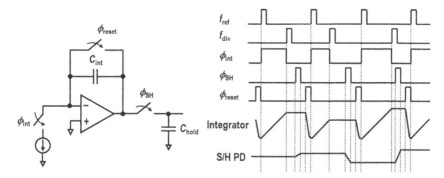

Figure 6.4 S/H PD.

Then, the integrator output is reset by a reset clock ϕ_{reset}. As a result, the output voltage at C_{hold} is proportional to a phase difference between f_{ref} and f_{div} for every reference clock period. When phase-locking occurs, f_{div} is in the middle of f_{ref} pulses, implying that a phase offset of 180° is formed like the flip-flop PD. Therefore, we have the PD gain as

$$K_d = \frac{V_{C\max}}{2\pi} \tag{6.5}$$

where V_{Cmax} is the maximum voltage range at C_{int} over 2π radian. Note that the S/H PD having an integrator does not create $1/s$ in the open-loop gain of a PLL. It is because the integrated voltage at C_{int} is reset for each reference period, thus not being propagated in the next clock periods. In other words, the S/H PD is also classified as a non-memory PD. As shown in Fig. 6.4, the output of the S/H PD is nearly constant if the reset time is negligible. Featuring such a quiet output, the S/H PD is able to achieve a low reference spur and was employed in high-performance signal source generators in the early days.

6.1.5 Sub-Sampling PD

A sub-sampling phase detector (SSPD) directly samples the waveform of a VCO with a reference clock. A functional block diagram of the SSPD with a charge pump is shown in Fig. 6.5(a). As the name implies, this PD is used for frequency multiplication without using a frequency divider. Like the multiplier PD, the gain of the SSPD depends on the amplitude or slope of an input waveform. The key advantage of the SSPD is a low-noise contribution at the PLL output since it directly samples the VCO phase. As illustrated in Fig. 6.5(b), the falling edge of a high-frequency VCO output is sub-sampled by the rising edge of a reference clock, which generates different output voltages depending on the sampling point of the VCO waveform. Since the slope of the VCO waveform is very steep, a high

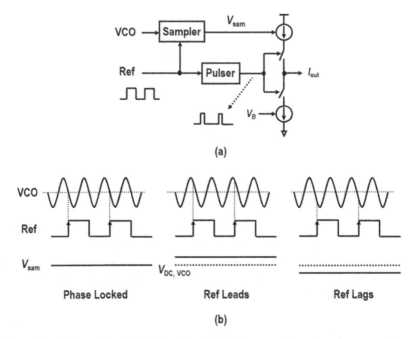

Figure 6.5 Sub-sampling PD: (a) functional block diagram with a charge pump and (b) operation principle.

PD gain could be obtained. When the rising edge of the reference clock is aligned to the zero-crossing point of the VCO waveform, a maximum PD gain is achieved. When the SSPD is combined with a charge pump, the noise contribution of the charge pump that operates at the reference frequency is deeply suppressed by the high PD gain. As a result, a low-jitter PLL can be designed with the SSPD.

The sub-sampling PLL, however, suffers from the narrow linear range in the PD characteristic and requires good frequency acquisition with additional circuitry. The reference spur performance could be problematic if there is a coupling or nonideal sampling effect when the VCO output is sub-sampled by the reference clock. More importantly, the SSPD exhibits high gain variation as well as nonlinearity since the PD gain depends on the slope of the VCO waveform. The poor linearity of the SSPD could be problematic for the in-band phase noise and spur performance in a fractional-N PLL, requiring additional compensation circuits.

6.2 Phase-Frequency Detector

A phase-frequency detector (PFD) combined with a charge pump is dominantly employed in the design of frequency synthesizers and clock generation circuits.

One of the most important features of the PFD is that it operates as a frequency acquisition aid circuit when a PLL is in the out-of-lock range.

6.2.1 Operation Principle

The PFD can memorize a previous state since its structure is based on a state machine. Figure 6.6(a) shows the common structure of the PFD that consists of two D-type flip-flops (DFFs) and one AND gate. Suppose that a feedback clock B comes earlier and that the DFF is triggered by the rising edge of the clock B. Since the data path of each DFF is connected to a supply voltage, the DFF in a lower path outputs a high signal when the rising edge of the clock B arrives. The output of the DFF in a lower path, namely a down signal D, stays high until the rising edge of a reference clock A arrives. As soon as the rising edge of the clock A is detected, the output of the DFF in an upper path, namely an up signal U, becomes high. As a result, the output of the AND gate goes to a high state, thus resetting both DFFs. Accordingly, the pulse width of the down signal D is determined by a phase offset between two clocks. When the clock A comes earlier than the clock B, the up signal U is generated with the pulse width equaling the phase error. When both clocks are synchronized, both U and D signals go to a high state but will be reset immediately as the AND output goes high. Therefore, during the

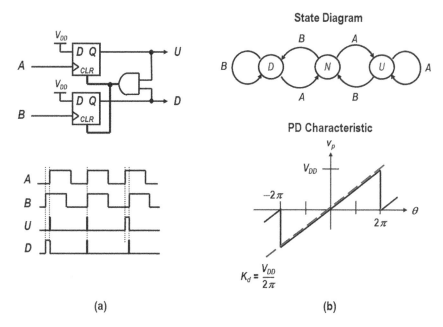

(a)   (b)

Figure 6.6 PFD: (a) block diagram and operation principle and (b) PD characteristic.

phase-locked state of the PLL, the PFD produces only a small logic glitch caused by the inherent logic delay in the feedback path. Since the pulse widths of both U and D signals are equal, a net error voltage of the PFD is considered zero. Unlike other PDs, the PFD has three states (up, down, neutral) and exhibits a phase offset of $0°$ at the phase-locking state. As depicted in Fig. 6.6, the linear range is extended to $\pm 2\pi$ radian since both U and D signals cover a detection range of 2π radian with different polarity.

Another important feature of the PFD is the frequency acquisition aid as mentioned previously. Figure 6.7 shows how the PFD itself can deliver frequency information to a PLL. Suppose that a PLL is out of a lock-in range. That is, the phase error is not bounded and varies over time with cycle slipping. Different from other PDs, the PFD generates an up signal U only during the cycle slipping when the frequency of clock A is higher than the frequency of clock B as depicted in Fig. 6.7(a). It is because two DFFs having a feedback path from a state machine

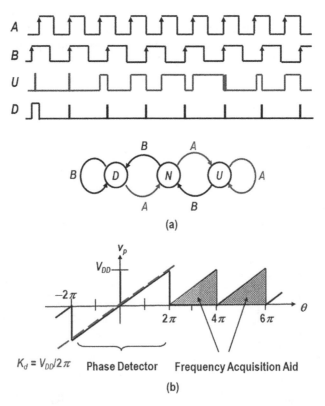

Figure 6.7 Frequency acquisition aid by the PFD: (a) up and down outputs with $f_A < f_B$ and (b) two different characteristics of the PFD.

and memorize a previous state. If the frequency of clock A is lower than the frequency of clock B, the PFD will generate a down signal D only during the cycle slipping. In that way, the PFD is able to help PLL acquire a desired frequency quickly. As learned in Chapter 2, any type 2 PLL can eventually acquire a desired frequency with a pull-in behavior. The pull-in time of the PLL, however, is too long to be useful. Hence, additional frequency acquisition aid circuits need to be employed when other PDs are used in the PLL.

It is worth mentioning that the PFD actually works as a *phase-or-frequency detector* rather than a *phase-and-frequency detector*. In other words, the PFD itself operates as either a phase detector or a frequency detector, depending on the boundary of the phase detection. The PFD operates as a normal PD when a PLL is within a lock-in range, while performing frequency detection by generating either a positive or negative pulse when the PLL is out of the lock-in range. Strictly speaking, the PFD becomes a frequency acquisition aid circuit rather than a frequency detector since the unipolar output of the PFD during the out-of-lock range is helpful for frequency acquisition rather than representing the amount of frequency difference. Figure 6.7(b) illustrates two characteristics of the PFD; phase detection for the PLL in a small-signal region and frequency acquisition aid for the PLL in a large-signal region. Compared with the flip-flop PD, the PFD has a wider linear range of 4π and the same PD gain of $V_{DD}/2\pi$.

6.2.2 Dead-Zone Problem

When a phase error is close to zero, the PFD generates a very narrow pulse whose width is too narrow to completely turn on a charge pump. As a result, the charge pump only produces a small portion of the full current, resulting in a weak PD gain. This phenomenon is called a PFD dead zone. The dead zone of the PFD significantly reduces a loop bandwidth. Such a reduced bandwidth degrades in-band noise performance since the phase noise of a VCO could not be suppressed much with the narrow bandwidth of the PLL. In the extreme case that the charge pump cannot be turned on at all, the PLL does not provide any feedback information to correct the VCO phase, making the in-band phase noise of the PLL close to the phase noise of an open-loop VCO.

The dead-zone effect in a CP-PLL can be classified into two kinds; one is a PFD-induced dead zone, and the other is a charge-pump-induced dead zone. Two cases are illustrated and compared in Fig. 6.8. The PFD-induced dead zone is caused by a weak driving capability of the output stage, typically an inverter, for a given load capacitance. Figure 6.8(a) shows the case that the PFD generates an insufficient voltage level of the U signal due to a very small phase error and does not turn on the up current of the charge pump. This problem can be overcome by having a proper transistor size at the output stage of the PFD. To the contrary,

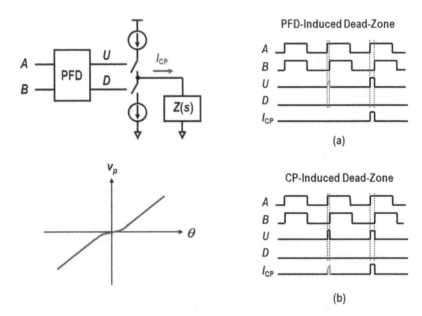

Figure 6.8 Dead-zone effect of the PFD: (a) PFD-induced and (b) charge-pump-induced.

the charge-pump-induced dead zone shown in Fig. 6.8(b) is more problematic. Even though a strong buffer is designed at the PFD output, the charge pump may not be completely turned on in a short time due to a limited switching time of the charge pump itself. It is mainly due to the parasitic capacitance and the turn-on resistance of a MOSFET switch. The charge-pump-induced dead zone generates some portion of the charge pump current, while the PFD-induced dead zone completely shuts down the charge pump current. Once the dead-zone problem is encountered, the variation of a PLL bandwidth is extremely sensitive to process, voltage, and temperature (PVT) variations. For instance, if the bandwidth changes by several times even with a slight change of a supply voltage in the testing of a PLL chip, it is very likely that the PLL has the dead-zone problem.

Fortunately, there is a simple way of solving the dead-zone problem of the PFD with logic gates only. By having the up and down pulses begin with a fixed minimum pulse width regardless of a phase error, we can make the charge pump turn on even with a zero phase error. This operation can be done by adding a delay cell between the AND gate and the reset input as shown in Fig. 6.9. The minimum turn-on time Δt_{on} of the PFD enables the charge pump to have flowing currents whenever the PFD generates U and D outputs. As illustrated in Fig. 6.9, both up and down currents, I_U and I_D, will be turned on even with the phase-locked condition. Since both currents have an equal pulse width, no net current will flow from the charge pump to a loop filter. The minimum turn-on time of the PFD is

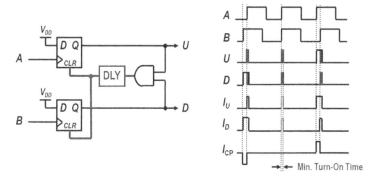

Figure 6.9 PFD without dead-zone.

an important parameter for PLL performance and should be carefully determined along with charge pump design as learned in Chapter 4.

6.2.3   Effect of the PFD Turn-On Time on PLL Settling

When a PLL is outside the lock-in range with cycle slipping, the PFD works as a frequency detector. If the ratio of T_{ref} to Δt_{on} is low, the minimum turn-on time Δt_{on} of the PFD can degrade the settling behavior of the PLL. Figure 6.10(a) shows how a reset delay affects the settling time of the PLL. If Δt_{on} takes a significant portion of a reference clock period T_{ref}, the PD gain inversion depicted in Fig. 6.10(a) occurs for a phase difference larger than $2\pi - 2\Delta\theta$, where $\Delta\theta = 2\pi\Delta t_{on}/T_{ref}$. The portion of the PD gain inversion could be substantial when the PLL uses high f_{ref}. In the extreme case of $\Delta\theta = \pi$, the PFD behaves as a flip-flop PD. Figure 6.10(b) shows the transient responses of a third-order type 2 PLL with different Δt_{on} values for the same T_{ref}. The $\Delta t_{on}/T_{ref}$ values of 4%, 8%, and 12% are used. As expected, the plot with the $\Delta t_{on}/T_{ref}$ value of 12% exhibits the lowest slew rate outside the lock-in range. Three plots clearly show that a large-signal settling time during frequency acquisition depends on the minimum turn-on time of the PFD.

6.2.4   Noise Performance of PFD

The thermal noise contribution of CMOS digital circuits operating with rail-to-rail signals is much less than that of analog circuits operating in a small-signal region. Accordingly, the phase noise contribution of a PFD to the PLL is considered negligible in comparison with that of a charge pump. Another concern would be supply noise coupling since the PFD consists of single-ended logic gates. The gate delay time of the logic gate is highly sensitive to supply voltage variation. Nonetheless, the PFD can offer power supply rejection to a certain degree since it operates like

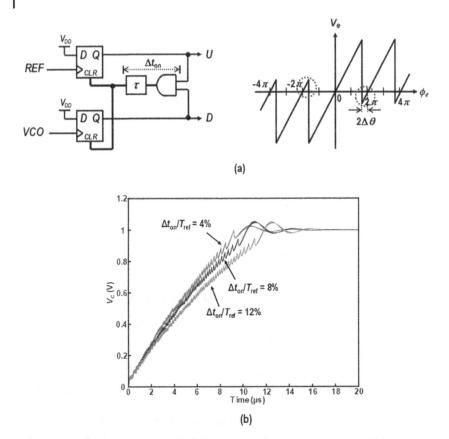

(a)

(b)

Figure 6.10 (a) PFD characteristic at high frequency and (b) Δt_{on} effect on PLL settling.

a pseudo-differential block in the PLL. For example, suppose that the DFFs and the delay cell shown in Fig. 6.9 are exposed to supply voltage variation. Since the delay modulation due to the supply voltage affects both U and D outputs nearly at the same time, a net error coming from the difference of U and D is negligible within the PLL.

To provide better immunity against the supply noise, we may think of a fully differential implementation even for the PFD. The fully differential PFD with a small-signal voltage swing, however, is not a good way of achieving high-resolution or low-power performance. Even though the PFD operates at a low-reference frequency, it still requires steep rising and falling slopes to achieve good timing resolution. If the PFD is designed with current-mode logic (CML) circuits, a huge bias current is required to have a similar gate delay time offered by CMOS standard logic circuits. In addition, a differential PFD with small-signal voltage swing gives a potential problem of increased timing jitter, which is mostly not a concern in the case of the single-ended PFD design.

6.3 Charge Pump

The use of a charge pump makes it possible to build a type 2 PLL with a passive loop filter. Since the charge pump is the important building block that determines the in-band phase noise and spur performance of the frequency synthesizer especially with a large division ratio, choosing a right circuit topology based on the design requirements of the PLL is important.

6.3.1 Circuit Design Considerations

Like other analog circuits, several parameters such as DC headroom, matching, noise, and linearity need to be addressed for the design of a charge pump. Below are key design factors to be considered for the charge pump:

- PFD dead-zone
- Operation frequency
- Output voltage compliance and DC headroom
- Matching between up and down currents
- Leakage current of a loop filter
- Linearity
- In-band phase noise contribution

As discussed previously, the minimum turn-on time of a PFD should consider the switching time of the charge pump. A fast-switching charge pump not only reduces the minimum turn-on time of the PFD but also enhances phase resolution and linearity. Especially when a single-ended charge pump is designed, minimizing the turn-on time of the charge pump is critical in achieving low spur and in-band noise contribution at the PLL output. The operation frequency of the charge pump is not a concern in the cases of frequency synthesizers and clock generators where low reference frequency is mostly used. However, when the charge pump is employed in the second stage of a cascaded PLL or a CDR system, the operation frequency becomes high, and the use of a fully differential charge pump is a must. If the charge pump offers a wide output-voltage range with good DC headroom, a low-gain VCO can be designed for a given tuning range requirement. When the VCO is designed with a lower gain, better phase noise performance can be achieved. Therefore, having a wide voltage compliance at the output of the charge pump is important to improve the phase noise of the PLL. The compliance voltage range of the charge pump should be determined based on the matching requirement of the up and down currents.

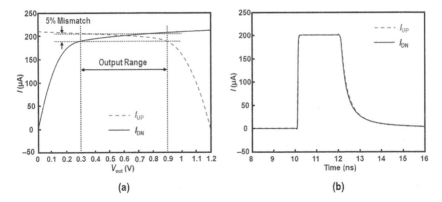

Figure 6.11 Charge pump design: (a) DC compliance voltage range under 5% matching requirement and (b) transient waveforms of the up and down currents with rising and falling edges matched.

6.3.1.1 Current Matching, Leakage Current, and Linearity

The current matching in the charge pump is important to achieve a good reference spur performance. The single-ended charge pump has difficulty in maintaining good current matching over a wide voltage range since the current matching between PMOS and NMOS current mirrors needs to be considered. Figure 6.11 shows the DC and transient simulation results of the matching performance between up and down currents in a single-ended charge pump. In Fig. 6.11(a), both up and down currents are plotted with the output voltage of the charge pump swept from 0 V to a supply voltage of 1.2 V. As the output voltage is reduced, the down current I_{DN} decreases due to the channel-length modulation of NMOS current-mirror transistors. When the output voltage gets close to 0 V, the NMOS current-mirror transistors operate in the linear region. When the output voltage is close to the supply voltage, a similar situation happens to the up current I_{UP} that is supplied by PMOS current-mirror transistors. The best matching point occurs when the output voltage is near the middle of the output voltage range. Based on the DC response of the up and down currents, the compliance-voltage range of the charge pump can be defined. In Fig. 6.11(a), the voltage range of 0.3–0.9 V can be considered the effective compliance-voltage range for a 1.2-V supply under a current mismatch of 5%. Not only the matching in the DC response but also the matching in the transient response is important since the DC response only shows the matching performance when both the up and down currents are settled. In other words, the current matching not only for the maximum amplitude but also for the rising and falling waveforms should be considered in the charge pump design since the transient mismatch can add a certain amount of net charge to a loop filter. Figure 6.11(b) shows the example of the transient matching between the up and down currents. Note that the worst-case performance of the transient matching must be examined not only with PVT variations but also at the highest and the lowest output voltages.

As discussed in Chapter 4, the mismatch between up and down currents causes a reference spur whose level in a second-order CP-PLL is approximately given by (4.28). It is written again here, that is,

$$P_{\text{spur}} \approx 20 \log \left(2\pi^2 N \frac{f_{\text{BW}}}{f_{\text{ref}}} \left(\frac{\Delta t_{\text{on}}}{T_{\text{ref}}} \right)^2 \frac{\Delta i}{I_{\text{CP}}} \right) \tag{6.6}$$

Equation (6.6) implies that, for the given values of the PLL bandwidth and the charge pump mismatch, the reference spur can still be reduced by having a shorter turn-on time of the PFD. Therefore, it is important to improve the current switching time of the charge pump. Similarly, the effect of a leakage current in the loop filter expressed in (4.24) is written again

$$P_{\text{spur}} \approx 20 \log \left(2\pi N \frac{f_{\text{BW}}}{f_{\text{ref}}} \frac{I_{\text{leak}}}{I_{\text{CP}}} \right) \tag{6.7}$$

Having a high charge pump current is good to mitigate the effect of the leakage current. Another way is to reduce the division ratio by having a higher reference frequency for the given output frequency.

The linearity of the charge pump is another important factor in the design of a fractional-N PLL, which will be discussed in Chapter 9. Note that the charge pump linearity performance should be separately considered from the matching performance. The current mismatch causes a static phase error, that is, the large-signal performance of a PLL, while the nonlinearity induces a noise folding effect, that is, the small-signal performance of a fractional-N PLL. Therefore, good matching does not necessarily guarantee high linearity since the charge pump having a current mismatch can still achieve good linearity even with a static phase error.

6.3.1.2 Phase Noise Contribution

If a stable reference frequency source such as a crystal oscillator is used in the PLL, then the charge pump usually becomes a dominant contributor of the in-band phase noise. To evaluate the phase noise contribution of the charge pump at the PLL output, it is important to consider an input-referred phase noise instead of an output-referred current noise. Otherwise, we will reach a wrong conclusion that higher charge pump current always results in worse-phase noise performance at the PLL output. When the charge pump current doubles, the PD gain also doubles, while current noise increases by the square root of two. As a result, there is a 3-dB noise improvement in the signal-to-noise ratio (SNR) performance of the charge pump.

In a single-ended charge pump, the turn-on time Δt_{on} of the PFD plays an important role. Since a tri-state charge pump has current flowing for Δt_{on} at the phase-locked state, having a small Δt_{on} is important to reduce the in-band phase noise contribution of the charge pump at the PLL output. When the reference frequency increases, the noise contribution of the charge pump is reduced with a decreased frequency division value. However, the ratio of $\Delta t_{on}/T_{ref}$ increases at the same time. If the reference frequency doubles, the frequency division ratio is reduced by half, improving phase noise by 6 dB at the PLL output, but the phase noise contribution of the single-ended charge pump is increased by 3 dB as the ratio of $\Delta t_{on}/T_{ref}$ doubles. As a result, only 3-dB noise improvement is achieved. To the contrary, Δt_{on} does not play a critical role to reduce the noise contribution in the case of a fully differential charge pump. Therefore, the inherent noise performance of the differential charge pump is worse than that of the singe-ended charge pump. As Δt_{on} does not affect the noise performance, the phase noise contribution of the differential charge pump increases by 6 dB at the PLL output if the frequency division ratio doubles.

6.3.1.3 Design Flow
Figure 6.12 shows a typical design flow for the charge pump. The first step is to set the value of the charge pump current by considering system parameters such as the overall PD gain, the in-band noise requirement, the leakage current of the loop filter, and so on. The next step is to choose a charge pump topology. Choosing a right architecture between differential and single-ended topologies is one of the

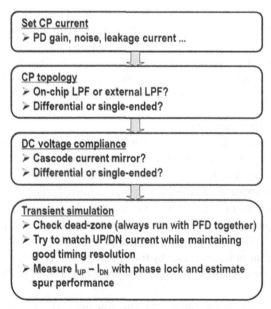

Figure 6.12 Design flow of the charge pump circuit.

important procedures in the design of the charge pump. When an on-chip loop filter is designed, designing a differential charge pump followed by a differential LPF is a good choice to mitigate the noise coupling and leakage current problems of the loop filter. The differential charge pump is also useful to achieve good current matching, to enhance charge pump linearity, and to improve output compliance voltage. When an external loop filter is used, a single-ended charge pump is a good choice to reduce a number of package pins. The single-ended charge pump is also useful to reduce power consumption and further improve the PLL noise performance with the tri-state operation.

The charge pump design should be completed with transient simulations. There are two reasons for running the transient simulation. One is to check the PFD dead zone, and the other is to match the transient waveforms of up and down currents as discussed previously. For the transient simulation, it is helpful to put a huge capacitor, for example 1 μF, at the charge pump output instead of an actual loop filter to have nearly a constant output voltage during the transient simulation. For checking the PFD dead zone, it is important to run the simulation with the PFD and the charge pump together. For example, with the simulation condition set to hot temperature or slow corner process, the turn-on time of the charge pump increases due to slow current switching. If the minimum turn-on time of the PFD is set by the worst-case performance of the charge pump, then it is possible that the minimum turn-on time of the PFD is set to a longer value than necessary, resulting in increased noise contribution of the charge pump. It is because the turn-on time of the PFD and the switching speed of the charge pump tend to move the same direction over temperature and process variations. For example, even though the minimum turn-on time of the PFD decreases with the simulation condition of the fast-corner process, the charge pump also operates faster with the fast-corner process. Indeed, there is no need of increasing the minimum turn-on time of the PFD to satisfy the worst-case requirement of the turn-on time of the charge pump.

When a fully differential charge pump is designed, the PFD dead zone is only caused by the PFD-induced dead zone rather than by the charge pump-induced dead zone. In that case, the delay time set by a delay line in the PFD can be reduced to a small value enough for the output stage to drive the gate capacitance of a charge pump switch. Lastly, it is important to check both the up and the down currents together instead of checking the difference of the up and the down currents. If the charge pump-induced dead zone occurs, the charge pump with incomplete switching still makes it possible to have no net current flow into the LPF. As a final step, we put two phase-locked signals to the PFD followed by the charge pump with an actual LPF. By measuring a voltage ripple in the high-order LPF, we can estimate the level of a reference spur caused by the charge pump mismatch as learned from Chapter 4.

6.3.2 Single-Ended Charge Pump Circuits

A single-ended charge pump provides low-power and low-noise performance with tri-state operation. In the conventional integer-N frequency synthesizer, a high charge pump current is used to achieve good reference spur performance over the leakage current of a loop filter and to have high SNR at the charge pump output. With the tri-state operation, the current consumption of the charge pump in a phase-locked condition is significantly reduced with the low duty ratio of the turn-on period.

A charge pump design with singled-ended operation can be categorized into three topologies according to the location of MOSFET switches for current-source transistors; at the drain, at the gate, and at the source, as shown in Fig. 6.13(a)–(c), respectively. The drain-switching charge pump shown in Fig. 6.13(a) has the straightforward structure and offers fast current switching. However, it suffers from a glitch current during the current switching. We consider the mechanism of the drain-current switching in three phases as illustrated in Fig. 6.14. When a

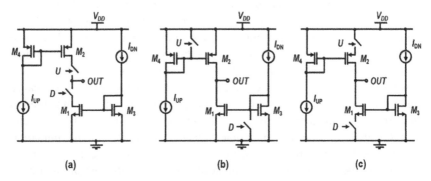

Figure 6.13 Circuit topologies of single-ended charge pumps: (a) with drain switching; (b) with gate switching; and (c) with source switching.

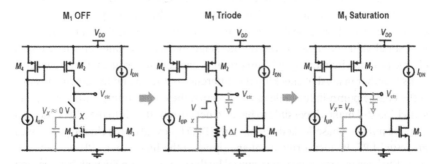

Figure 6.14 Current mirror with different modes in the drain-switching charge pump.

down switch D is turned off, the current pulls the drain of a transistor M_1 to the ground. After the switch is turned on, the voltage at the drain of M_1 increases from 0 V to the control voltage of a VCO. During the pull-up time of the drain voltage, M_1 has to be in a linear region until the drain voltage of M_1 is higher than the minimum saturation voltage $V_{DS1,sat}$. During this period, a high peak current is generated even when a clock feedthrough or charge coupling is not considered as illustrated in Fig. 6.14. The amount of the peak current is determined by $V_{ctr}/(R_1 + R_{SW1})$ where R_1 and R_{SW1} are effective turn-on resistors of the NMOS current mirror and a down switch respectively, and V_{ctr} is the output voltage of the charge pump. On the PMOS side, the same situation will occur with opposite polarity. Good matching for these peak currents from the NMOS and PMOS parts is difficult since the amount of each current varies over a large variation of V_{ctr} in the PLL.

The gate-switching charge pump shown in Fig. 6.13(b) has less current glitch during current switching than the drain-switching charge pump. Since the output voltage of the charge pump is fixed by a control voltage of the PLL and the gate voltage is set by diode transistors M_3 and M_4, the current mirrors are always in the saturation region. However, the operation speed of the gate-switching charge pump becomes slow due to large gate capacitance. Figure 6.15(a) shows the case that a large gate capacitance of $C_{ox}WL$ from M_1 transistor occurs when the current mirror transistors are designed with a high mirroring ratio to save the DC power. Then, the switching time is determined by a time constant of $(R_{DW} \| 1/g_{m3})(C_{G1} + C_{G3})$ where R_{DW} is the turn-on resistance of the D switch, g_{m3} is the transconductance of M_3, and C_{G1} and C_{G3} are the gate capacitance of M_1 and M_3, respectively.

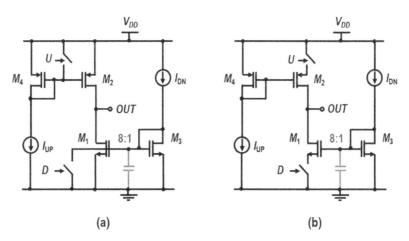

(a) (b)

Figure 6.15 Performance comparison with large gate capacitance for noise filtering: (a) gate-switching and (b) source-switching.

Since the gate capacitance affects the switching speed, it is difficult to add an extra capacitor at the gate to further improve the noise performance of the charge pump.

The third option is to have a switch at the source of the current mirror transistor as shown in Fig. 6.13(c). Like the gate-switching charge pump, output transistors are in the saturation region all the time. Unlike the gate-switching, neither the transconductance of M_3 or M_4 nor the gate capacitance of M_1 and M_2 affects the switching time. As a result, a low bias current can be used to generate a high output current with a high mirroring ratio. This architecture achieves a faster switching time than the gate-switching charge pump because of the lower parasitic capacitance seen by the switches. In addition, a large capacitor can be added at the gate of the current mirror for noise performance since the gate capacitance does not slow down the switching speed. In fact, the large gate capacitor improves the switching time since the source-switching charge pump suffers from a slow turn-off time due to fluctuation at the gate through capacitive coupling when the source node becomes a floating node. Based on the previous discussion, the source-switching charge pump is attractive by considering speed, power, and noise performance. Figure 6.16 shows a schematic example of a source-switching charge pump. Transistors M_4–M_{13} form current mirrors that are cascoded to have high output impedance and decouple the current mirror from variable control voltage. M_2 and M_{15} are used for replica biasing to give the same bias condition of switches, M_1 and M_{14}, when both up and down currents are on during phase-locked condition. A large capacitor of tens of pF is used at the gate of current mirrors to achieve fast switching as well as low noise performance.

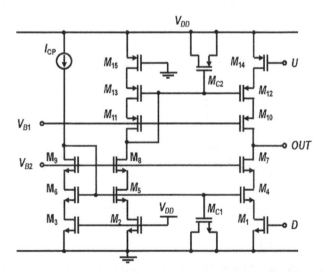

Figure 6.16 Schematic of a high-performance source-switching charge pump.

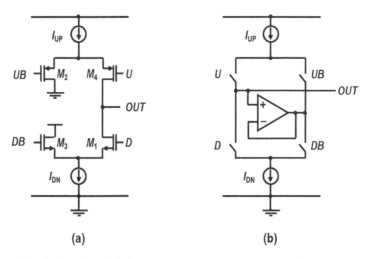

Figure 6.17 Semi-differential charge pumps: (a) current-steering switching and (b) current-steering switching with an active amplifier.

6.3.3 Semi- and Fully Differential Charge Pump Circuits

In addition to the typical configuration of the charge pump, some variations can be done to improve the performance of speed and matching. Figure 6.17(a) shows the charge pump that uses a current-steering switch instead of a voltage-mode switch. The current-steering charge pump avoids a floating node during switching by keeping a constant bias current at the cost of DC power, which results in fast switching. However, different parasitic capacitance and turn-on resistance values between NMOS and PMOS switches make it difficult to achieve a good matching. During a neutral state in which both U and D switches are turned off, the drain voltages of the NMOS and the PMOS current mirrors are set to supply and ground voltages, respectively, suffering from a large voltage fluctuation. As a result, a charge-sharing effect between the parasitic capacitor of a current mirror and a loop filter capacitor during a phase-locked state becomes serious especially when a small capacitor value is used in the loop filter. To mitigate the charge-sharing effect during the phase-locked state, the drain voltages of the NMOS and the PMOS current mirrors are always connected either by the up and down switches or by the UB and the DB switches as shown in Fig. 6.17(b), and a unity-gain amplifier is designed to make the drain voltage of the current mirror the same as the output voltage. The disadvantages of this topology would be that additional DC current is consumed by the amplifier and that the output compliance voltage of the charge pump is limited by that of the amplifier.

A fully differential charge pump has several advantages over the conventional single-ended charge pump or the semi-differential charge pump. Firstly, the matching requirement between NMOS and PMOS transistors in the single-ended charge pump is not necessary, and it is relaxed to the matching between NMOS transistors or between PMOS transistors. Secondly, as the differential charge pump uses either NMOS- or PMOS-only switches, a mismatch between up and down signals from a PFD is tolerable by the symmetric-input charge pump. Thirdly, a differential topology doubles the range of an output compliance voltage, which is useful to extend the tuning range of a VCO for a given supply voltage. Fourthly, the differential charge pump is less sensitive to a leakage current or the capacitor nonlinearity in a loop filter. Lastly, the differential topology offers good immunity against noise coupling. However, the differential charge pump has always-on currents, so it does not take advantage of the tri-state operation to mitigate the noise contribution at the PLL output. As to the matching requirement of the differential charge pump, even though the matching requirement is much relaxed compared with the single-ended charge pump, the current mismatch among PMOS or NMOS current mirrors behaves like the leakage current problem of the single-ended charge pump. Therefore, the current matching among current mirrors in both PMOS and NMOS parts is still critical.

Figure 6.18 shows a schematic example of the differential charge pump. NMOS transistors M_1–M_4 are current-steering switches for the up and down currents. Cascoded transistors M_5–M_8 provide high impedance output and isolate the current switches from a variable output voltage of the charge pump. The up and down current mirrors can also be cascoded with proper cascode biasing to further improve the charge pump performance at the cost of reduced DC headroom.

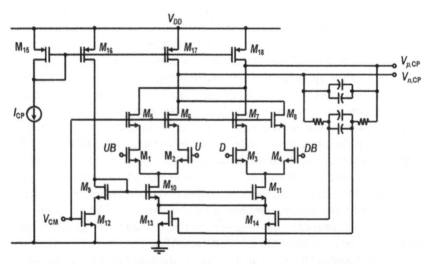

Figure 6.18 Schematic of a fully differential charge pump.

A common-mode-feedback (CMFB) circuit is implemented using linear-region transistors, M_{13} and M_{14}. The common-mode reference voltage V_{CM} is applied not only to the gate of M_{12} but also to the gates of M_5–M_8 for cascode biasing if the differential output swing of the charge pump is limited to $\pm 2V_{th}$ where V_{th} is the threshold voltage of M_5–M_8. To sample the CMFB information of the PLL, a control voltage across an integration capacitor is used instead of the differential output voltage, which is useful to avoid a high-frequency voltage ripple caused by resistors in the loop filter. When the current of PMOS current mirrors, M_{17} and M_{18}, is larger than that of NMOS current mirrors, M_{10} and M_{11}, both output voltages, $V_{p,CP}$ and $V_{n,CP}$, will increase. Then, the turn-on resistance of the linear-region transistors, M_{17} and M_{18}, decreases, making the gate-source voltage of M_{10} and M_{11} increase. Then, the output currents of M_{10} and M_{11} will increase and eventually have balanced up and down currents, performing the CMFB properly.

Figure 6.19 shows a simplified functional diagram of the differential charge pump. Two cases with the phase locked and the VCO lagged are shown in Fig. 6.19(a), (b), respectively. In the phase-locked mode, both the up and down switches, U and D, are turned on, which makes the charge pump current flow directly from PMOS current mirrors to NMOS current mirrors. As a result, there is no net current flowing to a loop filter. When the VCO phase is lagged, the up switch U and the complementary down switch DB are turned on as shown in Fig. 6.19(b). Accordingly, the complementary output node $V_{n,CP}$ will be pulled down with a sink current whose value equals I_{CP}, while the output node $V_{p,CP}$

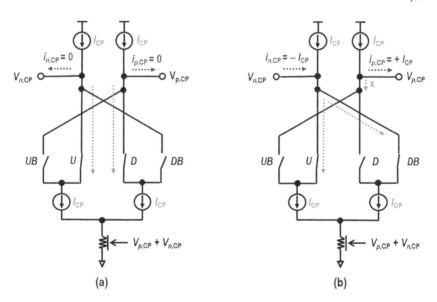

(a)   (b)

Figure 6.19 Functional diagram: (a) with phase locked and (b) with VCO phase lagged.

will be pulled up with a source current whose value equals I_{CP}. Therefore, the fully differential operation is achieved along with the CMFB circuit.

6.3.4 Design of Differential Loop Filter

When an on-chip LPF is designed, a capacitor in the integral path covers most of the loop filter area. Since a MOSFET capacitor has much higher capacitance density than a metal-to-metal capacitor, it is often employed to save the loop filter area in spite of the gate leakage current and poor linearity. The conventional MOSFET capacitor exhibits non-monotonicity in a voltage-to-capacitance transfer function curve as shown in Fig. 6.20(a). Those non-monotonicity and poor linearity characteristics are harmful for the design of the PLL. Having an NMOS transistor in an N-well, an *accumulation-mode varactor* offers a monotonic transfer function as shown in Fig. 6.20(b). If the accumulation-mode varactor is used as a differential low-pass filter (LPF), a symmetric configuration provides a small variation of capacitance over a wide voltage range as depicted in Fig. 6.21. If a nominal differential voltage ΔV_{diff} is set to 0 V as a middle point of the control voltage and the center frequency of a free-running VCO is tuned to a target frequency, a gate leakage current can be small. Compared with the

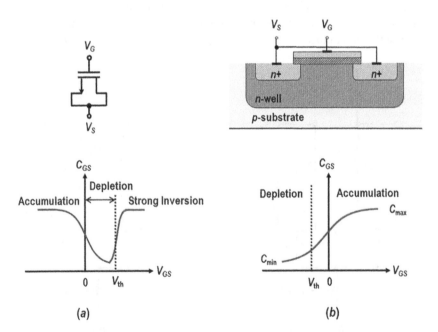

Figure 6.20 Comparison of the MOS capacitor and accumulation-mode varactor: (a) MOS capacitor and (b) accumulation-mode capacitor.

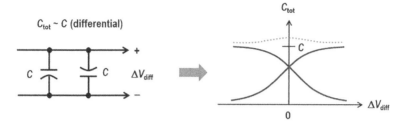

Figure 6.21 Effective capacitance of the symmetric accumulation-mode varactor.

metal-to-metal capacitor, the benefit of area reduction with a differential topology is mitigated for the accumulation-mode varactor since the effective capacitance of each capacitor at the middle of a tuning range is reduced by half as illustrated in Fig. 6.21.

To analyze the loop dynamics of the PLL having a differential loop filter, it is useful to convert the differential loop filter into the equivalent model of the conventional single-ended loop filter. To maintain the same loop gain as the differential model, a weighting factor of 2 must be carefully put in the equivalent single-ended model. As shown in Fig. 6.22, there are three parameters for which a factor of 2 can be considered in the equivalent singed-ended model. Those are the VCO gain K_v, the series resistor R_1, and the charge pump current I_{CP}. As a first case, if a factor of 2 is considered in K_v, we will have the same unity-gain frequency in the open-loop gain, but changing K_v gives confusion in consideration of the tuning range and the phase noise performance of the VCO. As a second case, the factor of 2 can be considered for R_1. Increasing the value of R_1, however, also affects the zero and pole frequencies of the open-loop gain. Lastly, let us consider

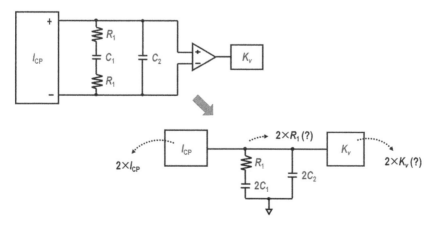

Figure 6.22 Conversion of a differential LPF to a single-ended LPF model.

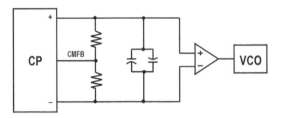

Figure 6.23 Second-order type 1 CP-PLL with the differential loop filter.

the factor of 2 in I_{CP}. Increasing I_{CP} by two only doubles the PD gain but does not affect the VCO parameter, the zero frequency, or the pole frequency. Therefore, using $2I_{CP}$ is the best choice in modeling the equivalent single-ended LPF from the differential LPF. Note that half the value of C_1 should be considered in Fig. 6.22 if an accumulation-mode varactor is used instead of a metal-to-metal capacitor.

With the use of a differential loop filter, designing a type 1 CP-PLL can be done without considering a DC voltage for the loop filter. Figure 6.23 shows an example of the loop filter configuration for a second-order type 1 CP-PLL. By having a differential resistor without a capacitor in series, an integral path is removed, resulting in the type 1 CP-PLL. A shunt capacitor is still needed to reduce a large voltage ripple, which prevents the output of the charge pump or the input of the VCO from being saturated as discussed in Chapter 2. If a high-order filtering is necessary, a third-order type 1 CP-PLL can also be designed by having an additional *RC* filter. Knowing that the main role of the integral path is to achieve frequency acquisition and can be considered the large-signal operation of the PLL, we consider implementing a separate frequency acquisition loop based on digital implementation. In that way, the use of a large area of the integration capacitor can be avoided. This is the basic operation principle of a so-called hybrid PLL, which will be further discussed in Chapter 10.

Example 6.1 *CP-PLL with other types of PDs*
It is often misunderstood that the charge pump can be used only with a PFD. As discussed, the PFD is just a kind of another PD once a PLL operates within a lock-in range. In fact, other logic-based PDs such as an XOR PD or a flip-flop PD can also work with the charge pump. Figure 6.24 shows the example of a CP-PLL using the XOR PD. Since the XOR PD has a single output, up and down outputs are generated by using the complementary output of the PD. Figure 6.25 shows the transient performance comparison of PLLs employing different types of PDs; the XOR PD, the flip-flop PD, and the PFD. Other loop parameters are unchanged in the transient simulations.

The CP-PLLs with the flip-flop PD and the PFD exhibit nearly the same transient settling performance as shown in Fig. 6.25(a). It is because the PD gain is the same

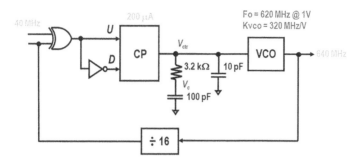

Figure 6.24 CP-PLL with an XOR PD.

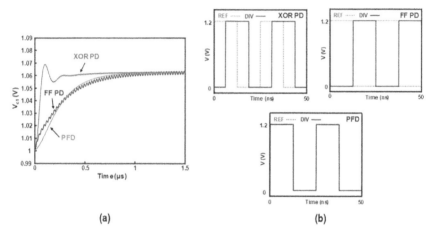

(a)   (b)

Figure 6.25 Simulation results of the CP-PLL with other types of PDs: (a) control voltage V_{C1} settling and (b) input waveforms of PDs.

for both PDs. The simulation result also verifies that the PFD works as a normal PD when the PLL is within a lock-in range. The different facts of the PFD are a wider range of linear phase detection and the 0° phase offset at the phase lock condition as discussed. As expected, the PLL with the XOR PD exhibits the 90° phase offset at the phase lock as shown in Fig. 6.25(b). It has faster settling than other PLLs since the XOR PD has higher PD gain than other PDs even though it has a narrower linear range. The bi-state PDs such as the XOR PD and the flip-flop PD always generate an output ripple with up and down states, while the tri-state PFD generates a pulse only when there is a phase error. Accordingly, the PLL using the PFD can achieve less voltage ripple, equivalently, lower reference spur than that with the XOR PD or the flip-flop PD. It also implies that the reference spur cannot be fully suppressed with the bi-state PD even when the perfect matching is achieved in the charge pump design.

References

1 F. M. Gardner, *Phaselock Techniques*, 3rd ed., Wiley, New York, 2005.

2 W. Egan, *Frequency Synthesis by Phase Lock*, 2nd ed., Wiley, New York, 2000.

3 D. H. Wolaver, *Phase-Locked Loop Circuit Design*, Prentice Hall, Englewood Cliffs, NJ, 1991.

4 B. Razavi, *Design of CMOS Phase-Locked Loops: From Circuit Level to Architecture Level*, Cambridge University Press, United Kingdom, 2020.

5 B. Razavi (ed.), *RF Microelectornics*, 2nd ed., Prentice Hall, Upper Saddle River, NJ, 2012.

6 K. Shu and E. Sanchez-Sinencio, *CMOS PLL Synthesizers; Analysis and Design*, Springer, 2005.

7 F. M. Gardner, "Charge-pump phase-locked loops," *IEEE Transactions on Communications*, vol. 28, pp. 1849–1858, 1980.

8 W. Rhee, "Design of high-performance CMOS charge pumps in phase-locked loops," in *Proceedings of IEEE International Symposium of Circuits and Systems*, June 1999, pp. 545–548.

9 X. Gao, E. Klumperink, M. Bohsali *et al.*, "A low noise sub-sampling PLL in which divider noise is eliminated and PD/CP noise is not multiplied by N^2," *IEEE Journal of Solid-State Circuits*, vol. 44, pp. 3253–3263, 2009.

10 K. Lee, B. Park, H. Lee *et al*, "Phase frequency detectors for fast frequency acquisition in zero-dead-zone CPPLLs for mobile communication systems," in *Proc. European Solid-State Circuits Conference*, 2003, pp. 525–528.

11 R. He, J. Li, W. Rhee *et al*, "Transient analysis of nonlinear settling behavior in charge-pump phase-locked loop design," in *Proc. IEEE International Symposium on Circuits and Systems*, May, 2009, pp. 469–472.

7

Voltage-Controlled Oscillator

For the design of an on-chip voltage-controlled oscillator (VCO), two types of VCOs are mostly considered. One is an LC VCO whose oscillation frequency is set by an inductor–capacitor (LC) tank. An ideal LC resonator oscillates by itself, but an actual LC tank cannot sustain oscillation due to an energy loss in the LC tank. The decay time of the oscillation depends on the quality factor of the LC tank. To maintain oscillation, an active circuit is required with DC power consumption. The LC resonator reinforced by the active circuit becomes an LC oscillator. As illustrated in Fig. 7.1(a), an active circuit creates an effective negative resistance to compensate for the energy loss in the LC resonator. On the other hand, a ring oscillator shown in Fig. 7.1(b) maintains oscillation with meta-stability of cascaded logic circuits. The ring oscillator was originally used to measure the gate delay time of an inverter to evaluate the performance of a newly developed CMOS technology by process engineers since the gate delay time of the inverter could be estimated by measuring the oscillation frequency of the ring oscillator. With the development of CMOS phase-locked loops (PLLs), the ring VCO became popular especially in wireline communication systems. The ring VCO features a wide tuning range, a compact area, and multi-phase generation, while the LC VCO offers a high maximum frequency and low phase noise. Another traditional VCO, namely a relaxation VCO shown in Fig. 7.1(c), was employed for low-frequency clock generation before the advent of the ring VCO.

We will learn the basic principle of the VCO based on the traditional LC VCO in the following sections. Then, the ring VCO will be discussed by focusing on different design aspects in comparison with the LC VCO. In the last part of this chapter, we will also briefly cover the relaxation VCO.

Phase-Locked Loops: System Perspectives and Circuit Design Aspects, First Edition.
Woogeun Rhee and Zhiping Yu.

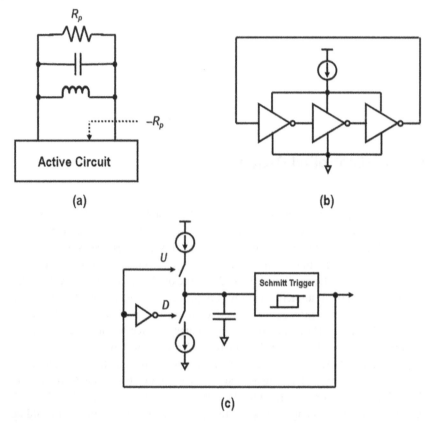

Figure 7.1 On-chip oscillators: (a) LC oscillator; (b) ring oscillator; and (c) relaxation oscillator.

7.1 Oscillator Basics

7.1.1 Oscillation Condition

The oscillator is basically a positive-feedback circuit with an open-loop gain of 1 at oscillation frequency. Figure 7.2(a) shows a linear model for startup condition. A system transfer function from an input signal V_i to an output signal V_o is given by

$$V_o(\omega) = \frac{G(j\omega)}{1 + G(j\omega)\beta(j\omega)} V_i(\omega) \tag{7.1}$$

where $G(j\omega)$ and $\beta(j\omega)$ are the feedforward and feedback gains, respectively. For oscillation at $\omega = \omega_0$, $V_o(\omega_0)$ should not be zero even with $V_i(\omega_0) = 0$. Hence, we

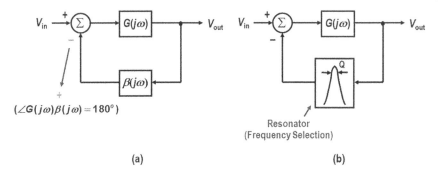

Figure 7.2 Oscillation principle: (a) positive-feedback condition and (b) frequency-selective gain boosting by an LC tank.

have the oscillation condition for some ω_0 as

$$1 + G(j\omega_0)\beta(j\omega_0) = 0 \qquad (7.2)$$

Nyquist criterion for the oscillation condition is defined by

$$|G(j\omega_0)\beta(j\omega_0)| = 1 \ \text{ and } \ \angle G(j\omega_0)\beta(j\omega_0) = 180° \qquad (7.3)$$

That is, the open-loop gain must exhibit a phase delay of 180° so that a negative feedback system is turned into a positive feedback system. Also, the magnitude of the open-loop gain must be unity at the oscillation frequency. Barkhausen's criterion is the same as the Nyquist criterion except that the phase delay of 360° is considered for a positive feedback model.

Now, let us consider the LC oscillator that consists of a passive LC resonator and an active circuit. Figure 7.2(b) shows the simplified functional model of the LC oscillator. The feedforward gain $G(j\omega)$ is set by the active circuit, while the feedback path is formed by the LC tank. $G(s)$ and the LC tank provide the total phase delay of 180° with the unity gain at the oscillation frequency. The LC tank resonates with a resonant frequency ω_0, enabling an LC oscillator to have the maximum open-loop gain at ω_0. Therefore, the LC tank is considered a frequency-selective network as it play as a band-pass filter at the resonant frequency. The quality of the LC tank as the frequency-selective network is quantitatively defined by a quality factor.

7.1.2 Quality Factor

Ideal LC resonator does not dissipate power and oscillates with a resonant frequency ω_0 given by

$$\omega_0 = \frac{1}{\sqrt{LC}} \qquad (7.4)$$

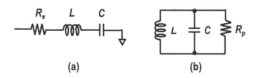

Figure 7.3 LC resonator: (a) series model and (b) parallel model.

In practice, any component has resistance, and a series LC resonator is modeled as shown in Fig. 7.3(a). The quality factor Q is defined by

$$Q = \omega \frac{E_{\text{stored}}}{P_{\text{avg}}} \tag{7.5}$$

where the stored energy E_{stored} in an ideal reactive device and the average dissipated power P_{avg} due to a series resistor R_s. For the series resonator, E_{stored} and P_{avg} are given by

$$E_{\text{stored}} = \frac{1}{2}LI_{\text{pk}}^2, \quad P_{\text{avg}} = \frac{1}{2}I_{\text{pk}}^2 R_s \tag{7.6}$$

where I_{pk} is the peak current across R_s. At a resonant frequency ω_0, Q is given by

$$Q = \omega_0 \frac{E_{\text{stored}}}{P_{\text{avg}}} = \omega_0 \frac{\frac{1}{2}LI_{\text{pk}}^2}{\frac{1}{2}I_{\text{pk}}^2 R_s} = \frac{\omega_0 L}{R_s} = \frac{1}{\omega_0 RC} \tag{7.7}$$

The Q value of a crystal oscillator exceeds 5,000, while the typical Q value of an integrated oscillator is less than 50.

For most circuits, a parallel LC resonator shown in Fig. 7.3(b) instead of the series LC resonator is used. It is because the parallel model is more straightforward to calculate the output impedance in the LC oscillator circuit. At a resonant frequency ω_0, Q is given by

$$Q = \frac{R_p}{\omega_0 L} = \omega_0 R_p C \tag{7.8}$$

where R_p is the resistance in parallel with the LC tank. In fact, any series LC resonator can be transformed into a parallel LC resonator with an equivalent model near ω_0, and it is called *passive impedance transformation*. The following example shows how to calculate equivalent values for the impedance transformation.

Example 7.1 *Passive impedance transformation*
At some frequency near the resonant frequency, we can transform a series LC resonator into a parallel LC resonator or vice versa. Let us consider the case of an inductor in Fig. 7.4(a). Suppose that we want to transform an inductor L_s in series

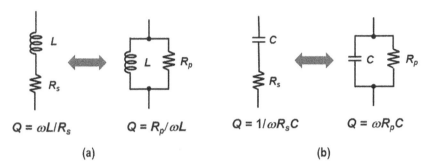

Figure 7.4 Impedance transformation: (a) inductor and (b) capacitor.

with a resistor R_s into an inductor L_p in parallel with a resistor R_p. To find the equivalent values of the parallel LC resonator, let

$$L_s s + R_s = \frac{L_p R_p s}{L_p s + R_p}$$

In steady-state, $s = j\omega$. Now, we have

$$(L_s R_p + L_p R_s)j\omega + R_s R_p - L_s L_p \omega^2 = L_p R_p j\omega$$

Then,

$$L_s R_p + L_p R_s = L_p R_p, \quad R_s R_p - L_s L_p \omega^2 = 0$$

From (7.7),

$$Q = \frac{\omega_0 L_s}{R_s}$$

For Q higher than 3, we can get approximated values of L_p and R_p for ω near ω_0, given by

$$L_p \approx L_s, \quad R_p \approx Q^2 R_s \tag{7.9}$$

Similarly, for a capacitor C_s in series with R_s shown in Fig. 7.4(b), we get

$$C_p \approx C_s, \quad R_p \approx Q^2 R_s \tag{7.10}$$

In the design of an LC oscillator, the LC tank is connected to other circuits. Accordingly, the quality factor of the LC tank in connection with other circuits becomes different from the original quality factor of the LC tank itself. The quality factor connected with other networks is called a *loaded Q* denoted by Q_L. Let us consider a parallel LC resonator connected with a source resistor R_{source} and a load resistor R_{load} as shown in Fig. 7.5. Then, the loaded Q is given by

$$Q_L = \omega_0 C(R_p \parallel R_{\text{source}} \parallel R_{\text{load}}) \tag{7.11}$$

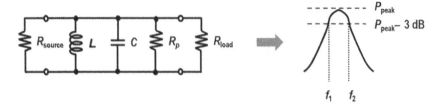

Figure 7.5 Loaded Q of the LC tank.

where R_{source} is the source resistor and R_{load} is the load resistor. In general, Q_L is given by

$$\frac{1}{Q_L} = \frac{1}{Q} + \frac{1}{Q_{ext}} \tag{7.12}$$

where Q_{ext} is the quality factor of an external component. In many cases, it is difficult to calculate the loaded Q. Another way of obtaining the loaded Q is observing frequency response. By measuring the 3-dB bandwidth from the band-pass filter characteristic of the LC network, we can obtain Q_L. That is,

$$Q_L = \frac{\omega_0}{2\omega_{BW}} \tag{7.13}$$

where ω_{BW} is the 3-dB bandwidth of a band-pass filter. When the LC resonator is used for the design of an oscillator, the loaded Q is important for the frequency stability of the oscillator, thus determining phase noise performance.

7.1.3 Frequency Stability

The quality factor Q of the oscillator indicates how stable the output frequency is against phase perturbation. The frequency stability of an LC oscillator is mainly determined by the quality factor of an LC resonator.[1] Figure 7.6(a) shows the model of an LC tank with an injected current I_n. The variation of an output voltage V_n due to I_n at the resonant frequency ω_0 is obtained by a transfer function $H(\omega)$

$$H(\omega) = \frac{V_n}{I_n} = \frac{R_p}{1 + jQ\left(\frac{\omega}{\omega_0} - \frac{\omega_0}{\omega}\right)} \tag{7.14}$$

where Q is the loaded quality factor of the LC resonator. A phase change θ_n from I_n to V_n is obtained by

$$\theta_n = \arg \frac{V_n}{I_n} = -\tan^{-1} Q\left(\frac{\omega}{\omega_0} - \frac{\omega_0}{\omega}\right) \tag{7.15}$$

---

1 When the LC resonator is used with other circuits, the loaded Q instead of the unloaded Q should be considered for accurate noise calculation. In this chapter, we will not treat them separately for the sake of simplicity.

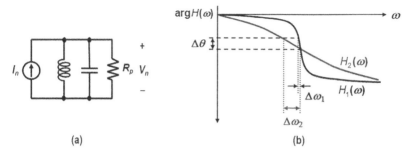

Figure 7.6 (a) LC tank with injected current noise and (b) frequency stability comparison of $H_1(\omega)$ and $H_2(\omega)$.

Then,

$$\left.\frac{d\theta_n}{d\omega}\right|_{\omega=\omega_o} = \frac{-2Q}{\omega_0} \equiv \frac{S_F}{\omega_0} \qquad (7.16)$$

where S_F is the frequency stability factor. Equation (7.16) implies that a high Q or S_F makes the frequency variation $\Delta\omega$ small in the oscillation system for the given phase variation $\Delta\theta$. That is, when we inject noise in the oscillation system, a small frequency change indicates high frequency stability or high quality factor. For example, in Fig. 7.6(b), the transfer function $H_1(\omega)$ has less frequency perturbation than $H_2(\omega)$ for a given $\Delta\theta$. A VCO with a high-frequency stability factor, however, suffers from limited tunability, having a fundamental trade-off between the phase noise and the tuning range.

7.1.4 Effect of Circuit Noise

Now, let us discuss how circuit noise within an oscillator affects phase noise based on a noise model shown in Fig. 7.7. Transistors in the oscillator generate current noise I_n to the LC load. The amount of phase variation $\Delta\theta_{n,\mathrm{LC}}$ at the LC tank is obtained from (7.15). When extra $\Delta\theta_{n,LC}$ is generated within the oscillator, the LC oscillator must maintain a phase shift of 360° by introducing $-\Delta\theta_{n,LC}$ in the loop. Otherwise, Barkhausen's criterion for oscillation cannot be met. Therefore, the oscillation frequency needs to be shifted slightly from the LC resonant frequency ω_0 to maintain oscillation. The required frequency shift $\Delta\omega_{n,\mathrm{osc}}$ in the oscillator is obtained from (7.16), that is,

$$\Delta\omega_{n,\mathrm{osc}} = \frac{\omega_0}{2Q}\Delta\theta_{n,\mathrm{LC}} \qquad (7.17)$$

For the given noise modulation frequency ω_m, the noise phase deviation $\Delta\theta_{n,\mathrm{osc}}$ of the oscillator can be obtained from (4.3), that is,

$$\Delta\theta_{n,\mathrm{osc}} = \frac{\Delta\omega_{n,\mathrm{osc}}}{\omega_m} = \frac{\omega_0}{2Q}\frac{\Delta\theta_{n,\mathrm{LC}}}{\omega_m} \qquad (7.18)$$

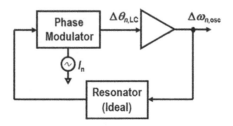

Figure 7.7 Noise model of oscillator.

The above equation is valid only if ω_m is within the bandwidth of the feedback system in Fig. 7.7. When ω_m is outside the bandwidth, the feedback gain is negligible, making $\Delta\theta_{n,\mathrm{osc}} = \Delta\theta_{n,\mathrm{LC}}$, which could be understood from Fig. 7.2(b). From (7.18), the corner frequency ω_c of the closed-loop phase noise and the open-loop phase noise can be obtained by having $\Delta\theta_{n,\mathrm{osc}} = \Delta\theta_{n,\mathrm{LC}}$

$$\omega_c = \frac{\omega_0}{2Q} \tag{7.19}$$

Figure 7.8 shows the pictorial view of how internal circuit noise is converted to phase noise at an oscillator output. At $\omega_m < \omega_c$, the phase noise of the oscillator exhibits a slope of $1/\omega_m$ from (7.18). Accordingly, the oscillator having a high-Q LC tank has low ω_c, achieving low phase noise as depicted in Fig. 7.8. Now we understand why we could always observe the slope of $-20\,\mathrm{dB/dec}$ in the phase noise plot of any oscillator. In integrated oscillators, the flicker noise of transistors contributes to phase noise as well. Unlike the thermal noise, the noise spectral density of flicker noise, namely $1/f$ noise, exhibits a slope of $-10\,\mathrm{dB/dec}$. As a result, the phase noise of an integrated oscillator shows a slope of $-30\,\mathrm{dB/dec}$ below the corner frequency $\omega_{1/f}$ of the flicker noise. Figure 7.9 shows the typical phase noise plots of high-Q and low-Q oscillators. In the case of a very high-Q oscillator, ω_c can be even lower than $\omega_{1/f}$. Hence, the phase noise has a slope of $-10\,\mathrm{dB/dec}$ at the

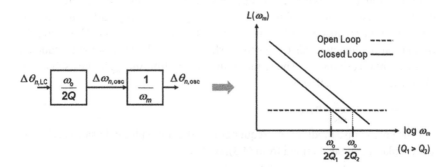

Figure 7.8 Effect of the internal noise and quality factor.

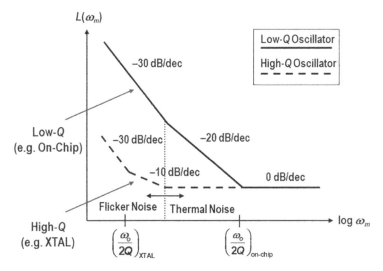

Figure 7.9 Typical phase noise performance of the VCO.

mid frequency and a slope of -30 dB/dec at the low frequency as shown in Fig. 7.9, which is the typical phase noise behavior of a crystal oscillator. On the other hand, most integrated oscillators show a slope of -20 dB/dec at mid frequency and a slope of -30 dB/dec at low frequency.

7.1.5 Leeson's Model and Figure-of-Merit

The phase noise behavior shown in Fig. 7.9 is generalized based on an empirical model and formulated by Leeson in 1966, which is known as Leeson's equation. The phase noise $L(\omega_m)$ at offset frequency ω_m is given by

$$L(\omega_m) = 10 \log \left[\frac{2FkT}{P_{avg}} \left(\frac{\omega_o}{2Q\omega_m} \right)^2 \right] \qquad (7.20)$$

where F is an empirical fitting factor, k is the Boltzman constant, and T is the absolute temperature. The first equation introduced by Leeson did not include the flicker noise contribution and was extended later as

$$L(\omega_m) = 10 \log \left[\frac{2FkT}{P_{avg}} \left\{ 1 + \left(\frac{\omega_o}{2Q\omega_m} \right)^2 \right\} \left(1 + \frac{\omega_{1/f}}{\omega_m} \right) \right] \qquad (7.21)$$

Note that the fitting factor F in (7.20) and (7.21) has nothing to do with a noise factor which is also denoted as F in RF circuits. The equations imply that having a high Q is the most efficient way of improving the phase noise performance for a given power budget.

Any oscillator has a design trade-off between power consumption and phase noise. To evaluate the oscillator performance with fair comparison, a figure-of-merit (FOM) can be defined based on Leeson's equation. It is expressed as

$$\text{FOM} = 10\log\left[\frac{kT}{P_{\text{avg}}}\left(\frac{f_o}{f_m}\right)^2\right] - L(f_m) \tag{7.22}$$

where the frequency is used instead of the angular frequency for convenience. A VCO is the oscillator that generates variable frequencies in response to a control voltage. Therefore, the VCO design needs to consider a fundamental trade-off between the phase noise and the tuning range, which is different from the design of a standalone oscillator. In that case, we replace f_o in (7.22) with the maximum output frequency $f_{o,\max}$ and the minimum output frequency $f_{o,\min}$ to define another FOM denoted as FOM_T given by

$$\text{FOM}_T = 10\log\left[\frac{kT}{P_{\text{avg}}}\left(\frac{f_{o,\max} - f_{o,\min}}{f_m}\right)^2\right] - L(f_m) \tag{7.23}$$

In addition to the quality factor, the VCO gain K_v is a key parameter for both the phase noise and the tuning range. When a multi-band LC VCO equipped with a capacitor array is designed, FOM_T is further improved since a low K_v can be used for the same tuning rage.

7.1.6 Effect of Noise Coupling

The effect of internal circuit noise on the phase noise also explains how noise coupling affects the oscillator performance. When there is noise coupling to an oscillator with a modulated frequency below ω_c, the coupled noise behaves like the internal noise of the oscillator. Hence, the coupled noise also causes a phase shift $\Delta\theta_{pk,LC}$ at the LC tank, resulting in a peak phase shift $\Delta\theta_{pk}$ at the oscillator output from (7.18). Assuming that the noise coupling is represented by a periodic modulation, $\Delta\theta_{pk}$ determines the level of spur at an offset frequency ω_m. If a spur is generated by periodic noise coupling and the modulation frequency is below ω_c, then the sideband sits on the top of the phase noise as illustrated in Fig. 7.10. It implies that a high-Q oscillator or a low-gain VCO offers not only low phase

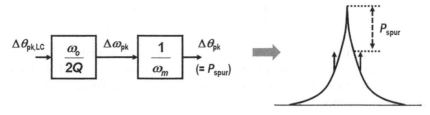

Figure 7.10 Effect of the external coupling noise.

noise but also low noise coupling. As the noise coupling behaves like the internal circuit noise, it cannot be suppressed by an external circuit once it happens. Therefore, it is important to lower the impedance of ground, supply, and local substrate to mitigate the noise coupling from digital circuits unless a substantial physical isolation is considered in the layout.

7.2 LC VCO

To control the frequency of an oscillator, we may consider controlling the phase-shifting based on (7.17), but the design of a phase-shifting circuit is complicated and does not provide wide frequency-tuning. For a very high-Q VCO such as the voltage-controlled crystal oscillator (VCXO) that uses an external crystal, a phase-injection method could be considered since the VCXO requires a huge voltage range even for a small frequency control due to an extremely low voltage-to-frequency gain. As to the design of an integrated LC VCO, a variable output frequency is achieved by having a variable capacitor, namely varactor. Using the MOSFET whose capacitance depends on a gate-source (or gate-drain) voltage is a typical way of implementing the varactor. Figure 7.11 shows the schematic of a conventional differential LC VCO with cross-coupled transistors. Even though single-ended VCOs such as Colpitts or Pierce oscillators can achieve a better FOM with less number of active transistors than the differential VCO, they suffer from on-chip supply noise. Therefore, a differential topology is dominantly used for the design of integrated LC VCOs. In this section, we will focus on the design of the differential LC VCO.

7.2.1 Design Considerations

There are three key parameters to be considered for the design of a VCO in general. Those are oscillation startup, tuning range, and phase noise. In addition, the

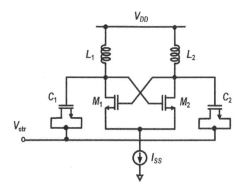

Figure 7.11 LC VCO with cross-couple transistors.

small-signal gain of transistors, the tunability of a varactor, the quality factor of an LC tank, and the flicker noise of a bias current should be considered in the design of an LC VCO.

7.2.1.1 Oscillation Startup

Since oscillators do not have any input source to generate an output signal except self-contained noise, we need a positive feedback for initial startup to have a small perturbed signal grow. When the amplitude of the growing signal with the positive feedback is limited by circuit nonlinearity and a supply voltage, a constant amplitude is obtained with a steady-state gain of 1. Therefore, it is important to make the initial gain of the feedback loop higher than 1 for process-voltage-temperature (PVT) variations.

Let us check the startup condition based on the conventional differential topology shown in Fig. 7.11. For the sake of simplicity, we consider an LC oscillator instead of the LC VCO and use a simplified model with a parallel LC tank as shown in Fig. 7.12(a). As discussed previously, an inductor with series resistance R_s can be transformed into an inductor with parallel resistance R_p if we are interested only in frequencies near the resonant frequency. We note that the schematic in Fig. 7.12(a) can be viewed as a two-stage amplifier with a positive feedback in such a way that an output signal is fed to an input of the first amplifier as shown in Fig. 7.12(b). When the frequency of an input signal is the same as the resonant frequency, the impedance of a parallel LC tank is approximated as R_p. Then, the required gain for oscillation startup is given by

$$g_{m1}R_{p1}g_{m2}R_{p2} > 1 \qquad (7.24)$$

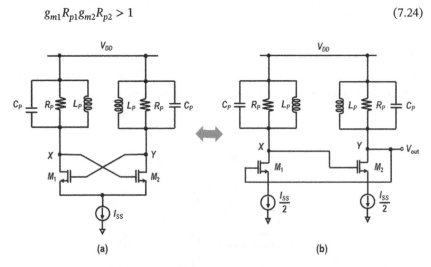

(a)   (b)

Figure 7.12 LC oscillator viewed as a two-stage LC-tuned amplifier with positive feedback: (a) LC oscillator with a parallel LC tank and (b) two-stage oscillator with an LC load.

where g_{m1}, g_{m2}, R_{p1}, and R_{p2} are transconductances and parallel resistances of first-stage and second-stage transistors and LC tanks, respectively. To have a symmetric circuit, we design an LC VCO with $g_{m1} = g_{m2} = g_m$ and $R_{p1} = R_{p2} = R_p$. Then, (7.24) is simplified as

$$g_m R_p > 1 \tag{7.25}$$

Example 7.2   *Amplifier with an LC load*
Suppose that we design an LC oscillator based on Fig. 7.13. Suppose that an LC tank has an inductor of 5 nH with a quality factor of 8 and a resonant frequency of 2 GHz. If the oscillator circuit is fully symmetric, what is the required g_m of both transistors for oscillation startup?

We need to get R_p to use (7.25). With $Q = 8$ at 2 GHz, we have $R_s = 7.85\,\Omega$ from (7.6) and $R_p = 502.4\,\Omega$ from (7.8). If we design an oscillator with a startup gain of 2, the required g_m should be about 2 mA/V from (7.25).

Example 7.2 shows why it is good to design an LC oscillator with a high-Q inductor. For the same power consumption, we can improve phase noise by having a large voltage swing with high R_p. For the same reason, we can reduce a bias current with high R_p for the given noise requirement. Accordingly, many researches have been done to improve the quality factor of an on-chip inductor. In the technology side, a thick metal is offered to reduce the series resistance of the on-chip inductor. In the circuit side, different shapes and shielding effects of inductors have been studied.

For the given R_p, it is important to define an optimum value of $g_m R_p$. High $g_m R_p$ gives a fast startup but causes a harmonic distortion since a large gain makes the output swing close to a rectangular waveform rather than a sinusoidal waveform. In addition, unnecessary power is consumed, degrading the FOM of the oscillator. As illustrated in Fig. 7.14, the internal voltage swing of an oscillator

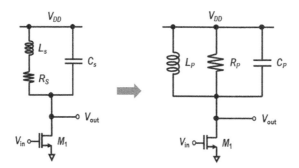

Figure 7.13   Amplifier with an LC-tuned load after impedance transformation.

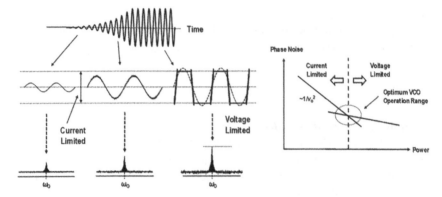

Figure 7.14 Oscillation with voltage- and current-limiting modes.

initially grows with a bias current. Since the output voltage swing is determined by the bias current, the oscillator operates in the current-limited region. Once the maximum amplitude is reached, the voltage swing does not increase with the bias current due to circuit nonlinearity and turns into a rectangular waveform. Therefore, the oscillator is in the voltage-limited region. Certainly, the cross point of the current-limited and the voltage-limited regions would be an optimum point for the best phase noise for a given power budget. However, the boundary of the current-limited region is exposed to the risk of not satisfying the startup condition when a large variation of g_m over process, voltage and temperature is considered. In practice, a startup gain in the range of 2–4 is typically set for the design of integrated oscillators, and a proper value should be determined by considering both technology and circuit aspects.

7.2.1.2 Tuning Range

Since the output frequency is tuned by a varactor in the conventional LC VCO, the capacitance range of the varactor is important to achieve a wide frequency range. As discussed in Chapter 6, the conventional MOS capacitor exhibits nonmonotonicity in a voltage-to-capacitance transfer function curve and severe nonlinearity when a control voltage is small. The nonmonotonicity or poor linearity is considered harmful for the PLL design. If we put an NMOS transistor in an n-well or a PMOS transistor in a p-well, the MOS transistor only has depletion and accumulation modes without an inversion mode.

Figure 7.15(a) depicts the accumulation-mode varactor consisting of an NMOS transistor in an n-well. The voltage-to-capacitance transfer function shows a decent linearity when a gate-source voltage V_{GS} is near the middle of an

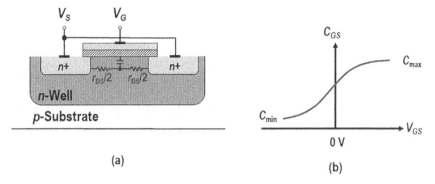

Figure 7.15 Accumulation-mode varactor: (a) physical structure and (b) voltage-to-capacitance characteristic.

entire range. The maximum capacitance C_{max} occurs when electrons are fully accumulated under the gate and is given by

$$C_{max} = C_{ox}WL + 2C_{ov}W \tag{7.26}$$

where C_{ox} is the oxide capacitance in units of F/μm^2, C_{ov} is the overlap capacitance in units of F/μm, W is the width of the NMOS transistor, and L is the length of the NMOS transistor. On the other hand, the minimum capacitance C_{min} occurs when the electrons under the gate are depleted by a negative gate voltage, having the capacitance only set by C_{ov} in both source and drain sides. That is,

$$C_{min} = 2C_{ov}W \tag{7.27}$$

From (7.26) and (7.27), the ratio of C_{max} to C_{min} becomes

$$\frac{C_{max}}{C_{min}} = \frac{C_{ox}WL + 2C_{ov}W}{2C_{ov}W} = 1 + \frac{C_{ox}}{2C_{ov}}L \approx \frac{C_{ox}}{2C_{ov}}L \tag{7.28}$$

The voltage-to-capacitance characteristic of the accumulation-mode varactor is depicted in Fig. 7.15(b). Since C_{ox} and C_{ov} are not design parameters but process parameters set by a given CMOS technology, the gate length is the only parameter to control the tuning range of the varactor. From (7.28), we see that a longer gate length is desirable to increase the tunability of the varactor.

A varactor with a large L, however, suffers from a degraded quality factor. As illustrated in Fig. 7.15(a), the resistance r_{DS} of an electron channel formed under the gate gives series resistance to the capacitor. The quality factor Q_{var} of the varactor is given by

$$Q_{var} = \frac{12k_p(V_{GS} - |V_T|)}{\omega_o C_{ox}L^2} \tag{7.29}$$

where k_p is the MOSFET transconductance parameter. Equations (7.28) and (7.29) imply that increasing the channel length L by k times improves the turning range by k times but degrades Q_{var} by k^2 times. Typically, L is chosen to be two or three times the minimum length of a transistor. The total quality factor Q_{tot} of an LC resonator is given by

$$\frac{1}{Q_{tot}} = \frac{1}{Q_{ind}} + \frac{1}{Q_{var}} \qquad (7.30)$$

In most cases, Q_{tot} is mainly determined by Q_{ind} since Q_{var} is higher than Q_{ind}. For the design of a >10-GHz VCO, Q_{var} also plays a critical role as the value of an inductor gets smaller.

7.2.1.3 Phase Noise

In general, an oscillator with a larger voltage swing achieves better phase noise since a fast transition at the zero crossing not only reduces timing uncertainty due to circuit noise but also minimizes AM-to-PM conversion. A high-Q inductor gives a large value of R_p, thus increasing a voltage swing for the given bias current. In addition to the loaded Q of an LC tank, there are other circuit design aspects to be considered for the design of a low-noise LC VCO. Razavi's books give a good explanation of those design aspects in an intuitive way, and key results are summarized here.

Cross-Coupled Transistors Cross-coupled transistors provide an active gain or transconductance g_m to maintain oscillation. Let us consider the case that one of the cross-coupled transistor, M_1, is fully on or off in an LC oscillator shown in Fig. 7.16. For the sake of simplicity, let us replace the LC tank with R_p by considering a small-signal model at resonant frequency and set the supply voltage to $(V_{DD} + 0.5I_{SS}R_P)$ as depicted in Fig. 7.16. Suppose that M_1 is turned and M_2 is turned off. If M_1 is in a saturation region, the noise contribution of M_1 will be degenerated by the high output impedance of the tail current source. If M_1 is turned off, there is no noise contribution. If both M_1 and M_2 transistors are turned on around crossing time, current noise will be injected by cross-coupled transistors. When white noise is gated by a periodic signal, the noise property of the periodically switched noise remains white as *cyclostationary* noise. The time duration of the conductance at the crossing point depends on the size of the cross-couple transistors. Since an output signal is generated by an oscillator itself, we do not consider an input-referred noise but an output noise for active transistors. Therefore, a small W/L ratio is desired to reduce the current noise generated by the cross-coupled transistors. On the other hand, a large W/L ratio can bring the fast rising and falling times of an output voltage, thus mitigating the noise contribution of the transistors at the zero crossing. Accordingly, the W/L

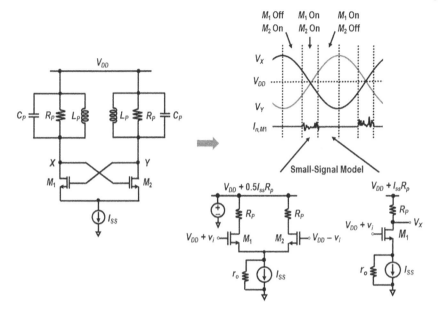

Figure 7.16 Noise contribution of the cross-coupled transistors.

ratio of the cross-coupled transistors has less contribution to the oscillator noise than other parameters if they operate in the saturation region.

However, it is difficult to have them operate in the saturation region for the whole cycle especially when a large output swing is desired for low-noise design. If M_1 is in a linear region, amplitude modulation occurs, causing AM–PM conversion. In addition, the effect of parasitic capacitance at the tail current source needs to be considered, which will be discussed in the following section.

Tail Capacitance Suppose that M_1 transistor in Fig. 7.16 partially operates in the linear region with an excessive swing at node Y. In practice, there exists the tail capacitance C_{tail} that comes from the parasitic capacitance of the cross-couple transistors and the tail current source, or it could be added to suppress the tail-current noise as a design parameter. Figure 7.17 shows the small-signal model of an LC oscillator when M_1 enters the linear region with turn-on resistance R_{on}. Then, the LC tank is connected to C_{tail} through R_{on}. The series connection of R_{on} and C_{tail} can be transformed to an equivalent parallel connection by (7.10). After the passive transformation, we have the parallel resistance R_{p2} and capacitance C_{p2} given by

$$R_{p2} = \frac{R_p R_{tail}}{R_p + R_{tail}}, \quad C_{p2} = C_p + C_{tail} \tag{7.31}$$

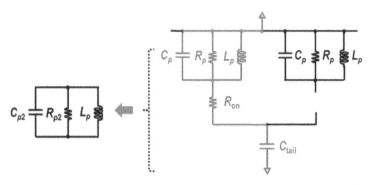

Figure 7.17 Tail capacitance effect for M_1 in the linear region.

where

$$R_{\text{tail}} = \frac{1}{R_{\text{on}} C_{\text{tail}}^2 \omega_0^2} \tag{7.32}$$

Equation (7.31) implies that increasing the tail current with the small W/L ratio of the cross-coupled transistors can degrade the phase noise performance since excessive tail current makes them operate in the linear region.

Another effect of the tail capacitance is an asymmetric waveform due to the second harmonic distortion. Even if the cross-coupled transistors do not enter the linear region, the voltage V_{tail} at C_{tail} exhibits a ripple at two times the oscillation frequency. It is because V_{tail} with both M_1 and M_2 turned on is driven by the common-mode voltage of a differential pair, while V_{tail} with only M_1 turned on is set by the input voltage of a source follower. Figure 7.18 illustrates how the voltage ripple at $2\omega_0$ distorts an output waveform, producing an asymmetric waveform. The asymmetric waveform causes AM–PM conversion and up-converts the flicker noise of a tail current, which will be discussed in the following sections.

Tail Current Even though the noise contribution of a tail current is mitigated by the common-mode rejection of a differential oscillator, it contributes more to the phase noise than the cross-coupled transistors. We consider the current noise contribution in three interesting frequencies: at DC or near DC (we only consider the flicker noise as it is the main design concern), at oscillation frequency ω_0, and at two times the oscillation frequency $2\omega_0$.

The flicker noise of the tail current contributes to the output phase noise when there is asymmetry or nonlinearity in the differential topology. As discussed, the tail capacitance is one of the main factors that cause the second harmonic distortion, thus resulting in an asymmetric waveform. When an output waveform is not symmetric, the common-mode fluctuation introduces phase noise via AM–PM conversion. The circuit nonlinearity can also make the common-mode

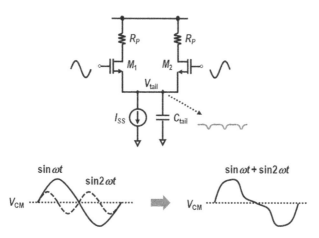

Figure 7.18 Tail capacitance effect on waveform asymmetry.

noise contribute to the phase noise. The nonlinear voltage-to-capacitance transfer function of a varactor could be one of the main sources to produce the AM–PM conversion. In addition to the asymmetry and the nonlinearity, imperfect common-mode rejection by the cross-coupled pair still modulates the output amplitude. This effect is weaker than the other two cases but may not be negligible especially when the cross-coupled transistors enter the linear region with a large output swing.

The thermal noise of the tail current at the oscillation frequency ω_0, and the second-harmonic frequency $2\omega_0$ also needs to be investigated. For that, it is good to note that the cross-coupled pair behaves like a single-sideband mixer to the modulated tail current. As illustrated in Fig. 7.19, the modulated current noise is analogous to an RF input, while the cross-coupled pair is analogous to a differential switching circuit gated by a local oscillator. As a result, the tail current noise at ω_0 is downconverted to DC and upconverted to $2\omega_0$, which does not affect the phase noise of the oscillator whose frequency is centered at ω_0. On the other hand, the tail current noise at $2\omega_0$ is downconverted to ω_0, affecting the phase noise performance. This implies that the flicker noise of the tail current can also be upconverted to ω_0 when the node V_{tail} of the tail capacitance C_{tail} is modulated by the common mode noise at $2\omega_0$, thus modulating the flicker noise of the tail current at $2\omega_0$ through channel length modulation. The upconverted flicker noise is added at the output phase noise of an oscillator, exhibiting a noise slope of $-30\,\text{dBc/Hz}$ as depicted in Fig. 7.9.

In summary, the flicker noise of the tail current affects the phase noise through direct and indirect ways. A slow-varying noise component directly affects the phase noise when there is asymmetry or nonlinearity. The asymmetric waveform

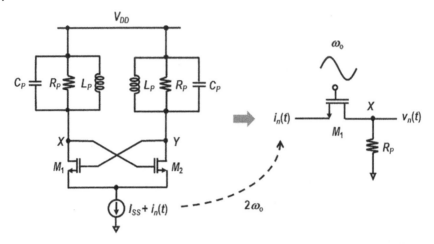

Figure 7.19 Up-conversion mechanism of the tail current noise by frequency mixing.

can be formed by the tail capacitance when the differential switching pair is fully on and off. Any even-order harmonic distortion or unbalanced circuit operation can also cause the asymmetric waveform in the oscillator. The importance of having the symmetric rising and falling times to mitigate the flicker noise was also analyzed in the time domain by introducing an impulse sensitivity function (ISF). As to the nonlinearity effect, the varactor nonlinearity could be one of the main factors in the VCO. When the flicker noise of the tail current is upconverted to $2\omega_0$ through the common-mode modulation, it affects the phase noise indirectly.

Based on the above discussion, there would be negligible flicker noise upconversion if a fully balanced class-A VCO is designed. In practice, designing such a VCO without the second-order harmonic distortion requires full operation in the current-limiting mode if an additional amplitude control circuit is not implemented, which will limit the maximum output swing or increase power consumption. There are other circuit techniques to mitigate the flicker noise upconversion, which will be discussed later.

7.2.2 LC VCO Circuit Topologies

7.2.2.1 Conventional VCO with Cross-Coupled Pair and Complementary Pair

An LC oscillator shown in Fig. 7.11 becomes an LC VCO after adding a varactor as shown in Fig. 7.20(a). Note that the gate of an MOS capacitor must be connected to an inductor with the source-drain node connected to a control voltage V_{cont}. It is because the source-drain node of the MOS capacitor is exposed to substrate noise coupling, while the gate node is well isolated. To improve supply noise rejection,

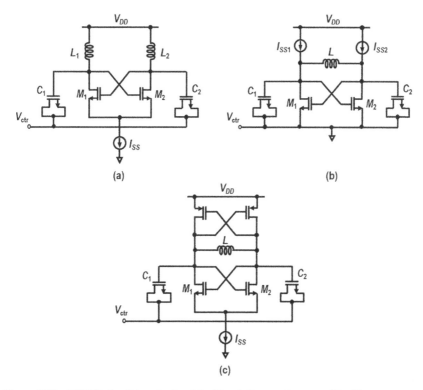

Figure 7.20 LC VCO circuit topologies: (a) with a tail current source; (b) with a top current source; and (c) with a complementary pair.

the LC VCO with a top bias current can be designed to have a ground-reference output as shown in Fig. 7.20(b).

Figure 7.20(c) shows an LC VCO with complementary cross-coupled pairs using both NMOS and PMOS transistors like a CMOS latch. There are some design aspects worth mentioning in comparison with the VCO using the NMOS pair only. With a symmetric configuration, the output of the VCO achieves better symmetricity with a common-mode voltage near half the supply voltage. As discussed previously, the symmetric waveform mitigates the effect of the flicker noise on the phase noise. With additional PMOS pair, better startup condition is obtained. For a given tail bias current, the VCO with the complementary pair can achieve a two-fold swing if the maximum swing is not limited by a supply voltage V_{DD}, thus achieving lower phase noise when the VCO is in the current-liming mode. However, the symmetric configuration limits the maximum swing amplitude to $V_{DD}/2$, while the VCO using the NMOS pair achieves a larger voltage swing close to V_{DD}. Moreover, the tuning range is worse due to additional parasitic capacitance from the PMOS pair. Accordingly, the FOM or FOM_T of the complementary-pair VCO is usually worse than the NMOS-pair VCO for the given tail current. In addition to those conventional architectures, several topologies to further enhance the phase noise or the tuning range will be discussed in the following sections.

7.2.2.2 Class-C VCO

The cross-coupled transistors degrade phase noise when they enter the linear region with the presence of the tail capacitance C_{tail}. The conventional LC VCO shown in Fig. 7.20(a) has difficulty in having them operate in the saturation region when an output swing is large with a high-Q inductor. In that case, the linear-region problem can offset the advantage of the high-Q inductor. To prevent the cross-coupled transistors from entering the linear region, the gate voltage of the transistor can be biased to have the transistors always operate in the saturation region. In that case, the cross-coupled transistors can generate the maximum current even higher than the tail current with a large g_m. Of course, the average current for each transistor should be the same as half the tail current. To keep the same total DC current, the conduction time of each transistor should be less than the half cycle of an oscillation period. Since the conduction angle of the transistor is less than 180°, the VCO is named as a class-C VCO. Considering that there is the period of no flowing current for both transistors, we need a large tail capacitance C_{tail} to have the node P stable. Figure 7.21 shows the schematic of the class-C VCO.

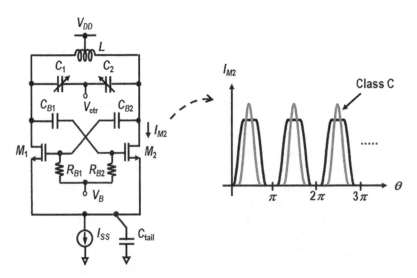

Figure 7.21 Class-C LC VCO.

Having a large peak amplitude with less conduction time of the transistor, the class-C VCO can achieve a better FOM than the conventional VCOs. However, the maximum peak current is determined by g_m of the cross-coupled transistors and can be varied a lot, resulting in different output swing amplitudes over PVT variations. As a result, the startup condition should be carefully examined in the design of the class-C VCO. Additional circuitry for amplitude control can be employed to have a robust startup condition.

7.2.2.3 Tail Noise Reduction with Common-Mode Noise Filtering

We learned that the tail current noise is upconverted to phase noise mainly because of the second-order harmonic component in the oscillator. In other words, if the tail current is not exposed to the second-order harmonic modulation, there is no flicker noise upconversion. Based on this observation, a noise-filtering technique to suppress the coupling of the second-order modulation is proposed. As shown in Fig. 7.22, an LC filter tuned at $2\omega_0$ is added at the drain of the tail transistor. A large capacitor in parallel with the current source shorts the current noise around $2\omega_0$, while an inductor along with total capacitance at the drain of the tail transistor provides high impedance at $2\omega_0$. As a result, the tail current is hardly modulated by the common-mode modulation, which helps mitigate the noise upconversion. With the noise filtering technique, the phase noise performance could be significantly improved.

The drawback of this method is to require a large area to have an auxiliary on-chip inductor just for the common-mode noise filtering. In addition, as the common-mode frequency is determined by the output frequency, the center

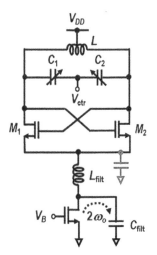

Figure 7.22 LC VCO with common-mode noise filtering.

frequency of the LC filtering circuit should be adjusted for different output frequencies, which requires an additional capacitor array to tune the resonant frequency of the auxiliary LC tank.

7.2.2.4 Wide Tuning with Discrete Capacitor Array

To increase the tuning range of the VCO, a large varactor is required. At the same time, the inductor value should be reduced to maintain the same resonant frequency. A small inductor reduces R_p in the parallel LC tank, resulting in degraded phase noise. The fundamental trade-off between the tuning range and the phase noise can be alleviated by employing a digitally programmable capacitor array as shown in Fig. 7.23. By having multiple coarse-tuning varactors with sufficient frequency overlaps, a relatively wide tuning range can be achieved without degrading phase noise. Each tuning curve must cover temperature variation and should have a sufficient overlap with adjacent curves. A metal-to-metal capacitor shows good linearity but suffers from a large minimum capacitance on the order of 10 fF. In practice, both the metal-to-metal capacitor array and the MOS capacitor array are used for coarse- and fine-tuning controls, respectively.

One problem of using a band-switching LC VCO is the possibility of having a large gain variation over temperature. The center frequency shifting over process variation can be calibrated during system initialization, but the center frequency drift over temperature needs to be continuously covered by the PLL without switching bands. Otherwise, the system clock will have an abrupt frequency change during the normal operation. As illustrated in Fig. 7.23, discrete coarse tuning curves can cause a significant gain reduction at the edge of a single tuning curve with high temperature, requiring a careful VCO design for each band.

7.2.2.5 Seamless Wide Tuning with Single-Input Dual-Path Control

Instead of the discrete capacitor array, a dual-path VCO can be considered to provide not only a seamless wide tuning but also a constant VCO gain over

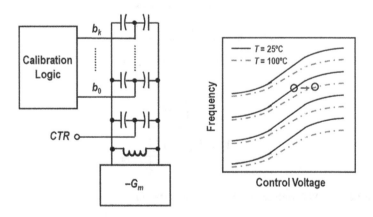

Figure 7.23 LC VCO with capacitor array and its tuning range.

temperature and process variations. Figure 7.24 shows a single-input dual-path LC VCO. By having a separate large varactor controlled by a narrowband loop control, a wide frequency acquisition is achieved without increasing the small-signal gain of the VCO in the loop. With the DC gain of a narrowband linear amplifier, the control voltage range of the VCO is significantly reduced, thus enhancing the linearity of the VCO in the small-signal path. Figure 7.24 illustrates the relationship between the coarse-tuning range and the fine-tuning range. The solid line represents the tuning curve of a coarse-tuning varactor, and the dotted line represents the tuning curve of a fine-tuning varactor. When the gain of the coarse-tuning linear amplifier is set to k and greater than 1, the actual voltage range of a loop filter in the PLL (or, equivalently, the output compliance voltage of the charge pump) is narrower than the output voltage range of the coarse-tuning linear amplifier by a factor k. Therefore, the dual-path VCO effectively minimizes the VCO gain variation over a wide tuning range. The reduced voltage range also improves the charge pump design as it relaxes the current mismatching between up and down currents. This kind of LC VCO could be useful in some wireline applications where a wide-range LC VCO is needed to replace the ring VCO that suffers from poor jitter and noise coupling.

The stability problem in the PLL can be solved by designing a linear amplifier to have a much narrower bandwidth in the coarse-tuning path. Figure 7.25 shows the linear model and the open-loop gain of a PLL with the dual-path VCO, where M and N represent the coarse-tuning gain factor and the division ratio, respectively. The coarse-tuning gain factor M is the ratio of the coarse-tuning path gain to the fine-tuning path gain, which includes the linear amplifier gain ratio as well as the varactor size ratio. Compared to the conventional Type 2 PLL, the dual-path PLL has additional pole and zero located at $s = 1/RC$ and $(M + 1)/RC$, respectively.

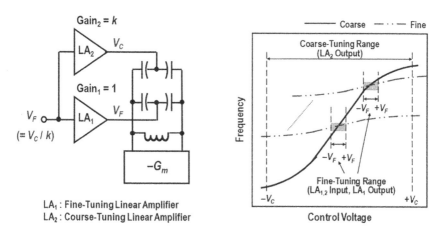

Figure 7.24 Single-input dual-path VCO and its tuning range.

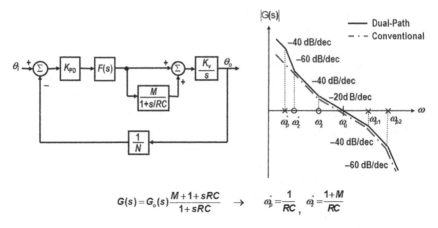

$$G(s) = G_o(s)\frac{M+1+sRC}{1+sRC} \quad \rightarrow \quad \omega_p = \frac{1}{RC}, \quad \omega_z = \frac{1+M}{RC}$$

Figure 7.25 Loop dynamics of the PLL with the single-input dual-path VCO.

A high value of the coarse-tuning gain factor M degrades the phase margin even with narrow coarse-tuning bandwidth. Hence, it is important to keep additional pole and zero at much lower frequencies than the unity-gain frequency for stability. Note that a high-value resistor can be used to reduce the area of a capacitor for a low-pass filter at the output of the linear amplifier since the thermal noise contribution of the resistor is suppressed by the band-pass characteristic of the PLL and the low-pass filter characteristic of the *RC* filter. In the layout, it is important to minimize substrate noise coupling at the output of the linear amplifier in the coarse-tuning path that controls a high-gain varactor.

7.3 RING VCO

If three inverters are connected in series by having the output of the last inverter fed back to the input of the first inverter as shown in Fig. 7.26(a), the output of each inverter is toggling between logic 1 and logic 0 due to meta-stability. Even though the oscillation mechanism of a ring oscillator could be understood as a meta-stable feedback logic that produces a toggling output, Nyquist criteria or Barkhausen criteria can still be applied to a ring oscillator by considering the small-signal gain of an inverter as well as the amount of a phase delay for each stage. In general, if the number of inverters is odd and not less than 3, then the oscillation frequency f_{osc} is given by

$$f_{osc} = \frac{1}{2N\tau_d} \tag{7.33}$$

where N is the number of the inverters and τ_d is the gate delay time of the inverter. Equation (7.33) implies that the gate delay time can be estimated simply

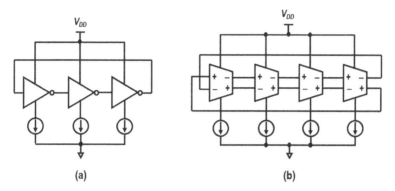

Figure 7.26 Ring oscillator: (a) single-ended and (b) differential topologies.

by measuring the oscillation frequency of the inverter-based oscillator, namely a ring oscillator. For that reason, the ring oscillator has been a popular testing vehicle to measure the gate delay time of an inverter or other digital gates.

CMOS PLLs using a ring VCO became popular when they were employed for clock de-skewing purpose in high-speed I/O interfaces. With the development of the differential ring VCO that significantly improves the power supply rejection performance of a PLL, the PLL expanded its applications to clock generation as well as clock recovery systems. When a differential delay cell is designed, the ring oscillator can work with the even number of differential inverters by swapping the polarity of the connection between the last-stage and the first-stage inverters as illustrated in Fig. 7.26(b). If the number of stages is a multiple of four, the differential ring VCO can generate quadrature phases of 0° and 90°, which is a highly useful feature for clocking and clock-and-data recovery (CDR) applications.

7.3.1 Design Aspects

Choosing the right VCO topology between an LC VCO and a ring VCO is critical for the design of a PLL and should be determined based on system design aspects in various applications. Having an LC tank, an LC VCO achieves lower phase noise and much higher frequencies than a ring VCO under the same power. However, the LC VCO with an on-chip inductor occupies a large area and suffers from a narrow tuning range. Most wireline systems are implemented not with a mixed-mode or RF CMOS technology but with a standard digital CMOS technology where a thick metal option is not available. In addition, the standard CMOS technology does not offer good model-to-hardware correlation for on-chip inductors or varactors. Therefore, the choice of the inductor-less ring VCO could also come from cost and design flexibility, which is sometimes more important than the noise performance.

Since the frequency is not determined by a narrowband LC resonator but based on the gate delay time, the ring VCO shows a wide tuning range, thus not requiring an accurate control of a free-running frequency. The poor phase noise of the ring VCO can be suppressed by the open-loop gain of the wideband PLL that has the high-pass filter characteristic to the VCO noise. Another strong feature of the ring VCO is multi-phase generation. One popular application that utilizes the multi-phase feature is data sampling in the CDR system to relax the speed requirement of a PLL. For instance, a 1-GHz four-phase clock can be used to sample 4-Gb/s NRZ data by using four phases of a ring VCO if a CDR-PLL is designed with a 4-stage differential ring VCO. Note that the use of the multi-phase clock with lower frequency does not relax the random jitter requirement at a target data rate and that a phase mismatch gives another source of deterministic jitter.

The ring VCO, however, is highly sensitive to PVT variations since the oscillation frequency is directly related to the delay time of each inverter. The use of a wideband PLL can improve the phase noise, but an integrated phase error is still much higher than that of an LC VCO. It is because the noise bandwidth is also increased when the wide bandwidth PLL is designed. As the data rate exceeds 5 Gb/s, the jitter requirement in serial I/O links becomes tight and the LC VCO becomes a natural choice for high-frequency clock generation systems. In the wireless system, not only the in-band phase noise but also the out-of-band phase noise is important to meet the spectrum mask or mitigate the reciprocal mixing effect as discussed in Chapter 5. Accordingly, the use of the ring VCO is limited in the wireless systems.

7.3.2 Phase Noise

7.3.2.1 Single-Ended Ring Oscillator

Unlike the LC VCO, the ring VCO operates with a strong voltage-limiting mode, having a rail-to-rail swing for a single-ended circuit and a hard-limit swing for a differential circuit. Since the swing amplitude is set by an active circuit rather than by the quality factor of a passive resonator, the phase noise performance is mainly determined by a DC power that determines the slope of rising and falling edges at zero crossing. Interestingly, the number of stages does not play a critical role for the phase noise. For the given power, a large number of inverters boosts an open-loop gain but also increases the noise contribution of active transistors. Accordingly, increasing the average supply current or the supply voltage is an effective way to reduce the phase noise.

As the oscillation frequency is determined by the delay time of an inverter, the phase noise of a ring oscillator is severely affected by supply or substrate noise coupling. In addition, the noise contribution from a control voltage could be substantial. Overall, the ring VCO has more difficulty in having good model-to-hardware correlation than the LC VCO especially when the ring VCO is embedded in a large

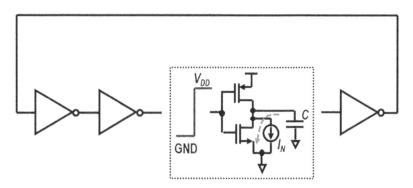

Figure 7.27 Noise analysis of a single-ended ring oscillator.

digital system. Hence, only considering the thermal or flicker noise contribution of active transistors may not be so meaningful to define a best or optimum circuit design or architecture. Among several phase noise equations in the literature, we consider rather simple one based on a square-law FET model with hardware validation based on the Abidi's paper in addition to the Razavi's book.

To derive the timing uncertainty of an inverter in the ring oscillator, we apply a step input voltage at the input of the inverter and measure a voltage ramp at the output. We define the propagation delay τ_d by the time for the output ramp to cross $V_{DD}/2$. We also define the period jitter by the standard deviation of the discrete sequence of periods for every cycle. Figure 7.27 shows the case that a PFET is off with a step input voltage at V_{DD}. Assuming that the mean value of the current noise i_n is zero, the average propagation delay τ_{dN} with an NMOS transistor only is given by [2]

$$\tau_{dN} = \frac{CV_{DD}}{2I_N} \tag{7.34}$$

where C is the load capacitor at the inverter output and I_N is the drain current of the NMOS in a saturation region. Even if the NMOS transistor enters a triode region, the value of I_N does not change substantially as the output node gradually reaches $V_{DD}/2$. Then, the variance of $\sigma^2_{\tau dN}$ is expressed as

$$\sigma^2_{\tau dN} = \frac{4kT\gamma_N\tau_{dN}}{I_N(V_{DD} - V_{tN})} + \frac{kTC}{I_N^2} \tag{7.35}$$

where γ_N is the body effect coefficient of the NMOS and V_{tN} is the threshold voltage of the NMOS transistor. By including another case with a PMOS transistor

---

2 τ_{dN} is actually a random variable, but we will treat it as the same as the mean value for the sake of simplicity.

turned on, we have the variance of a period jitter $\sigma^2{}_\tau$ given by

$$\sigma_\tau^2 = N \left(\sigma_{\tau dN}^2 + \sigma_{\tau dP}^2 \right) \tag{7.36}$$

where N is the number of inverters. For simplicity, let $\tau_{dN} = \tau_{dP} = \tau_d$, $V_{tN} = V_{tP} = V_t$, and $I_N = I_P = I$. Then, we have

$$\sigma_\tau^2 = N \left[\frac{4kT(\gamma_N + \gamma_P)\tau_d}{V_{DD} - V_t} + \frac{2kTC}{I^2} \right] \tag{7.37}$$

Using (7.34), the output frequency f_o of an N-stage ring oscillator is given by

$$f_o = \frac{1}{2N\tau_d} = \frac{I}{NCV_{DD}} \tag{7.38}$$

From (7.37) and (7.38), σ_τ^2 is expressed as

$$\sigma_\tau^2 = \frac{kT}{If_o} \left[\frac{2(\gamma_N + \gamma_P)}{V_{DD} - V_t} + \frac{2}{V_{DD}} \right] \tag{7.39}$$

For the given variance of the period jitter, the single-sideband phase noise $L(f)$ is expressed as

$$L(f) = \sigma_\tau^2 \frac{f_o^3}{f^2} \tag{7.40}$$

Using (7.39), the phase noise is given by

$$L(f) = \frac{2kT}{I} \left(\frac{\gamma_N + \gamma_P}{V_{DD} - V_t} + \frac{1}{V_{DD}} \right) \left(\frac{f_o}{f} \right)^2 \tag{7.41}$$

The above equation implies that the number of inverters does not affect the phase noise based on a simple model and that increasing power or supply voltage is the only option for the designer to lower the phase noise. When flicker noise is added, the phase noise equation becomes

$$L(f) = \frac{C'_{ox}}{8NI} \left(\frac{\mu_N + K_{fN}}{L_N^2} + \frac{\mu_P + K_{fP}}{L_P^2} \right) \frac{f_o^2}{f^3} \tag{7.42}$$

where C'_{ox} is the gate-oxide capacitance, μ_N (μ_P) is the mobility of the NMOS (PMOS), K_{fN} (K_{fP}) is the transconductance parameter of the NMOS (PMOS), and L_N (L_P) is the gate length of the NMOS (PMOS). Like other circuits, using a longer gate length is good to reduce the flicker noise but will slow down the slope of rising and falling edges due to increased load capacitance.

7.3.2.2 Differential Ring Oscillator

Differential ring VCOs are dominantly used in the design of an integrated PLL circuit since an inverter-based ring VCO is too sensitive to supply noise. For jitter analysis, we consider the conventional differential ring oscillator based

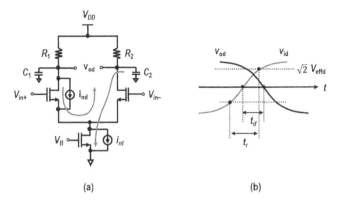

Figure 7.28 Noise analysis of a differential ring oscillator: (a) noise sources and (b) transition time and propagation delay.

on a differential amplifier having a resistor load. In fact, additional circuitry is required to maintain a constant output amplitude over different bias currents. Otherwise, the output frequency of a differential ring oscillator suffers from a very narrow tuning range since an increased output swing with the increased bias current can decrease the oscillation frequency. For that reason, the phase noise performance of an actual ring VCO is worse than that of a current-controlled differential ring oscillator with additional circuits such as a voltage-to-current converter and an amplitude control circuit. In this section, we will consider a conventional differential ring oscillator only and also assume a constant output swing for the sake of simplicity.

Figure 7.28 shows a differential delay circuit where a constant differential amplitude V_{op} is assumed, that is, the product of a bias current I and a load resistor R is fixed ($V_{op} = I \times R$). With the RC load, the proportional delay τ_d governed by exponential decay is expressed as

$$\tau_d = \frac{CV_{op}\ln 2}{I} = RC\ln 2 \tag{7.43}$$

The variance of a period jitter σ_τ^2 is given by

$$\sigma_\tau^2 = \frac{2kT}{If_o\ln 2}\left[\gamma\left(\frac{3}{4V_{effd}} + \frac{1}{V_{efft}}\right) + \frac{1}{V_{op}}\right] \tag{7.44}$$

where V_{effd} is the effective gate voltage of the differential pair and V_{efft} is the effective gate voltage of the tail transistor. To fully steer the current of the differential input pair in the next stage, V_{op} needs to satisfy

$$V_{op} \gg \sqrt{2}V_{effd} \tag{7.45}$$

From (7.40), the single-sideband phase noise from the variance of the period jitter is given by

$$L(f) = \sigma_\tau^2 \frac{f_o^3}{f^2} = \frac{2kT}{I \ln 2} \left[\gamma \left(\frac{3}{4V_{\text{effd}}} + \frac{1}{V_{\text{efft}}} \right) + \frac{1}{V_{op}} \right] \left(\frac{f_o}{f} \right)^2 \tag{7.46}$$

It is also shown that the effect of flicker noise on the phase noise is expressed as

$$L(f) = A \frac{K_f}{WLC'_{ox}f} \left(\frac{1}{V_{\text{efft}}^2} \right) \frac{f_o^2}{f^3} \tag{7.47}$$

The above equations imply that white noise from the differential pair and flicker noise from the tail transistor contribute to overall phase noise. With actual design parameters, it is shown that the flicker noise is more dominant than the white noise for the phase noise of the differential ring oscillator. For both single-ended and differential ring VCOs, increasing the average power is the most effective way to reduce the phase noise.

7.3.3 Circuit Implementation

Since the performance of a ring VCO is significantly affected by the circuit topology, various architectures could be considered based on different design requirements including phase noise, short-term jitter, power consumption, operation voltage, power supply rejection, multi-phase generation, operation speed, and so on. A voltage-controlled delay line (VCDL) is the key building block for the design of a ring VCO, and basic circuit topologies will be discussed in the following sections.

7.3.3.1 Single-Ended VCDL

Figure 7.29 shows three basic circuits of an inverter-based VCDL where the delay time of an inverter is controlled by current, load capacitance, or supply voltage. If the supply current of an inverter is reduced, the gate delay time of the inverter increases. Accordingly, the frequency of the ring VCO decreases from the maximum frequency with the reduced current, and this kind of VCDL shown in Fig. 7.29(a) is called a current-starved VCDL. In practice, the tail currents are controlled through a current mirror instead of a gate voltage, and a voltage-to-current converter is required. Instead of the current control, a variable capacitive load can also be used for the design of a VCDL. In Fig. 7.29(b), a MOS capacitor in series with a triode-region NMOS is used at the output of an inverter. As a control voltage increases, the resistance of the triode-region NMOS decreases, making an effective load capacitance increase. Therefore, the delay time of each stage increases with the increased control voltage. Instead of the triode-region NMOS and the fixed load capacitor, a varactor can also be used as a

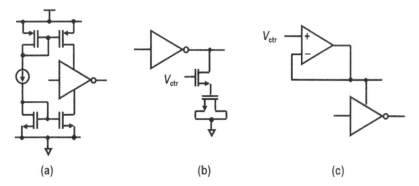

(a)   (b)   (c)

Figure 7.29   Single-ended VCDL: (a) using a current-starved inverter; (b) using a variable capacitive load; and (c) using a variable supply.

voltage-controlled load capacitor. Another option for controlling the delay time of an inverter is directly changing the supply voltage of the inverter. As shown in Fig. 7.29(c), a source follower is used to control the supply of the inverter. Instead of the source follower, a digitally-controlled resistor array can also be considered to improve the DC headroom of the VCDL. In practice, the single-ended VCOs are seldom employed in VLSI systems because of their high sensitivity to supply voltage variation.

7.3.3.2   Fully Differential VCDL

A differential ring VCO with a resistor load suffers from a very narrow tuning range since the amplitude of an output swing is nearly proportional to the value of a bias current. To achieve a linear frequency control, the voltage swing of the VCO needs to be unchanged regardless of the bias current, which is the case for the single-ended ring VCO. To limit the voltage swing, an active load is used. Figure 7.30(a) shows a straightforward way to limit the voltage swing by employing a PMOS diode as a load. Based on the square law of MOSFET, moderate swing regulation is achieved. However, the gate-source voltage heavily depends on process and temperature variations. Moreover, it is difficult to have a linear voltage-to-frequency control due to the inherent square-law property of the MOSFET.

Figure 7.30(b) shows the differential VCDL that offers a linear voltage-to-frequency control by using a linear-region PMOS and a replica biasing control. The linear-region PMOS behaves as a voltage-controlled resistor since the resistance can be changed by a gate voltage. Suppose that a ring oscillator has a hard-limit swing. When M_1 is on and M_2 is off, output voltages V_X and V_Y at nodes X and Y, respectively, are given by

$$V_X = V_{DD} - I_{\text{tail}}R_{L3}, \quad V_Y = V_{DD} \tag{7.48}$$

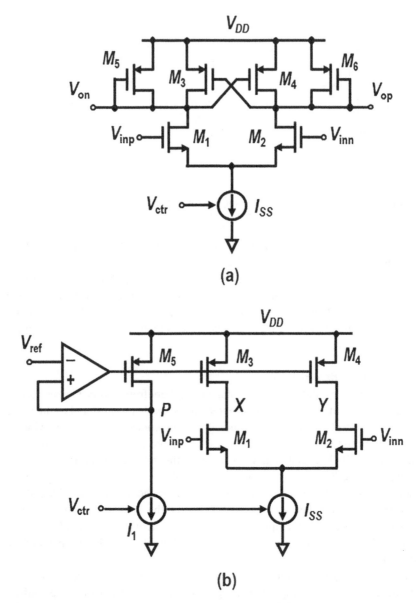

Figure 7.30 Differential VCDL with amplitude control: (a) using a diode load and (b) using a triode-region transistor load.

where the effective load resistance R_{L3} of the linear-mode M_3 is given by

$$R_{L3} = \frac{1}{\mu_p C_{ox}(V_{GS3} - V_{t3})\left(\frac{W}{L}\right)_3} \tag{7.49}$$

A replica circuit in the left shows the same condition that M_1 is fully on and sets V_P to a reference voltage V_{ref} by a feedback circuit. If I_{tail} increases, the resistance of M_1 decreases by feedback to have V_P unchanged. With the replica circuit, V_X is the same as V_P when M_1 is fully on. Therefore, the ring oscillator has an output swing from $(V_{DD} - V_{ref})$ to V_{DD}. Then, the oscillation frequency is expressed as

$$f_o = \left(\frac{\alpha}{CV_{ref}}\right) I_{tail} \tag{7.50}$$

where α is a scaling parameter to consider the effect of a circuit delay. Equation (7.50) shows that the output frequency is linearly controlled by a tail current. To improve power supply rejection, instead of the NMOS input pair, a PMOS input pair with NMOS loads can be designed to have the output swing of the ring VCO referenced to the ground.

The differential VCDL with an active load has difficulty in achieving fast speed as the active load adds substantial parasitic capacitance. In addition, the rising and falling times of an output swing get slewed when a ring VCO operates at a very high frequency. As a result, controlling the output frequency by changing the resistance of the active load is difficult. Figure 7.31 shows another type of a differential VCDL for high-frequency operation. When an original sinusoidal signal and a delayed sinusoidal signal are summed on average, an interpolated phase is generated. The amplitude of a summed signal remains the same if the total current is unchanged. By changing the weighting ratio of two signal paths, namely the fast path and the slow path, a phase control based on the phase interpolation is achieved as illustrated in Fig. 7.31(a). The amount of the phase delay in a slow path should not exceed 90° of a fast-path signal for smooth phase interpolation. Since the variable delay is achieved in the voltage domain by using current-steering amplifiers with a resistor load as shown in Fig. 7.31(b), a linear frequency control can be achieved at high frequency. The phase-interpolated VCDL, however, suffers from a narrow tuning range unless a multi-phase input is available.

7.3.3.3 Pseudo-Differential VCDL

The use of a bias current makes it difficult for the differential ring VCO to achieve a compact, low-power or low-voltage design. If a supply voltage can be regulated to a certain degree, a pseudo-differential VCDL shown in Fig. 7.32 is another choice to have a compact area and low power for moderate-speed operation. With a rail-to-rail swing and a positive-feedback latch, fast rising and falling times are obtained, achieving low-noise performance. In addition, the pseudo-differential

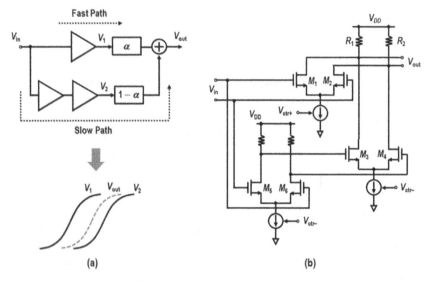

Figure 7.31 Differential VCDL based on the phase-interpolated delay: (a) delay control by phase interpolation and (b) circuit schematic.

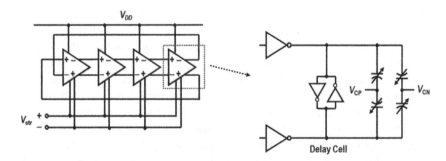

Figure 7.32 Pseudo-differential ring VCO with a latch and a varactor.

topology makes it possible to have even-number delay cells as illustrated in Fig. 7.32. A variable delay is achieved by having a variable capacitive load. Like the capacitor array in the design of an LC VCO, a switching capacitor array at the output node of each delay cell can also be employed to provide coarse tuning, while fine tuning is done by a control voltage through a varactor. Another way of controlling the delay time is to have a digitally-controlled resistor array at the supply node of inverters since the gate delay time highly depends on the supply voltage even for the pseudo-differential inverter.

7.4 Relaxation VCO

The relaxation oscillator generates a periodic signal by periodically charging and discharging a capacitor with source and sink currents. Unlike the ring oscillator, the relaxation oscillator generates a triangular waveform based on large-signal operation. In addition, the relaxation oscillator has less dependency on the g_m variation of transistors since an output frequency is determined mainly by a tail current and a timing capacitor. Especially for low-frequency operation, the use of a large timing capacitor makes the oscillator insensitive to parasitic capacitance and offers a constant frequency-tuning gain. Accordingly, the relaxation oscillator shows good noise performance for low-frequency clock generation by having the minimum number of active transistors. Multi-phase generation is also possible if multistage cores are designed with feedback. There are two kinds of conventional relaxation oscillators; one is with a ground capacitor, and the other is with a floating capacitor.

7.4.1 Relaxation Oscillator with Ground Capacitor

Figure 7.33 shows a relaxation oscillator with a ground capacitor. It simply charges and discharges the capacitor when the capacitor voltage exceeds the threshold of a comparator that is typically designed with a Schmitt trigger. At high speed, however, the delay time of the comparator becomes substantial over the oscillation period, making the performance sensitive to PVT variations. Due to a single-ended

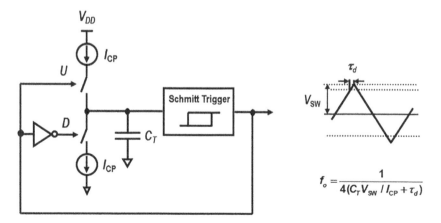

$$f_o = \frac{1}{4(C_T V_{SW} / I_{CP} + \tau_d)}$$

Figure 7.33 Relaxation oscillator with a ground capacitor.

structure, it is difficult to achieve a 50% duty cycle and good supply noise rejection. The oscillation frequency is given by

$$f_o = \frac{1}{4(CV_{sw}/I + \tau_d)} \tag{7.51}$$

where V_{sw} is the peak amplitude of a capacitor voltage and τ_d is the comparator delay. A semi-differential architecture can be designed to have the 50% duty cycle with symmetric operation and relax the comparator design by employing two single-ended VCOs, but it consumes a large area with two timing capacitors and increased power consumption.

7.4.2 Relaxation Oscillator with Floating Capacitor

A relaxation oscillator with a cross-couple pair and a floating capacitor shown in Fig. 7.34 has a fully balanced operation and provides decent supply noise rejection. Therefore, the symmetric relaxation oscillator with the floating capacitor is a good choice for the design of an integrated oscillator. However, the main limitation comes from nonlinear amplitude-dependency on the tail current. When a small floating capacitor is designed for high-speed low-power operation, parasitic capacitance also degrades both tuning range and phase noise.

Let us see how the NMOS loads affect the oscillator performance. We consider a relaxation oscillator for low-frequency operation and assume that parasitic capacitors at nodes A, B, X, and Y in Fig. 7.34 are negligible. We begin with an initial condition that M_1 is on and M_2 is off. Then,

$$V_A = V_{DD} - V_{GS3} = V_{DD} - V_{t3} - \Delta_3, \quad V_B = V_{DD} - V_{t4} \tag{7.52}$$

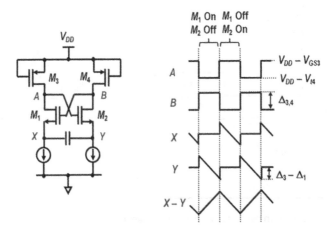

Figure 7.34 Relaxation oscillator with a floating capacitor.

where $V_{t,i}$ is the threshold voltage and Δ_i is the minimum drain-source voltage of the M_i transistor for saturation region ($i = 1, 2, 3$, and 4). Then, V_X is set to

$$V_X = V_B - V_{GS1} = V_{DD} - V_{t4} - V_{t1} - \Delta_1 \tag{7.53}$$

Since current I is flowing from X to Y, V_Y at the floating node Y keeps decreasing until M_2 is turned on, that is, $V_A - V_Y > V_{t2}$. Then, the turn-on voltage $V_{Y,on}$ at the node Y becomes

$$V_{Y,on} = V_A - V_{t2} = (V_{DD} - V_{t3} - \Delta_3) - V_{t2} \tag{7.54}$$

For simplicity, assume $V_{t,1} = V_{t,2}$ and $V_{t,3} = V_{t,4}$. Then, the peak swing voltage V_{SW} is given by

$$V_{SW} = V_X - V_{Y,on} = (V_{DD} - V_{t4} - V_{t1} - \Delta_1) - (V_{DD} - V_{t3} - \Delta_3 - V_{t2}) = \Delta_3 - \Delta_1 \tag{7.55}$$

From Fig. 7.34, the oscillation period T_0 is expressed as

$$T_0 = 2 \times \frac{C}{I} \times 2(\Delta_3 - \Delta_1) \tag{7.56}$$

Then, the oscillation frequency is given by

$$f_0 = \frac{I}{4C(\Delta_3 - \Delta_1)} \approx \frac{I}{4C\Delta_3} \text{ if } \Delta_1 \ll \Delta_3 \tag{7.57}$$

Equation (7.57) implies that the (W/L) ratio of M_1 and M_2 should be much larger than the (W/L) ratio of M_3 and M_4 to mitigate the effect of the cross-coupled transistors. To minimize the flicker noise, a long channel length is desirable for load transistors M_3 and M_4. However, Δ_3 is highly dependent on process and temperature variations, which makes it difficult to generate a stable frequency. When a relaxation VCO is designed, the nonlinear characteristic of the active load over different bias currents makes it difficult to achieve a linear voltage-to-frequency transfer function.

To overcome the nonlinear characteristic of the load transistor, a similar technique employed in a differential ring oscillator could be considered to achieve a constant output swing over different bias currents. As shown in Fig. 7.35, the load transistors M_3 and M_4 operate in the triode region and work as voltage-controlled resistors. The effective load resistances of M_3 and M_4, denoted by R_{M3} and R_{M4}, are assumed to be equal and denoted as R_{load}. Then,

$$R_{load} = R_{M3} = R_{M4} = \frac{1}{\mu_p C_{ox}(V_{GS3,4} - V_{t3,4})\left(\frac{W}{L}\right)_{3,4}} \tag{7.58}$$

Suppose that M_1 is turned off and M_2 is turned on as an initial state. Then, the voltages at nodes A and B are given by

$$V_A = V_{DD}, \quad V_B = V_{DD} - 2IR_{load} \tag{7.59}$$

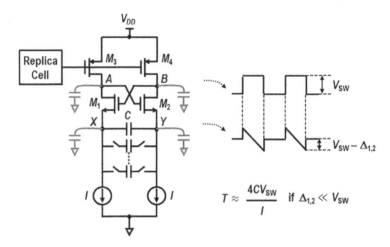

Figure 7.35 Relaxation oscillator with constant amplitude swing.

The voltage jump depicted in Fig. 7.35 is given by

$$V_Y - V_X = (V_{DD} - V_{t2} - \Delta_2) - (V_{DD} - 2IR_{load} - V_{t1}) = 2IR_{load} \qquad (7.60)$$

where the threshold voltages of M_1 and M_2 are assumed to be equal. During a half cycle, the capacitor voltage changes by $2(V_Y - V_X)$ and the oscillation period T_{osc} is determined as follows:

$$T_o = 2\left(\frac{C \times 2(2IR_{load} - \Delta_2)}{I}\right)$$

$$= \frac{4CV_{SW}}{I}\left(1 - \frac{\Delta_2}{V_{SW}}\right) \text{ where } V_{SW} = 2IR_{load} \qquad (7.61)$$

$$\approx \frac{4CV_{SW}}{I} \text{ if } \Delta_2 \ll V_{SW} \qquad (7.62)$$

Therefore, the oscillation period is fully characterized by the capacitor value, the fixed amplitude and the tail current if $\Delta_{1,2}$ is much less than V_{SW}. Since V_{SW} is constant regardless of the tail current, the oscillation frequency is linearly controlled by the tail current. Equation (7.62) also implies that the VCO gain can be scaled linearly with different capacitor values. Accordingly, using a capacitor array as depicted in Fig. 7.35 significantly improves the tuning range of the relaxation VCO. A set of switchable reference voltages can also be used to achieve fine-tuning control.

References

1 W. Egan, *Frequency Synthesis by Phase Lock*, 2nd ed., Wiley, New York, 2000.

2 U. L. Rohde, *Microwave and Wireless Frequency Synthesizers: Theory and Design*, Wiley, New York, 1997.

3 V. Manassewitsch, *Frequency Synthesizers, Theory and Design*, Wiley, New York, 1987.

4 D.B. Leeson, "A simple model of feedback oscillator noise spectrum," *Proceedings of the IEEE*, 1966, pp. 329–330.

5 U. L. Rohde, J. Whitaker and T. N. N. Bucher, *Communication Receivers*, 2nd ed., McGraw Hill, New York, 1997.

6 T. Lee, *The Design of CMOS Radio-Frequency Integrated Circuits*, Cambridge University Press, United Kingdom, 1997.

7 B. Razavi, *Design of CMOS Phase-Locked Loops: From Circuit Level to Architecture Level*, Cambridge University Press, United Kingdom, 2020.

8 B. Razavi (ed.), *RF Microelectornics*, 2nd ed., Prentice Hall, Upper Saddle River, NJ, 2012.

9 A. Hajimiri and T. H. Lee, "A general theory of phase noise in electrical oscillators," *IEEE Journal of Solid-State Circuits*, vol. 33, pp. 179–194, Feb. 1998.

10 C. Samori, A. L. Lacaita, F. Villa *et al.*, "Spectrum folding and phase noise in LC tuned oscillators," *IEEE Trans. Circuits Syst. II*, vol. 45, pp. 781–790, July 1998.

11 B. De Muer, M. Borremans, M. Steyaert *et al.*, "A 2-GHz low-phase-noise integrated LC-VCO set with flicker-noise upconversion minimization," *IEEE Journal of Solid-State Circuits*, vol. 35, pp. 1034–1038, July 2000.

12 P. Andreani and A. Fard, "More on the 1/f phase noise performance of CMOS differential-pair LC-tank oscillators," *IEEE Journal of Solid-State Circuits*, vol. 41, pp. 2703–2712, Dec. 2006.

13 J.R. Long and M.A. Copeland, "Modeling of monolithic inductors and transformers for silicon RFIC design," in *Proc. IEEE MTT-S Int. Symp. Tech. Wireless Appl.*, Feb. 1995, pp. 129–134.

14 M. Soyuer, J. N. Burghartz, K. A. Jenkins *et al.*, "Multi-level monolithic inductors in silicon technology," *Electron. Letters*, vol. 31, no. 5, pp. 359–360, Mar. 1995.

15 C. P. Yue and S. S. Wong, "On-chip spiral inductors with patterned ground shields for Si-based RF ICs," *IEEE Journal of Solid-State Circuits*, vol. 33, pp. 743–752, May 1998.

16 A. Niknejad and R. G. Meyer, "Analysis, design, and optimization of spiral inductors and transformers for Si RF ICs," *IEEE Journal of Solid-State Circuits*, vol. 33, pp. 1470–1481, Oct. 1998.

17 A. Zolfaghari, A. Y. Chan and B. Razavi, "Stacked inductors and transformers in CMOS technology," *IEEE Journal of Solid-State Circuits*, vol. 36, pp. 620–628, Apr. 2001.

18 F. Svelto, S. Deantoni and R. Castello, "A 1.3 GHz low-phase noise fully tunable CMOS LC VCO," *IEEE Journal of Solid-State Circuits*, vol. 35, pp. 356–361, Mar. 2000.

19 A. Mazzanti and P. Andreani, "Class-C harmonic CMOS VCOs, with a general result on phase noise," *IEEE Journal of Solid-State Circuits*, vol. 43, no. 12, pp. 2716–2729, Dec. 2008.

20 E. Hegazi, H. Sjöland and A. A. Abidi, "A filtering technique to lower *LC* oscillator phase noise," *IEEE Journal of Solid-State Circuits*, vol. 12, pp. 1921–1930, Dec. 2001.

21 D. Young and B. Boser, "A micromachined based RF low noise voltage controlled oscillator," in *Proc. IEEE Custom Integrated Circuits Conference*, May 1997, pp. 431–434.

22 A. Dec and K. Suyama, "A 1.9-GHz CMOS VCO with micromachined electromechanically tunable capacitors," *IEEE Journal Solid-State Circuits*, vol. 35, pp. 1231–1237, Aug. 2000.

23 W. Rhee, H. Ainspan, D.J. Friedman *et al.*, "A uniform bandwidth PLL using a continuously tunable single-input dual-path LC VCO for 5Gb/s PCI Express Gen2 application," in *Proc. IEEE Asian Solid-State Circuits Conference*, Nov. 2007, pp. 63–66.

24 D.-K. Jeong, G. Borriello, D. A. Hodges *et al.*, "Design of PLL-based clock generation circuits," *IEEE Journal of Solid-State Circuits*, vol. SC-22, pp. 255–261, Apr. 1987.

25 M. Horowitz, A. Chan, J. Cobrunson *et al.*, "PLL design for a 500 MB/s interface," in *Proc. IEEE International Solid-State Circuits Conference*, Feb. 1993, pp. 160–161.

26 T. Lee, K. Donnelly, J. Ho *et al.*, "A 2.5 V CMOS delay-locked loop for an 18 Mbit, 500 MB/s DRAM," *IEEE Journal of Solid-State Circuits*, vol. 29, pp. 1491–1496, Dec. 1994.

27 B. Kim, D. N. Helman and P. R. Gray, "A 30-MHz hybrid analog/digital clock recovery circuit in 2-pm CMOS," *IEEE Journal of Solid-State Circuits*, vol. 25, pp. 1385–1394, Dec. 1990.

28 I. A. Young, J. K. Greason and K. L. Wong, "A PLL clock generator with 5 to 110 MHz of lock range for microprocessors," *IEEE Journal of Solid-State Circuits*, vol. 27, pp. 1599–1607, Nov. 1992.

29 T.C. Weigandt, B. Kim, and P.R. Gray, "Timing jitter analysis for high-frequency, low-power CMOS ring oscillator design," in *Proc. IEEE International Symposium on Circuits and Systems*, June 1994, pp. 27–30.

30 J. A. McNeil, "Jitter in ring oscillators," *IEEE Journal of Solid-State Circuits*, vol. 32, pp. 870–879, June 1997.

31 A. A. Abidi, "Phase noise and jitter in CMOS ring oscillators," *IEEE Journal of Solid-State Circuits*, vol. 41, pp. 1803–1816, Aug. 2006.

32 A. Hajimiri, S. Limotyrakis and T. H. Lee, "Jitter and phase noise in ring oscillators," *IEEE Journal of Solid-State Circuits*, vol. 34, pp. 790–804, June 1999.

33 P. R. Gray and R. G. Meyer, *Analysis and Design of Analog Integrated Circuits*, Wiley, New York, 1993.

34 A. Abidi and R. G. Meyer, "Noise in relaxation oscillators," *IEEE Journal of Solid-State Circuits*, vol. sc-18, pp. 794–802, Dec. 1983.

35 W. Rhee, "A low power, wide linear-range CMOS voltage-controlled oscillator," in *Proc. IEEE International Symposium on Circuits and Systems*, May, 1998, pp. 85–88.

36 T. Sawlati and H. Shakiba, "A 20-800 MHz relaxation oscillator with automatic swing control," in *Proc. IEEE International Solid-State Circuits Conference*, Feb. 1998, pp. 222–223.

37 H. Lv, B. Zhou, W. Rhee, Y. Li, and Z. Wang, "A relaxation oscillator with multi-phase triangular waveform generation," in *Proc. IEEE International Symposium on Circuits and Systems*, May, 2011, pp. 2837–2840.

8

Frequency Divider

The use of a frequency divider in a feedback path enables a phase-locked loop (PLL) to perform frequency multiplication. When a programmable frequency divider is employed, the PLL becomes the frequency synthesizer that generates multiple frequencies for different channels in wireless systems. In digital systems, the PLL with a fixed or variable frequency division ratio acts as a clock multiplier unit. In terms of circuit design, the frequency divider is an important block for the low-power design of frequency synthesizers as it brings a voltage-controlled oscillator (VCO) frequency down to a PD frequency. According to the operation principle, the frequency divider can be categorized into three types: an integer divider (digital), a fractional divider (mixed-mode), and an injection-locked divider (analog). The integer divider generates an output signal whose period is an integer multiple of the period of an input signal. It often employs a dual-modulus divider (DMD) followed by programmable counters. The fractional divider generates an output signal whose average period can be a non-integer multiple of the period of an input signal. There are two common ways of implementing the fractional divider. One is using a multi-phase input and performing phase selection. The other is based on oversampled digital modulation. By the frequency-locking property of a PLL, the oversampled modulation effectively realizes the fractional division ratio on average, while quantization noise is filtered by the low-pass filter (LPF) characteristic of the PLL. The injection-locked divider performs frequency division not by counting the rising or falling edge of a period signal but by using the harmonic-locking property of a regenerative circuit.

8.1 Basic Operation

8.1.1 Frequency Division with Prescaler

Since a frequency divider is the most power-hungry block in the frequency synthesizer, the design of a programmable frequency divider is not based on the

Phase-Locked Loops: System Perspectives and Circuit Design Aspects, First Edition.
Woogeun Rhee and Zhiping Yu.

straightforward use of synchronous counters. For low power, the programmable frequency divider based on a prescaling method is designed to reduce the number of synchronous flip-flops. A DMD is a frequency divider that provides two division modes, typically with consecutive division values. In the fractional-N PLL discussed in Chapter 5, the DMD is used in a feedback path to realize a fractional division ratio with oversampled modulation. To the contrary, the prescaler is the DMD that should be followed by a counter, providing a prescaled value of the total division ratio. The operation principle and benefit of the prescaler can be well understood with the following examples.

Figure 8.1 shows how a divide-by-9 frequency divider, DIV_9, can be designed by using a 2/3 prescaler and a divide-by-4 frequency divider, DIV_4. A division ratio of 9 is realized by the product of a prescaling factor of 2.25 and a multiplication factor of 4, that is, $9 = 2.25 \times 4$. An effective division ratio of 2.25 is achieved by the 2/3 prescaler that performs the periodic operation of three divide-by-2 and one divide-by-3 for every four-clock period. A modulus control (MC) signal for the 2/3 prescaler comes from the trigger circuit that is enabled by the DIV_4 circuit as illustrated in Fig. 8.1. Note that the output clock f_o of the DIV_9 circuit always has a constant clock period since a fractional clock period produced at the output f_{pre} of the 2/3 prescaler is multiplied by a following divider with an inverse ratio. This example shows that various frequency division values can be achieved if the fixed divider-by-4 and the trigger circuits are replaced with a programmable counter and a programmable trigger circuits, respectively. The function of a standard programmable frequency divider will be discussed in detail in the next section.

We may wonder why the prescaler-based counter having feedback achieves lower power than a generic programmable counter. Figure 8.2 shows a good example of the comparison of a prescaler-based counter and a synchronous counter for the design of a programmable frequency divider at 2 GHz. In Fig. 8.2, the prescaler-based counter consisting of a 4/5 prescaler, an 8-bit counter, and a 2-bit counter performs 10-bit programmable frequency division at 2 GHz. The 8-bit main counter provides various division ratios after the 4/5 prescaler, while

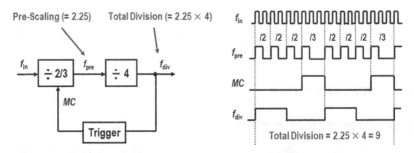

Figure 8.1 Frequency division based on the prescaling technique.

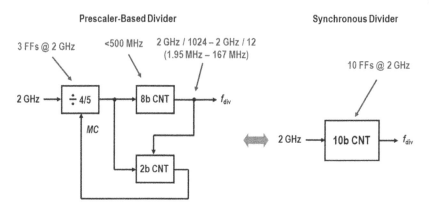

Figure 8.2 Prescaler-based counter versus synchronous counter.

the 2-bit auxiliary counter generates the MC signal for the 4/5 prescaler. How to calculate the overall division ratio will be learned in the next section. In this topology, only three D-type flip-flops (DFFs) operate at 2 GHz and the other two synchronous counters operate at <500 MHz. If a 10-bit synchronous counter is used, at least ten DFFs must operate at 2 GHz, consuming huge power. The prescaler-based counter also gives another advantage of reduced switching noise. Since the 4/5 prescaler requires only three DFFs, the current-mode logic (CML) circuit with bias currents can be considered for the DFF to minimize the kickback noise to a VCO by avoiding a rail-to-rail swing. In practice, the 4/5 prescaler cannot be placed too far from the VCO to minimize the parasitic capacitance due to the metal connection in the layout. Two 8-bit and 4-bit synchronous counters operating at low frequency can be implemented with standard CMOS logic and placed away from the VCO.

Example 8.1 *Fixed frequency division with multiple prescalers*
Let us take a heuristic example to see how the prescaling technique can be useful for power saving. Figure 8.3 shows a 2-GHz frequency divider with a fixed division ratio of 1,001. Since 1,001 is 11 times 91, we first use two frequency dividers with the fixed division ratios of 11 and 91. Then, the divide-by-11 divider is designed with a 3/4 prescaler followed by a divide-by-3 divider with a periodic sequence {3, 4, 4, 3, 4, 4, ...}. That is, the 3/4 prescaler generates an averaged division ratio of 11/3, and the following divide-by-3 divider generates the divide-by-11 output clock. With the prescaler-based divider topology, only three DFFs operate at 2 GHz, while the divide-by-3 divider operates at <670 MHz. Similarly, the divide-by-91 divider can be implemented with an 8/9 prescaler and a divide-by-11 divider based on the equation: $91 = 8 \times 8 + 3 \times 9$. That is, the 8/9 prescaler will

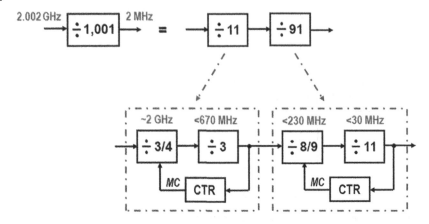

Figure 8.3 Low-power frequency divider for a fixed division ratio of 1,001.

perform eight divide-by-8 functions and three divide-by-9 functions for every 11 clock periods.

8.1.2 Standard Configuration of Prescaler-based Frequency Divider

The prescaler-based frequency divider not only reduces power consumption but also offers flexible programing. Figure 8.4 shows the standard configuration of a prescaler-based frequency divider. It consists of a $P/(P+1)$ prescaler, a main counter (M-CNT), and an auxiliary counter (A-CNT). The A-CNT with a counting value of A makes the MC signal set to logic high for A clock cycles. The M-CNT with a counting value of M resets the MC signal after M clock cycles. That is, if $(P+1)$ division is selected with MC = 1 in the $P/(P+1)$ prescaler design, then $(P+1)$ division is performed for A clock cycles, while P division with $(M-A)$ clock cycles. Therefore, the overall division ratio N is given by

$$N = A \times (P+1) + (M-A) \times P = A + M \times P \qquad (8.1)$$

For example, if the values of M and A are set to 8 and 3, respectively, then the use of a 4/5 prescaler gives the total division ratio given by

$$N = 3 \times 5 + (8-3) \times 4 = 15 + 20 = 35$$

Since M and A are set by the programmable counters of the M-CNT and the A-CNT, a fully programmable division ratio can be set based on the prescaler-based frequency divider. One caution of using the prescaler-based frequency divider is that a continuous set of division ratios is not guaranteed if a programmed division ratio is less than a certain value. The minimum

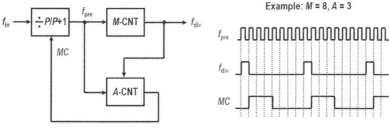

$$N = A \times (P + 1) + (M - A) \times P = A + M \times P$$

For continuous division ratio, $0 \leq A \leq P - 1$

Then, $N_{min} = P(P - 1) = P^2 - P$

Figure 8.4   General frequency division with the prescaler.

division ratio N_{min} for continuous programming should be set by

$$N_{min} = P(P - 1) = P^2 - P \tag{8.2}$$

For instance, the use of a 4/5 prescaler sets a minimum division ratio of 12, while the use of a 16/17 prescaler has a minimum division ratio of 240. Apparently, the use of a small value of P is better for the $P/(P + 1)$ prescaler to have a small N_{min}, but the high output frequency f_p of the prescaler gives a tight timing control for the MC signal. More discussion will be made later in this chapter.

Even though calculating the division value of the prescaler-based frequency divider is straightforward, it could be tedious to calculate the values of M and A and to verify all the division values with different programming control words. Fortunately, we can directly program the total division ratio without considering the values of M and A if we keep the standard configuration shown in Fig. 8.4 with the following rules:

- P value must be the power of 2. For example, we can use an 8/9 prescaler but cannot use a 7/8 prescaler.
- The maximum number of bits in the A-CNT must be the same as $\log_2 P$. That is, we must use a 3-bit A-CNT if an 8/9 prescaler is employed.
- Total division ratio cannot be smaller than $P^2 - P \, (= N_{min})$.

Figure 8.5 shows an example of how to program a division ratio of 4,197 with a 16/17 prescaler, a 10-bit M-CNT, and a 4-bit A-CNT. From (8.1), we obtain $M = 262$ and $A = 5$. That is,

$$N = 5 \times 17 + (262 - 5) \times 16 = 5 + 262 \times 16 = 4{,}197$$

Now, we will learn another way of programming the M-CNT and the A-CNT for the same case. What we need to do is just to convert the value of the target division ratio to a binary number. The division ratio of 4,197 is converted to a binary

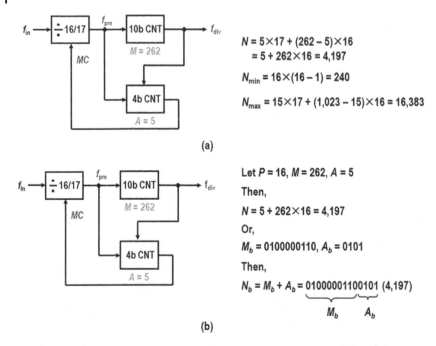

$$N = 5 \times 17 + (262 - 5) \times 16$$
$$= 5 + 262 \times 16 = 4{,}197$$

$$N_{min} = 16 \times (16 - 1) = 240$$

$$N_{max} = 15 \times 17 + (1{,}023 - 15) \times 16 = 16{,}383$$

(a)

Let $P = 16$, $M = 262$, $A = 5$

Then,

$$N = 5 + 262 \times 16 = 4{,}197$$

Or,

$$M_b = 0100000110, \; A_b = 0101$$

Then,

$$N_b = M_b + A_b = 01000001100101 \; (4{,}197)$$

$$\underbrace{}_{M_b} \underbrace{}_{A_b}$$

(b)

Figure 8.5 General frequency division with the prescaler: (a) calculation of M and A and (b) direct programming.

division value N_b as

$$N_b = 01\,0000\,0110\,0101\,(= 4{,}197)$$

Then, we take the last four bits from the least significant bit (LSB) and assign them as the input of the 4-bit A-CNT as illustrated in Fig. 8.5(b). That is, the binary input A_b of the A-CNT is

$$A_b = 0101\,(= 5)$$

Then, the rest of the remaining bits are assigned as the input of the 10-bit M-CNT and the binary input M_b of the M-CNT is

$$M_b = 01\,0000\,0110\,(= 262)$$

The example above shows that we do not need to calculate the input bits of the M-CNT and the A-CNT if the standard configuration is employed. All we need is to convert the desired division ratio to binary bits and decompose them into two control words for the M-CNT and the A-CNT. Note that the standard programming is valid when the desired division ratio is higher or equal to the minimum division ratio N_{min} set by (8.2). In Fig. 8.5, N_{min} is given by $16 \times (16 - 1) = 240$. As to the maximum division ratio N_{max}, we put all "1"s for N_b, so N_{max} is 16,383 ($= 2^{14} - 1$). It can also be calculated from (8.2) by having M of 1,023 ($= 2^{10} - 1$) and A of 15 ($= 2^4 - 1$)

$$N_{max} = 15 \times 17 + (1023 - 15) \times 16 = 15 + 1023 \times 16 = 16{,}383$$

Indeed, there is no limit for the maximum division ratio if the maximum number of bits of the M-CNT is not limited. To increase the maximum division ratio, we just need to increase the number of bits of the M-CNT.

8.1.3 Operation Principle of Dual-Modulus Divider

The operation principle of the DMD is based on a pulse swallowing method. Figure 8.6(a) shows the logic block diagram of a 2/3 DMD that consists of two DFFs, an AND gate, and an OR gate. When a modulus control bit MC is set to high, the output of the OR gate will be high regardless of the output of the first DFF. Then, the output of the AND gate is the same as the output of the complementary output of the second DFF. As a result, the whole block becomes

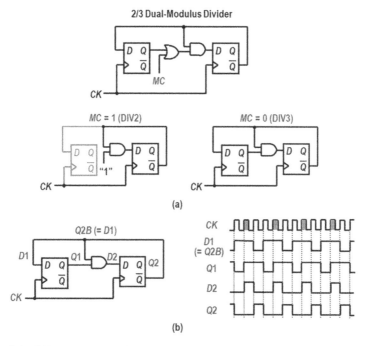

(a)

(b)

Figure 8.6 2/3 dual-modulus divider based on the pulse-swallowing technique: (a) functional block diagram and (b) divide-by-3 mode.

a toggle flip-flop (TFF) and works as a divide-by-2 circuit. When *MC* is set to logic low, the output of the OR gate depends on the output of the first DFF *Q*1. An equivalent block diagram for the divide-by-3 mode is drawn in Fig. 8.6(b). Compared with the divider-by-2 mode, one input pulse is removed (or swallowed) by the AND function of *Q*1 and *D*1 in the middle of two DFFs. Otherwise, the divide-by-2 function would have been performed. With the absence of one pulse at the input of the TFF, the output period of the TFF becomes three times the input period. This kind of pulse-swallowing technique is commonly employed in the design of a high-speed DMD since it is more power efficient than the standard synchronous counting method. As shown in Fig. 8.6, both the divide-by-2 and divide-by-3 functions are achieved with two DFFs and one AND gate only.

Even though the 2/3 DMD itself has the minimum number of logic gates and DFFs, it suffers from a tight MC timing control when used as a prescaler in the design of a standard frequency divider. The reason is that the MC signal is not static when the 2/3 DMD is used as a prescaler and that an updated MC signal must arrive within two periods of an input clock. For example, if a 1-GHz 2/3 prescaler is designed with main and auxiliary synchronous counters, an updated MC signal must arrive in 2 ns to warrant the right function of the prescaler for the dynamic modulus control. For that reason, a 4/5 prescaler or higher-division-ratio prescaler is employed in the design of standard frequency dividers in practice. Figure 8.7 shows the logic block diagram of a 4/5 DMD composed of three DFFs, one AND gate, and one OR gate. For the division ratio of 4, the DMD equals to a frequency divider based on cascaded TFFs, thus performing the divide-by-4 function. In the case of the divide-by-5 function, the pulse swallowing technique is used with a few

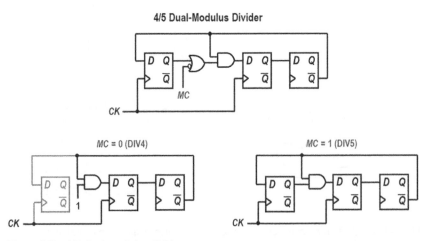

Figure 8.7 4/5 dual-modulus divider.

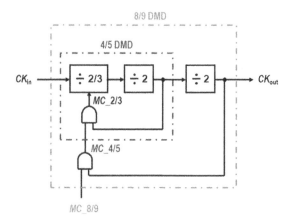

Figure 8.8 Dual-modulus divider with nested prescalers.

control logic gates, which is similar to the case of the divide-by-3 in Fig. 8.6(b). Apparently, the 4/5 DMD will consume more power than the 2/3 DMD, but the MC timing is relaxed from two periods of an input clock to four periods.

With the knowledge of the pulse-swallowing principle, it is worth noting that the 4/5 DMD or higher-division-ratio DMD can also be implemented by the extensive use of prescalers and logic gates to reduce the number of synchronous DFFs. Figure 8.8 illustrates how an 8/9 DMD is designed with a nested-prescaler structure. To begin with, let us consider the design of a 4/5 DMD by using a 2/3 prescaler. As shown in Fig. 8.8, the 4/5 DMD can be implemented by having the cascaded configuration of the 2/3 prescaler and a TFF. For the division ratio of 4, the MC input of the 4/5 DMD, denoted by $MC\_4/5$ in Fig. 8.8, is set to low, so that the AND output is always low. Then, the 2/3 prescaler becomes a divide-by-2 circuit, generating the division ratio of 4 after the following divide-by-2 circuit. For the division ratio of 5, $MC\_4/5$ is set to high, having the AND output $MC\_2/3$ determined by the output of the second-stage divide-by-2 circuit. The 2/3 prescaler performs both divide-by-2 and divide-by-3 functions with an equal duty ratio, resulting in an effective division ratio of 2.5. Therefore, the following divide-by-2 circuit generates an output clock whose period is equal to five periods of the input clock. From this example, we can see that the 4/5 DMD is realized by using the 2/3 prescaler, the TFF, and one AND gate. In a similar way, an 8/9 DMD can be designed by using the 4/5 DMD, the TFF, and the AND gate. Figure 8.8 shows a complete functional diagram of the 8/9 DMD by using one 2/3 prescaler, two TFFs, and two AND gates. As discussed, the 4/5 DMD is usually designed with three DFFs and control logic circuits instead of using the 2/3 prescaler in practice due to the tight timing margin of the modulus control.

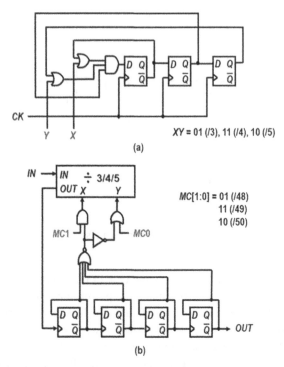

Figure 8.9 Multi-modulus dividers: (a) 3/4/5 prescaler and (b) 48/49/50 modulus divider.

Example 8.2 *Design of multi-modulus divider*

The operation principle of the pulse swallowing method is skipping one period of an input clock. In fact, more than one clock period can be skipped with additional control logic. For example, by swallowing two clock periods, the division ratio can be extended by two. Therefore, we can also build a multi-modulus divider with the pulse swallowing method. Figure 8.9(a) shows the block diagram of a 3/4/5 modulus divider. Divide-by-5 function is accomplished by skipping two periods of an input clock, while divide-by-4 is done by skipping one clock. As shown in Fig. 8.9(b), the 3/4/5 divider is used as a prescaler to build a 48/49/50 modulus divider with a 2-bit modulus control word. This kind of multi-modulus divider is useful in the design of a customized divider for a specific division range but is seldom used as a standard configuration. Note that the multi-modulus divider can also be realized by a standard prescaler-based frequency divider with a dynamic control word. However, the standard prescaler-based frequency divider must satisfy the requirement of the minimum division ratio as discussed. When a specific division range is considered, a customized architecture could be useful for optimum power and speed performance.

8.2 Circuit Design Considerations

If a type 2 PLL is designed with an underdamped loop, frequency overshooting during transient settling makes the output frequency of a VCO exceed the target frequency of the PLL. For that reason, the maximum operation frequency of a frequency divider should be higher than the maximum locking range of the PLL. In practice, the frequency range of the divider is designed to be wider than the tuning range of the VCO. Depending on the maximum speed, either a single-ended logic circuit with rail-to-rail swing inputs or a CML circuit with differential swing inputs could be considered. Even if a prescaler designed with single-ended latches meets the maximum speed, a CML-based prescaler is preferred since switching noise generated by high-frequency digital circuits can affect the noise performance of a VCO through substrate noise coupling. If there is no concern about the noise coupling or maximum speed, a prescaler with single-ended logic circuits offers compact area and low power. As the power and speed performance of a frequency divider is mostly determined by a front-end prescaler whose operation is based on a digital latch and logic circuits, the main design effort lies in the low-power design of a high-speed latch.

8.2.1 Frequency Divider with Standard Logic Circuits

When a latch is designed with CMOS logic circuits, either static-logic or dynamic-logic circuits are considered. Figure 8.10(a), (b) shows the schematics of a conventional static latch and a dynamic latch, respectively. The dynamic latch is

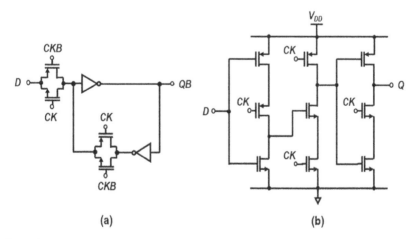

Figure 8.10 Single-ended DFF: (a) with the static latch and (b) with the TSPC dynamic latch.

based on a true single-phase clock (TSPC) logic. Even though the dynamic latch offers faster speed and lower power than the static latch, it is highly sensitive to process, voltage, and temperature (PVT) variations. Another critical issue of the dynamic latch is the minimum operation frequency. Unlike the static latch, a dynamically charged node in the dynamic latch cannot hold a voltage for a long time, resulting in malfunction with a low-frequency clock. Since both the maximum and the minimum operating frequencies heavily depend on PVT variations, the use of a dynamic latch is not recommended when a general-purpose PLL is designed.

Both static and dynamic latches with rail-to-rail inputs are more dependent on a physical layout than the CML latch, requiring careful post-layout simulations. If the distance between a VCO and a prescaler is too far, the rising and falling edges at the input of the prescaler slow down or could be slewed due to increased parasitic capacitance, which brings a situation that a positive-feedback latch could not flip the previous state properly. When the VCO and the prescaler are close, the VCO may be affected by substrate noise coupling as the prescaler operates at high speed, let us say, several GHz with a rail-to-rail swing. Nonetheless, the dynamic latch is frequently employed for low power, compact area, and bias-current-free operations especially when advanced CMOS technology is available. Figure 8.11 shows an example of the 2/3 prescaler designed with the TSPC logic. The TSPC logic is popular as it does not require a complementary clock and a non-overlapping clock generation circuit. Many other types of dynamic logics such as clocked CMOS (C^2MOS) logic, enhanced true single-phase clock (E-TSPC), and pseudo-differential dynamic latches could also be designed.

8.2.2 Frequency Divider with Current-Mode Logic Circuits

The CML latch with a differential input offers high-speed operation, reduced switching noise, and good immunity against supply voltage variation. Figure 8.12 shows a conventional divide-by-2 circuit using the master–slave CML latch that

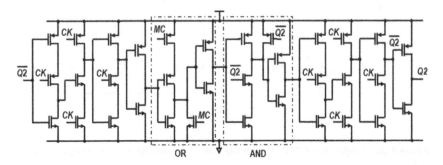

Figure 8.11 2/3 prescaler with the TSPC logic circuits.

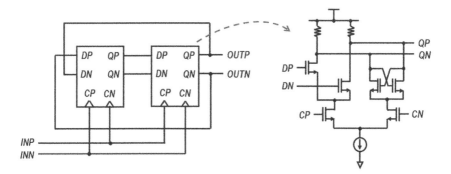

Figure 8.12 Divide-by-2 circuit with a master-slave CML latch.

generates an output at the rising edge of a clock. The master–slave CML latch forms a TFF by having an inverted output of the second latch connected to the data input of the first latch. The CML latch can also work without a tail bias current, but the performance is highly sensitive to supply voltage unless a large voltage swing is assumed like a pseudo-differential logic. In this section, we consider the CML latch having the tail bias current only. Since current biasing and transistor sizing affect the speed and the input sensitivity of the CML latch, the design procedure of the CML latch needs to consider several design factors as listed below:

- Input sensitivity ($\Delta V_{in,min}$)
- Maximum frequency (f_{max})
- Power consumption (P_{DC})
- Voltage amplitude (V_{SW})
- DC headroom
- Small-signal gain and noise

Figure 8.13 shows the schematic of a CML TFF with a comprehensive diagram illustrating several design aspects. A single CML latch is analyzed as a two-stage circuit whose function is similar to the comparator that consists of a preamplifier and a positive-feedback latch. Input sensitivity is an important parameter to warrant the robust function of the CML divider. If the gain of a first-stage amplifier is not high enough, a cross-coupled latch at the second stage cannot flip the state properly. The input sensitivity of a common differential amplifier depends on the bias current and the transistor size. In the conventional differential amplifier, a minimum differential input voltage $\Delta V_{in,min}$ to have a sufficient current steering is approximately given by

$$\Delta V_{in,min} = \sqrt{2}V_{DS,sat} = \sqrt{\frac{2I_B}{\mu_n C_{ox}\frac{W}{L}}} \tag{8.3}$$

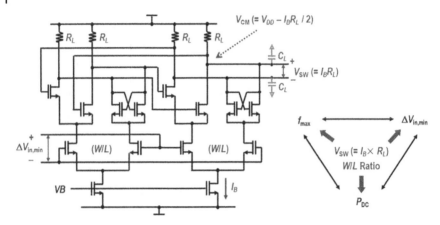

Figure 8.13 Design aspects of the CML divider.

where $V_{DS,sat}$ is the overdrive voltage of input transistors, I_B is the tail current, μ_n is the mobility of the NMOS transistor, and C_{ox} is the oxide capacitance. Equation (8.3) implies that designing a CML latch with a small load resistor and a large tail current for high-speed operation may encounter a potential problem of poor input sensitivity especially when all the transistors are designed with a small W/L ratio to minimize the parasitic capacitance. Simply increasing the bias current to have a large voltage swing does not solve the input sensitivity problem since it increases $\Delta V_{in,min}$ as well. If a high W/L ratio of the differential-input transistors is used to improve the input sensitivity, the high W/L ratio increases both the parasitic capacitance at the output node of the divider itself and the load capacitance to the previous stage, thus significantly reducing the maximum speed of the divider for the given bias current.

Another caution is that not only the amplitude but also the slope of an input waveform is important for input sensitivity. It is because an input waveform with slow rising or falling edge causes a long period of the simultaneous turn-on overlap of input-pair transistors. For instance, when an input signal is slewed like a triangular waveform, even a large input waveform can also cause an insufficient current steering. The slewed waveform happens because of large load capacitance, which is often the case when a VCO and a frequency divider are far apart. Therefore, maintaining the fast rising and falling times of an input waveform is important for the robust function of the CML divider at the cost of increased power.

The maximum speed is mainly determined by an RC time constant at the output node. The resistance is set by a load resistor R_L, while the load capacitance C_L is given by the sum of all parasitic capacitances at the output node including the gate-oxide capacitance of input transistors in the following stage. If we do not change the transistor size and the voltage swing, the only way to increase the speed is to reduce the resistance and increase the bias current. As discussed, increasing the bias current without changing the transistor size degrades the input sensitivity. Therefore, there is a fundamental trade-off between the maximum operation frequency and the input sensitivity. In the design of a CML frequency divider, the worst-case condition for the input sensitivity could happen with fast-corner process and hot temperature. It is because the fast-corner process has a low value of R_L, while the hot temperature slows down the mobility of the transistor, giving the worst condition for a small-signal gain. On the other hand, the worst-case condition for the maximum speed occurs with slow-corner process and hot temperature since the slow-corner process gives a large RC time constant at the output node, while hot temperature makes transistors slow down. Therefore, both worst-case conditions, that is, the fast- and slow-corner processes under hot temperature, must be run for the proper design of the CML frequency divider.

A voltage swing amplitude V_{SW} is an important design parameter in determining the speed, power, DC headroom, and robust CML operation. It should be sufficiently larger than $\Delta V_{in,min}$ in (8.3) for robust operation. The small-signal gain of a CML latch must be sufficiently higher than 1 for proper differential logic operation. Being a switching circuit rather than an amplifying circuit, the CML divider with sufficient V_{SW} has negligible noise contribution at the PLL output compared with a VCO or a charge pump. In the CML divider, the voltage swing sets the output common-mode voltage V_{CM} as well. Accordingly, if the voltage swing is too large, the input common-mode voltage of the following latch is too low to make tail transistors operate in a saturation region with enough DC headroom.

When cascaded CML dividers are used for a divider chain, any modification of a latch can affect the previous- or next-stage latch performance. Figure 8.14 shows

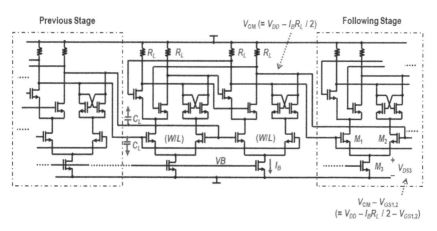

$$V_{CM} - V_{GS1,2}$$
$$(= V_{DD} - I_B R_L / 2 - V_{GS1,2})$$

Figure 8.14 Design of cascaded divide-by-2 circuits.

an example of how the change of the second-stage divider affects both the first- and the third-stage dividers in the design of a divide-by-8 circuit that consists of three cascaded divide-by-2 circuits. Suppose that the first-stage divide-by-2 circuit is designed with relatively a small output swing to meet the maximum speed with low power. As a result, we need to compensate for the weak output of the first-stage divide-by-2 circuit by designing a second-stage divide-by-2 circuit with good input sensitivity. In addition, a slightly bigger output swing can be considered since the operation frequency of the second-stage divide-by-2 circuit is half the frequency of the first-stage divide-by-2 circuit. To have a bigger output swing, we need to increase either the size of an input-stage transistors for high transconductance or the bias current. The former case significantly increases the output load capacitance of the first-stage divider since the dominant part of the output load capacitance is given by $C_{ox}WL$. Accordingly, the first-stage divider circuit needs to be redesigned as the output load capacitance is increased. In the latter case, increasing the bias current lowers the output common-mode voltage. If the following-stage divider has a tight DC headroom, then the value of an output load resistor needs to be reduced. Otherwise, the current mirror in the following stage may enter the linear region due to a limited drain-source voltage. In addition, increasing the bias current without increasing the transistor size will degrade the input sensitivity based on (8.3). Therefore, all design parameters should be considered with entire latches together in the divider chain.

Example 8.3 *Voltage swing variation over process and temperature*
As the voltage swing amplitude is an important parameter not only for the speed but also for the robust operation, it is important to consider the variation of the voltage swing amplitude over process and temperature. Figure 8.15 illustrates how an on-chip reference voltage and an on-chip resistor are used to control the voltage swing of a CML divider. A bandgap reference voltage V_{ref} and an on-chip resistor R_{ref} are used to generate the bias current I_B of the CML divider. Then, the voltage swing V_{sw} is given by

$$V_{sw} = \frac{I_B R_L}{2} = \frac{1}{2}\left(\frac{V_{ref}}{R_{ref}}\right)R_L = \frac{V_{ref}}{2}\left(\frac{R_L}{R_{ref}}\right)$$

Therefore, a constant voltage swing is achieved regardless of process and temperature variations if the same kind of resistor is used for both the current generation circuit and the CML divider. This approach is useful to control a tight DC headroom and avoid speed degradation in slow corner process or hot temperature. When a low-voltage swing is designed for low-power and high-speed operation, the fixed-voltage swing may cause insufficient current steering or input sensitivity problem to the next stage at slow corner or hot temperature. In that

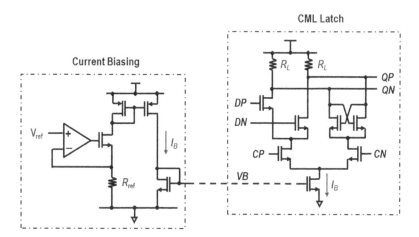

Figure 8.15 CML latch with an on-chip current biasing circuit.

case, a fixed current with an external resistor or a programmable reference voltage instead of the bandgap reference voltage could be considered. To further increase the voltage swing to compensate for degraded sensitivity at high temperature, a proportional-to-absolute-temperature (PTAT) voltage can also be considered at the cost of reduced maximum frequency.

Figure 8.16 shows a typical topology of a programmable frequency divider, consisting of a CML preclear, a CML-to-CMOS converter, a counter, and a modulus controller. The CML circuit is designed for high speed and low switching noise, while the low-speed counter is designed with standard CMOS logic for low power and compact area. The CML-to-CMOS converter generates a single-ended rail-to-rail output from a differential input. In the next section, we will discuss the timing issue of the modulus controller, which sometimes is more critical than the speed of a front-end prescaler for the overall performance of a frequency divider.

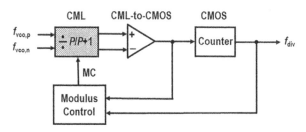

Figure 8.16 Programmable divider with the CML prescaler and standard counters.

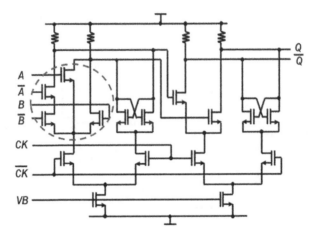

Figure 8.17 CML DFF with an embedded OR gate.

The current-steering operation of the CML divider makes it easy to have an embedded logic gate to achieve low-power prescaler design. Figure 8.17 shows a DFF circuit with an embedded OR gate that consists of NMOS transistors. You may wonder how all transistors are guaranteed to be in the saturation region when a reasonable voltage swing is assumed. In fact, it is difficult for all the transistors to operate in the saturation region during the entire voltage switching period. Except current mirrors, other transistors may operate in the linear region when the voltage swing becomes maximum or minimum. CMOS transistors are fast in the transition between linear and saturation regions and do not slow down the speed because of the transition. For comparison, bipolar transistors are slow in recovering the linear region from a saturation region,[1] so an emitter follower should be added at the output to provide a proper DC level shifting, which is called an emitter-coupled logic (ECL). To the contrary, a source follower is not required in the CMOS CML design due to a significant gain loss.

8.2.3 Critical Path of Modulus Control

The prescaler is the most critical building block in the design of a high-speed low-power frequency divider. However, a feedback path for the modulus control should not be neglected since the timing margin of the MC signal could be a bottleneck for the design of a high-speed frequency divider. Figure 8.18 shows an example of how the maximum speed of a 32/33 DMD could be limited by the MC signal path instead of a prescaler. The MC signal of the 32/33 DMD is generated

---

1 The saturation region for bipolar transistors is analogous to the linear region for CMOS transistors, while the linear region with bipolar corresponds to the saturation region with CMOS.

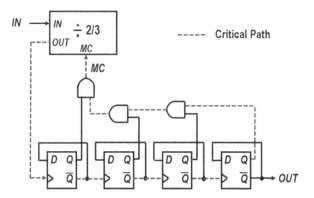

Figure 8.18 32/33 dual-modulus divider with a wrong MC signal path.

by four cascaded TFFs and a few logic gates to control a 2/3 prescaler. As depicted in Fig. 8.18, the critical path of the MC signal is set by the total delay of four TFFs and three AND gates. If the input frequency of the 2/3 prescaler is 2 GHz, the time margin of the MC signal path for the 2/3 prescaler should be less than 1 ns since the minimum clock period at the output of the 2/3 prescaler is 1 ns. In that case, minimizing the MC signal delay is more challenging than designing a 2-GHz 2/3 prescaler, making the delay of the MC signal path be a bottleneck for the design of a low-power DMD.

The MC timing issue can be significantly relaxed by judiciously connecting the TFFs and logic gates in the prescaler-based frequency divider. Figure 8.19

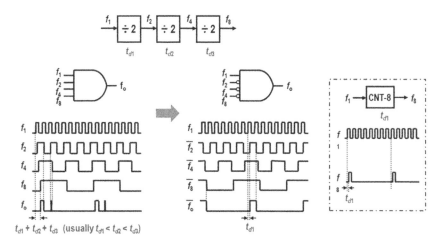

Figure 8.19 Generation of synchronous MC delay with asynchronous counters.

illustrates how to generate the output of a divide-by-8 circuit like a synchronous counter even if the divide-by-8 circuit is composed of three cascaded TFFs. The total delay of the asynchronous divide-by-8 counter is given by $(t_{d1} + t_{d2} + t_{d3})$, where t_{d1}, t_{d2}, and t_{d3} are the delay times of the first, the second, and the third TFFs, respectively. Therefore, the output of the AND gate exhibits an accumulated time delay from the cascaded TFFs. Moreover, a glitch waveform due to an overlap between the output f_2 of the first TFF and the output f_4 of the second TFF could occur as shown in Fig. 8.19. Now, let us see how the situation changes if we use the complementary outputs of the TFFs for the MC signal. As illustrated in Fig. 8.19, the output of the AND gate has only one gate delay time of the first TFF if the complementary outputs of the TFFs are used. In addition, the output of the AND gate does not have a glitch waveform since $\overline{f_2}$ is fully embedded within the pulse widths of $\overline{f_4}$ and $\overline{f_8}$. The result implies that the gate delay of the following TFFs is not important since the critical path of the MC timing is determined by the gate delay of the first TFF. Therefore, we could achieve the same delay as that of a synchronous counter even with the asynchronous counters, taking the full benefit of the prescaler-based topology.

Based on what we discussed, let us get back to the design of the 32/33 DMD. In Fig. 8.20, the complementary outputs $\overline{Q}$ of the TFFs are used for the modulus control so that the critical path of the MC timing is given by the gate delay time of the first TFF and one AND gate. In this way, the critical path of the MC signal path of the 64/65 or the 128/129 DMD can be nearly the same as that of the 32/33 DMD. Accordingly, there is no need of keeping the same power for low-frequency TFFs. For example, the total power consumption of the 128/129 DMD is comparable to that of the 32/33 DMD since the extra two TFFs can consume only one-tenth of the power of the first high-frequency TFF (why?).

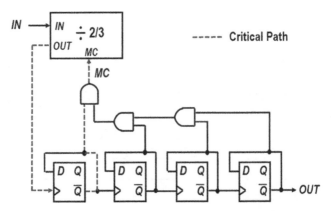

Figure 8.20 32/33 dual-modulus divider with an improved MC signal path.

8.3 Other Topologies

Even though the prescaler-based frequency divider provides a versatile function as a standard programmable divider, other structures could also be considered for various applications. In the following sections, several different topologies of frequency dividers are discussed.

8.3.1 Phase-Selection Divider

The operation principle of a DMD comes from the pulse swallowing technique. Hence, the same function can be achieved if a phase jump is performed with a phase step equal to a single period of a VCO. For example, if a divide-by-4 circuit provides quadrature outputs with a phase resolution of 90°, a single phase-jump after every divide-by-4 output results in divide-by-5 function since each phase-jump equals to skipping one VCO clock period. If two-phase steps are jumped, it is equivalent to swallowing two periods of the VCO, resulting in divide-by-6 operation. Therefore, the pulse-swallowing 4/5 prescaler can also be designed with a phase-selection circuit after generating quadrature phases from a divide-by-4 divider.

Figure 8.21 shows the comparison example of two 16/17 DMDs based on different topologies: (a) the conventional 4/5 prescaler and (b) the phase-selection 4/5 prescaler. The conventional 4/5 prescaler uses three synchronous DFFs, which increases power consumption not only for the prescaler itself but also

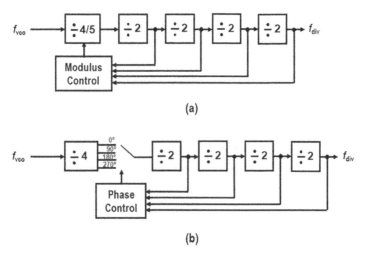

Figure 8.21 16/17 dual-modulus divider: (a) with the conventional 4/5 prescaler and (b) with the phase-selection 4/5 prescaler.

for the VCO output buffer that sees a large capacitive load due to three DFFs. In the phase-selection 4/5 prescaler, the VCO output buffer sees a single TFF if the divide-by-4 divider is implemented with two cascaded TFFs. Therefore, significant power reduction is achieved with the phase-selection 4/5 prescaler. Note that quadrature phases can easily be generated by cascaded TFFs and that the design of a multi-phase VCO is not needed for the phase-selection divider. The phase-selection divider, like the phase-interpolated fractional-N divider, requires careful design of the phase control circuits to minimize phase mismatch and potential glitch problems in the phase selection among multiple phases. Note that the phase mismatch at the output of the divide-by-4 divider causes a reference spur in a frequency synthesizer since the phase-selection rate is as low as the reference rate.

8.3.2 Phase-Interpolated Fractional-N Divider

If multiple phases are available from a VCO, a frequency divider with a non-integer division ratio, namely a fractional-N divider, can be implemented by interpolating a phase at the output of an integer-N divider. Figure 8.22 shows an example of building a 4/4.25 prescaler with a 4-stage differential ring VCO and a phase control circuit. Eight phases from the 4-stage differential ring VCO are used to have the phase resolution of one-fourth of the VCO period T_{vco}. As shown in a timing diagram in Fig. 8.22, when we have a phase jump from ϕ_k to ϕ_{k+1}, an extra phase of one-fourth of the VCO period is interpolated. The interpolated time Δt equals to

$$\Delta t = \frac{T_{vco}}{4} \tag{8.4}$$

Accordingly, a single-phase jump is equal to one-fourth of the VCO period. If the MC signal is always set to high, the phase jump occurs for every reference clock

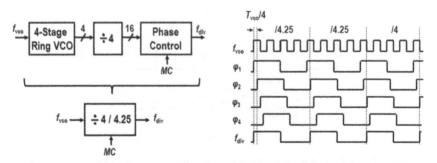

Figure 8.22 Phase-interpolated fractional divider.

period, producing the output period T_{div} of the DMD given by

$$T_{\text{div}} = \left(4 + \frac{1}{4}\right) T_{\text{vco}} \tag{8.5}$$

This kind of a *fractional-modulo* prescaler shown in Fig. 8.22 is also considered a pulse- swallowing prescaler since the phase interpolation of the one-fourth of the T_{vco} is analogous to swallowing the sub-period of the VCO. Even though the phase-interpolated fractional-N divider enables the PLL to have a high reference frequency for a given frequency resolution, it requires a careful design to achieve good matching and linearity among the multiple phases. A better way of realizing the fractional division ratio will be discussed in Chapter 9.

8.3.3 $(2^k + M)$ Multi-Modulus Divider

The conventional prescaler-based frequency divider has the drawback of having a large minimum division ratio set by (8.2). Fractional-N frequency synthesizers can use a high reference frequency regardless of frequency resolution, requiring a multi-modulus divider with a low minimum division ratio. To have the low-division-ratio multi-modulus divider, the use of cascaded 2/3 prescalers, namely a $(2^k + M)$ multi-modulus divider, can be considered. Figure 8.23 shows the operation principle of the $(2^k + M)$ divider with a timing diagram. By employing the pulse-swallowing method for each 2/3 prescaler, the cascaded 2/3 prescalers achieve the division ratio that is equal to the number of pulse swallowing cycles plus 2^k where k is the number of 2/3 prescalers. For example,

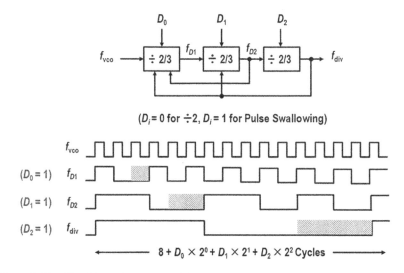

Figure 8.23 $(2^k + M)$ multi-modulus divider.

as illustrated in Fig. 8.23, swallowing one pulse in the third-stage 2/3 prescaler equals four cycles of an input clock. With a binary-weighted modulus control, various division values can be programmed with a minimum division ratio of 2^k, and increasing the division ratio is done by swallowing clock cycles successively with a binary-weighted function. That is, the division ratio N in Fig. 8.23 is given by

$$N = 8 + D_0 + 2D_1 + 4D_2 \tag{8.6}$$

where D_0, D_1, and D_2 are modulus control bits for the first, the second, and the third 2/3 prescalers, respectively. Since the 2/3 prescaler is used as a first-stage prescaler, this topology can have lower power than the standard topology of the prescaler-based frequency divider. However, having the cascaded 2/3 prescalers gives a harsh timing margin for the modulus control.

8.3.4 Regenerative Divider

8.3.4.1 Miller Divider
Unlike the counter-based frequency divider, the regenerative frequency divider employed in early-stage frequency synthesizers is a feedback-based circuit consisting of a mixer, a frequency multiplier, and a low-pass filter. Such an analog-oriented divider, namely a Miller divider can operate at higher frequencies than the conventional digital divider. As shown in Fig. 8.24, the regenerative divider performs frequency division by having a frequency multiplier in the feedback path and a mixer to generate a beat tone. Suppose that the frequency multiplier with a multiplication factor of $(N - 1)$ is used. Then, the output frequency of the Miller divider is obtained by

$$f_{out} = f_{in} - f_{fb} = f_{in} - (N - 1)f_{out}$$

or

$$f_{out} = \frac{f_{in}}{N} \tag{8.7}$$

Note that we can also obtain a fractional division ratio if a multiplication factor of N instead of $(N - 1)$ is used in the feedback path. The Miller divider does not

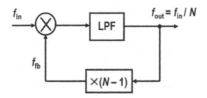

Figure 8.24 Miller divider.

perform frequency division based on digitally counting the clock edge but uses the mixer and the multiplier with small-signal waveforms, so it can operate at a very high frequency. Despite high-speed operation, the Miller divider adds significant design complexity for the control of the loop gain and locking range, suffering from high power and large area.

8.3.4.2 Injection-Locked Divider

Another class of the regenerative divider for low-power design is an injection-locked frequency divider (ILFD). The ILFD employs the oscillator that is injection locked by an input signal whose frequency is close to the integer multiple of the free-running frequency of an oscillator. Figure 8.25 shows typical differential ILFDs that perform divide-by-2 by using an LC oscillator or a ring oscillator. Since the common-source connection of a differential input pair oscillates at twice the output frequency, an input signal whose frequency is twice the oscillation frequency is injected to the common-mode node. Then, the oscillator will be injection locked with a desired output frequency, that is, half the input frequency. As a result, the injection-locked oscillator functions as a divide-by-2 ILFD. The ring-oscillator-based ILFD offers a wider locking range, compact area, and multi-phase generation capability but suffers from limited speed and poor noise performance.

To have a free-running oscillator locked by an injected signal, the injected frequency should be within the locking range of the ILFD. A strong injected signal

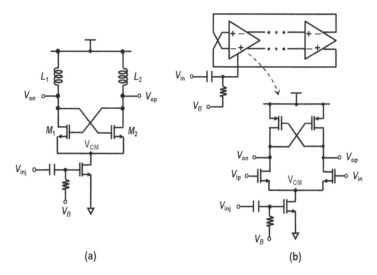

Figure 8.25 Injection-locked dividers: (a) based on the LC oscillator and (b) based on the ring oscillator.

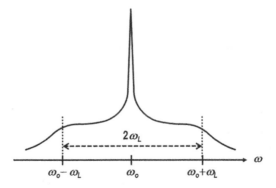

Figure 8.26 ILO spectrum with injection-lock range ω_L.

gives a large phase perturbation in the oscillator, which will cause frequency deviation whose amount depends on the frequency stability factor or the quality factor of the oscillator as discussed in Chapter 7. High-quality factor gives less frequency change for a given phase perturbation based on (7.16). The locking range ω_L of an injection-locked oscillator (ILO) is approximated as

$$\omega_L = \frac{\omega_0}{2Q} \frac{I_{\text{inj}}}{I_{\text{osc}}} \tag{8.8}$$

where I_{inj} and I_{osc} are the amplitudes of the injection current and the tail current, respectively. Figure 8.26 depicts the typical output spectrum of the ILO. In fact, the locking behavior of the ILO is similar to that of a type 1 PLL. After the injection locking, that is phase/frequency locking, the phase noise of an open-loop oscillator is suppressed within the locking range of the ILO, which is analogous to the high-pass filter characteristic of the PLL to the VCO noise. Also, the input jitter from an injection signal is low-pass filtered at the ILO output. That is, the ILO can be viewed as a band-pass filter with an effective Q of $\omega_0/2\omega_L$. Unlike the PLL, the ILO does not have the frequency acquisition aid by a phase detector, resulting in a limited lock-in range. Like other regenerative dividers, the main drawback of the ILFD using an LC oscillator is a narrow locking range, while the drawback of the ILFD using a ring oscillator is its high sensitivity to process, voltage, and temperature variations.

References

1 W. Egan, *Frequency Synthesis by Phase Lock*, 2nd ed., Wiley, New York, 2000.
2 U. L. Rohde, *Microwave and Wireless Frequency Synthesizers: Theory and Design*, Wiley, New York, 1997.

3 V. Manassewitsch, *Frequency Synthesizers, Theory and Design,* Wiley, New York, 1987.

4 B. Razavi, *Design of CMOS Phase-Locked Loops: From Circuit Level to Architecture Level,* Cambridge University Press, United Kingdom, 2020.

5 B. Razavi (ed.), *RF Microelectornics,* 2nd ed., Prentice Hall, Upper Saddle River, NJ, 2012.

6 Y. Kado, M. Suzuki, K. Koike *et al.,* "A 1-GHz/0.9-mW CMOS/SIMOX Divide-by- 128/129 Dual-Modulus Prescaler Using a Divide-by-2/3 Synchronous Counter," *IEEE Journal of Solid-State Circuits,* vol. 28, pp. 513–517, Apr. 1993.

7 J. Craninckx and M. Steyaert, "A 1.75-GHz/3-V dual modulus divide-by-128/129 prescaler in 0.7 μm CMOS," *IEEE Journal of Solid-State Circuits,* vol. 31, pp. 890–897, July 1996.

8 M. H. Perrott, T. L. Tewksbury III, and C. G. Sodini, "A 27-mW CMOS fractional-N synthesizer using digital compensation for 2.5-Mb/s GFSK modulation," *IEEE Journal of Solid-State Circuits,* vol. 32, pp. 2048–2060, Dec. 1997.

9 C. S. Vaucher, I. Ferencic, M. Locher *et al.,* "A Family of low-power truly modular programmable dividers in standard 0.35-μm CMOS technology," *IEEE Journal of Solid-State Circuits,* vol. 35, pp. 1039–1045, July 2000.

10 W. Rhee, "Design of low jitter 1-GHz phase-locked loops for digital clock generation," in *Proc. IEEE International Symposium on Circuits and Systems,* May, 1999, pp. 520–523.

11 R. L. Miller, "Fractional-frequency generators utilizing regenerative modulation," *Proc. Institute of Radio Engineering (IRE),* vol. 27, pp. 446–456, July 1939.

12 R. Rategh and T. H. Lee, "Superharmonic injection-locked frequency dividers," *IEEE Journal of Solid-State Circuits,* vol. 34, pp. 813–821, June 1999.

13 R.J. Betancourt-Zamora, S. Verma, and T.H. Lee, "1-GHz and 2.8-GHz CMOS Injection-locked Ring Oscillator Prescalers," in *Proc. Symposium on VLSI Circuits,* June 2001, pp. 49–50.

14 S. Verma, H. R. Rategh and T. H. Lee, "A unified model for injection-locked frequency dividers," *IEEE Journal of Solid-State Circuits,* vol. 38, no. 6, pp. 1015–1027, June 2003.

15 B. Razavi, "A study of injection locking and pulling in oscillators," *IEEE Journal of Solid-State Circuits,* vol. 39, pp. 1415–1424, Sep. 2004.

Part IV

PLL Architectures

9

Fractional-N PLL

9.1 Fractional-N Frequency Synthesis

In the conventional phase-locked loop (PLL)-based frequency synthesizer, a phase detector (PD) frequency determines the minimum step frequency. Therefore, the PD frequency is normally set by the channel spacing requirement of wireless systems. As discussed in Chapter 5, designing a high-performance PLL is quite challenging with a large division ratio. The fractional-N PLL was developed to use high PD frequencies for the same frequency resolution so that the frequency division ratio can be significantly reduced. The fractional-N frequency synthesis is done based on a digital phase-control method, which was originally called a *digiphase* technique and named later as a fractional-N technique for commercial products. The development of the fractional-N frequency synthesizer greatly improved the performance of modern transceiver systems not only as a low-noise local oscillator but also as a direct-digital frequency modulator.

9.1.1 Basic Operation

The fractional division ratio could be obtained by modulating the control input of a dual-modulus divider, which is basically frequency interpolation. Figure 9.1 shows an example of how to achieve a fractional division ratio of $(N + 1/4)$ with an $N/(N + 1)$ dual-modulus frequency divider. To have the division ratio of $(N + 1/4)$, a division ratio of $(N + 1)$ is done for every three N divisions. A 2-bit accumulator is used to generate a carry sequence of $\{\ldots 000100010001 \ldots\}$ where the carry output "1" corresponds to the $(N + 1)$ division. An output period T_{out} and the reference period T_{ref} are related by

$$T_{\text{ref}} = T_{\text{out}} \left(N + \frac{1}{4} \right) = NT_{\text{out}} + \frac{T_{\text{out}}}{4} \tag{9.1}$$

Phase-Locked Loops: System Perspectives and Circuit Design Aspects, First Edition.
Woogeun Rhee and Zhiping Yu.

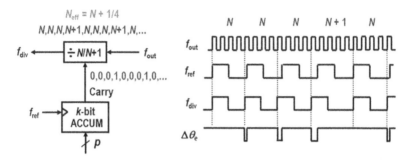

Figure 9.1 Interpolative frequency division by oversampling.

Then, an instantaneous phase error due to the divide-by-N Δt_N is given by

$$\Delta t_N = T_{\text{ref}} - N T_{\text{out}} = \frac{T_{\text{out}}}{4} \tag{9.2}$$

Similarly, the instantaneous phase error due to the divide-by-$(N+1)$ is given by

$$\Delta t_{N+1} = T_{\text{ref}} - (N+1) T_{\text{out}} = \frac{3}{4} T_{\text{out}} \tag{9.3}$$

Therefore, the instantaneous phase error for each reference period gives the sequence of $\{\ldots, +\frac{T_{\text{out}}}{4}, +\frac{T_{\text{out}}}{4}, +\frac{T_{\text{out}}}{4}, -\frac{3T_{\text{out}}}{4}, \ldots\}$ for the fractional division ratio of $(N + 1/4)$. Similarly, the phase error sequences are $\{\ldots, +\frac{T_{\text{out}}}{2}, -\frac{T_{\text{out}}}{2}, \ldots\}$ and $\{\ldots, +\frac{3T_{\text{out}}}{4}, -\frac{T_{\text{out}}}{4}, -\frac{T_{\text{out}}}{4}, -\frac{T_{\text{out}}}{4}, \ldots\}$ for the division ratio of $(N + 1/2)$ and $(N + 3/4)$, respectively.

Figure 9.2 shows the block diagram of a traditional fractional-N frequency synthesizer. The frequency resolution Δf with a k-bit accumulator is simply given by

$$f_{\text{out}} = \left(N + \frac{p}{2^k}\right) f_{\text{ref}} = N f_{\text{ref}} + \frac{p}{2^k} f_{\text{ref}} \tag{9.4}$$

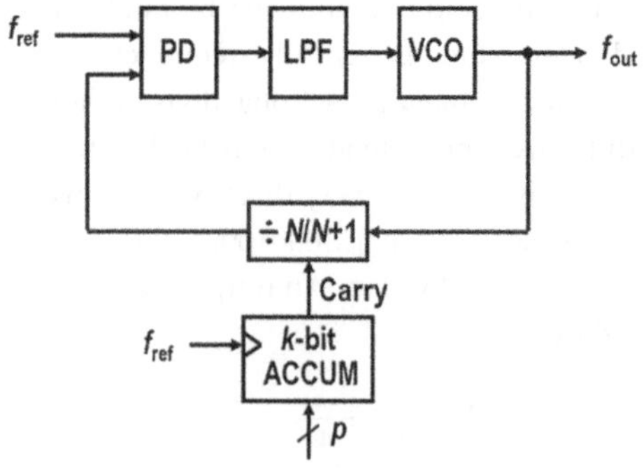

Figure 9.2 Traditional fractional-N PLL.

where p is the number of the carry "1" over 2^k clock periods. Then, the frequency resolution Δf with a k-bit accumulator is given by

$$\Delta f = \frac{p}{2^k} f_{\text{ref}} \qquad (9.5)$$

By increasing the number of bits of an accumulator, finer frequency resolution is achieved. Since the PD frequency is higher than the frequency resolution of the fractional-N PLL, the loop bandwidth is not limited by the frequency resolution.

The fractional-N PLL generates an averaged output frequency by periodically modulating a dual-modulus divider. Since the modulation period corresponds to the fraction of the reference frequency, a spur will be generated based on narrowband frequency modulation (FM). As a result, a close-in spur, namely a fractional spur, between the carrier and the reference spur is generated. Unlike the reference spur that depends on the circuit mismatch or leakage current, the fractional spur due to a digital phase control is considered a systematic modulation, and its level can be estimated with the knowledge of PLL parameters. The following example shows how to calculate the level of the fractional spurs for the given loop parameters of a PLL.

Example 9.1 *Calculating the level of fractional spurs*
The level of a spur generated by the periodic modulation of a dual-modulus divider can be estimated based on the narrowband FM assumption as done for the reference spur calculation in Chapter 4. Figure 9.3 shows the block diagram of a fractional-N PLL using an 8/9 dual-modulus divider and a 2-bit accumulator to

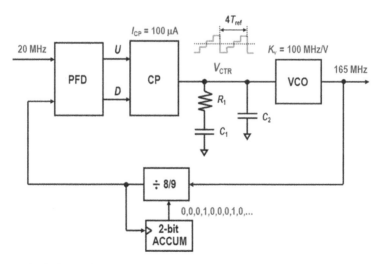

Figure 9.3 4-Modulo fractional-N PLL with a 2-bit accumulator.

generate an output frequency of 165 MHz from a reference frequency of 20 MHz. Therefore, an effective division ratio of 8.25 is required, and a fractional spur will appear at 160 and 170 MHz, which are 5-MHz offset frequencies from the carrier frequency. The output of the 8/9 dual-modulus divider is used as the clock of the 2-bit accumulator instead of the reference clock, which has the same effect when the PLL is locked. A charge pump current of 100 μA and a voltage-controlled oscillator (VCO) gain of 100 MHz/V are set, and an overdamped loop with a loop bandwidth of 1 MHz is designed by having $R_1 = 5.2\,\mathrm{k\Omega}$, $C_1 = 150\,\mathrm{pF}$, and $C_2 = 15\,\mathrm{pF}$.

Let us estimate the level of the fractional spur at 5-MHz offset frequency. From (9.2) and (9.3), the maximum phase deviation is given by

$$\Delta\theta_{\mathrm{pk}} = \frac{1}{2} \times \frac{3}{4} T_{\mathrm{out}} \times \frac{2\pi}{T_{\mathrm{ref}}} = \frac{3\pi}{4}\frac{T_{\mathrm{out}}}{T_{\mathrm{ref}}} = \frac{3\pi}{4}\frac{1}{N_{\mathrm{eff}}}$$

where N_{eff} is an effective fractional division ratio. The peak frequency deviation becomes

$$\Delta f_{\mathrm{pk}} = a_1 \times \Delta\theta_{\mathrm{pk}} \times \frac{I_{\mathrm{CP}}R_1}{2\pi}K_v = a_1 \times \frac{3}{8}\frac{I_{\mathrm{CP}}R_1 K_v}{N_{\mathrm{eff}}}$$

where a constant a_1 represents the coefficient of a fundamental tone. For a sawtooth waveform generated by the 4-modulo operation of the dual-modulus divider, a value of $2/\pi$ is set for a_1. The modulation frequency would be $f_{\mathrm{ref}}/4$. For the sake of simplicity, let us neglect a shunt capacitor C_2 to consider a second-order PLL first. Then, the spur level is given by

$$P_{\mathrm{spur}} = 20\log\left(\frac{\Delta f_{\mathrm{pk}}}{2f_m}\right) = 20\log\left(a_1\frac{3}{4}\frac{I_{\mathrm{CP}}R_1 K_v}{f_{\mathrm{ref}}N_{\mathrm{eff}}}\right) = -16.4\,[\mathrm{dBc}]$$

For an overdamped loop, the above equation can also be expressed as

$$P_{\mathrm{spur}} \approx 20\log\left(3\frac{f_{\mathrm{BW}}}{f_{\mathrm{ref}}}\right)$$

where the loop bandwidth f_{BW} is approximated as

$$f_{\mathrm{BW}} \approx \frac{I_{\mathrm{CP}}R_1 K_v}{2\pi N_{\mathrm{eff}}}$$

Note that the fractional spur level does not depend on the division ratio for the given ratio of f_{BW} to f_{ref}, which is different from the reference spur case as shown in (4.24). It is because the fractional division is done by the dual-modulus divider that swallows a pulse whose period is the same as that of a VCO.

Now, let us consider a third-order PLL by including C_2 shown in Fig. 9.3. The spur level becomes

$$P_{\mathrm{spur}} = 20\log\left(a_1\frac{3}{4}\frac{I_{\mathrm{CP}}R_1 K_v}{f_{\mathrm{ref}}N_{\mathrm{eff}}}\right) - 20\log\left(2\pi\frac{f_{\mathrm{ref}}}{4}R_1 C_2\right)$$

$$= 20\log\left(a_1\frac{3}{2}\frac{I_{\mathrm{CP}}K_v}{\pi f_{\mathrm{ref}}^2 N_{\mathrm{eff}} C_2}\right)$$

Then, the spur level at 5-MHz offset frequency is given by

$$P_{\text{spur}} = 20 \log \left(a_1 \frac{3}{2} \frac{I_{\text{CP}} K_v}{\pi f_{\text{ref}}^2 C_2} \right) = -24.2 \, [\text{dBc}]$$

In the case of the third-order PLL, the waveform with C_2 becomes triangular, but the same value of $2/\pi$ can be used for Fourier coefficient a_1. Note that the fractional spur suppression by the high-order pole becomes less effective as the control bits of the accumulator increases, that is with finer frequency resolution.

The example shows that the fractional-N synthesis is not useful in practice unless the fractional spurs are removed or suppressed. We can also deduce that the spur level becomes higher when an accumulator with a larger number of bits is used for finer frequency resolution. Therefore, additional circuitry must be added to suppress the fractional spurs. Let us discuss several spur-reduction techniques.

9.1.2 Spur Reduction Methods

9.1.2.1 Phase Compensation by a DAC

In theory, the fractional spur can be suppressed completely since the amount of an instantaneous phase error for each reference period is predictable for the given control word of an accumulator in a fractional-N PLL. A straightforward way for spur reduction is to employ a digital-to-analog converter (DAC) and compensate for the instantaneous phase error in the voltage domain as illustrated in Fig. 9.4. This kind of analog method suffers from mismatching and DAC non-linearity. Note that this method does not remove the fractional spur completely

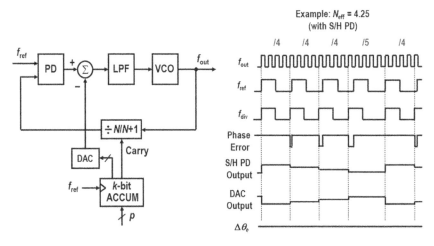

Figure 9.4 Phase compensation by the DAC and a 4-modulo example.

even with the perfect matching and an ideal DAC when a phase-frequency detector (PFD) is used. It is because the PFD generates a phase error based on pulse-width modulation. The amount of voltage or charge to compensate for the pulse-width-modulated pulse can balance the amount of charge for each reference period but does not fully remove the voltage ripple. The DAC-based compensation method is more effective when a sample-and-hold (S/H) PD is used in the fractional-N PLL. A timing diagram in Fig. 9.4 shows that the periodic tone can be fully suppressed after the DAC compensation when the S/H PD is used in the fractional-N PLL.

9.1.2.2 Phase Compensation by a DTC

The phase compensation by the DAC in the voltage domain is not so effective to suppress fractional spurs in the conventional CP-PLL having a PFD. Instead of the DAC, a digital-to-time converter (DTC) could be employed for phase compensation in the time domain. Figure 9.5 shows the 4-modulo fractional-N PLL that employs the DTC-based phase compensation method. Unlike the DAC-based method that compensates for a phase error at the output of a charge pump, the phase compensation is done right after a frequency divider, that is, before a PFD. Even though the amount of a compensated phase for the given control word of an accumulator could be predicted from (9.2) and (9.3), the value depends on the output frequency. To properly compensate for the frequency-dependent phase, a delay-locked loop (DLL) is employed to adaptively control the delay cell in the DTC. The DLL acts as a replica bias cell by providing a process-voltage-temperature (PVT)-insensitive control voltage V_{ctr} to the DTC as depicted in Fig. 9.5.

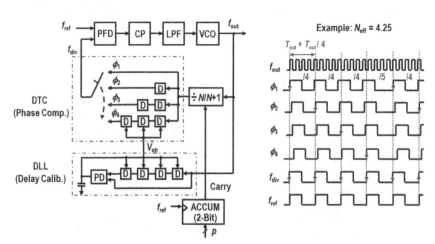

Figure 9.5 Phase compensation by the DTC.

For example, if a fractional modulo of 4 is chosen, the DLL consists of a voltage-controlled delay line (VCDL) having four delay cells can be designed. In that case, the total delay of the VCDL becomes T_{out} when the DLL is locked, making the delay of each delay cell the same as $T_{out}/4$. By performing sequential phase selection among multiple phases from the VCDL, the instantaneous phase error due to a modulated dual-modulus divider is cancelled after the DTC. The bandwidth of the DLL should be much wider than that of the PLL so that the loop dynamics of the PLL should not be affected by the DLL. If a multi-phase ring VCO is designed in the PLL, the input frequency of the DLL or the number of the delay cells can be reduced by judiciously using multiple phases.

9.1.2.3 Multi-Phase Fractional-N Division

If a multi-phase VCO such as a ring VCO is available, the multiple phases could be utilized to realize a *true* fractional-N divider without phase compensation or modulation. Figure 9.6 shows the basic operation principle of a multi-phase fractional-N divider with the timing diagram of a 4-modulo case. A frequency divider with a fixed division ratio of N is followed by the phase selector that effectively stretches a clock period with a fixed phase amount of $T_{out}/4$ at the divider output. By selecting the phase edge of the multiple phases in sequence from ϕ_1 to ϕ_4, the selected phase is locked to a reference clock without generating an instantaneous phase error as illustrated in Fig. 9.6. The basic operation principle is the same as the DTC-based method, but the amount of phase interpolation is automatically tuned to an output frequency since the multiple phases come directly from the ring VCO. To further improve frequency resolution, a phase interpolator is designed to generate more number of phases than the number of phases from the ring VCO. Like other multi-phase-based circuits, the

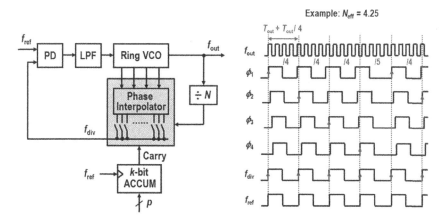

Figure 9.6 Multi-phase fractional-N division and a 4-modulo example.

mismatch and nonlinearity among the multiple phases limit the performance of the fractional-N PLL. Performing a glitch-less phase selection at high frequency is quite a challenging task for this architecture. Another drawback of this method is the requirement of a multi-phase ring VCO. For most wireless applications, frequency synthesizers employ an LC VCO to satisfy the phase noise requirement, and the design of a multi-phase LC VCO substantially increases hardware cost.

9.1.2.4 Pseudo-Random Modulation Method

As the output frequency of a fractional-N PLL increases, the DAC- or DTC-based compensation method makes it difficult for the fractional-N PLL to achieve fine resolution since the required compensation phase given by (9.2) and (9.3) becomes very small. Knowing that the fractional spur originates from a periodic pattern in the dual-modulus control, we consider dithering the modulus control in the digital domain. Figure 9.7 shows a fractional-N frequency synthesizer based on a pseudo-random modulation (PRM) method. At every output of the divider, a random number generator produces a random word P_n that is compared with the frequency word K. If P_n is less than K, the division by N is performed. If P_n is greater than K, the division by $(N+1)$ is performed. Hence, the frequency word K determines the average occurrence ratio of N and $(N+1)$, which gives an effective control on the fractional division value. Compared with the previous methods that rely on the phase compensation or the multi-phase selection in the analog domain, the PRM method can provide a very fine frequency resolution regardless of the output frequency or the division ratio. It is because the frequency resolution is not limited by the mismatch or nonlinearity of analog circuits but determined by clock frequency and the number of bits of the random number generator. Therefore, the PRM-based spur reduction is considered an all-digital modulation method.

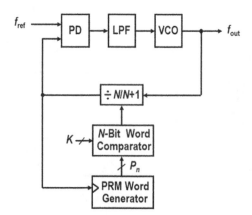

Figure 9.7 Fractional-N division with pseudo-random modulation.

In the PRM-based fractional-N division, a white-noise jitter will be produced due to the random modulation of a dual-modulus divider. A loop filter in the PLL suppresses the injected jitter from the divider since the PLL plays as a low-pass filter to the phase noise of the divider. This method, however, suffers from $1/f^2$ phase noise near the carrier, which cannot be suppressed by the loop filter of the PLL. It is because the dual-modulus divider modulated by random bits generates white noise in the frequency domain, resulting in $1/f^2$ noise in the phase domain due to an integration factor in the frequency-to-phase conversion.

9.1.2.5 Delta-Sigma Modulation Method

When the random number generator in the PRM-based fractional-N frequency synthesis is replaced with a delta-sigma ($\Delta\Sigma$) modulator, not only the random modulation property but also the noise shaping property of the $\Delta\Sigma$ modulator can be obtained for the fractional-N frequency synthesis. Figure 9.8 shows the block diagram of a fractional-N PLL with $\Delta\Sigma$ modulation, or namely $\Delta\Sigma$ fractional-N PLL. A dual-modulus divider controlled by a digital $\Delta\Sigma$ modulator interpolates a fractional frequency by oversampling. The operation principle of the $\Delta\Sigma$ fractional-N PLL is similar to that of a 1-bit $\Delta\Sigma$ analog-to-digital converter (ADC), while the dual-modulus divider is analogous to a 1-bit quantizer as depicted in Fig. 9.8. Like the PRM method, the operation of the $\Delta\Sigma$ modulator method is based on fully digital modulation. Hence, finer frequency resolution can be obtained simply by increasing the number of bits of the digital modulator. The second- or higher-order $\Delta\Sigma$ modulators, in theory, do not generate fixed tones for DC inputs, thus enabling a spur-free fractional-N frequency synthesis.

Thanks to the noise shaping property of the $\Delta\Sigma$ modulator that pushes quantization noise to high frequencies like a high-pass filter, the $\Delta\Sigma$ fractional-N PLL does not exhibit $1/f^2$ noise near the carrier. Figure 9.9 illustrates the difference between the PRM-based and the $\Delta\Sigma$ modulator-based fractional-N PLLs in the frequency domain. As discussed, the PRM-based fractional divider generates white noise in

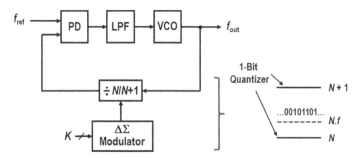

Figure 9.8 Fractional-N division with $\Delta\Sigma$ modulation.

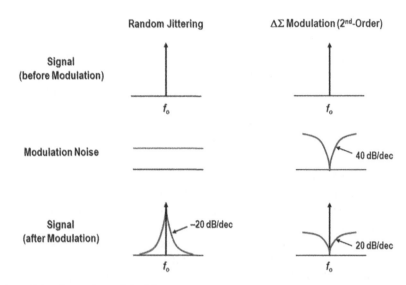

Figure 9.9 Comparison of the PRM and the $\Delta\Sigma$ modulation.

the frequency domain, which results in the noise slope of -20 dB/dec in the phase domain. In the case of a second-order $\Delta\Sigma$ modulator, the noise slope of $+40$ dB/dec is shown in the frequency domain, exhibiting the noise slope of $+20$ dB/dec near the carrier in the phase domain. Therefore, the phase noise near the carrier is typically much lower than the phase noise from other circuits such as a VCO or a charge pump, while the phase noise in high frequencies due to the noise-shaping effect of the $\Delta\Sigma$ modulator is suppressed by the loop filter of the PLL. We will discuss the basic operation and analysis of the $\Delta\Sigma$ fractional-N PLL in detail in Section 9.2.

9.1.3 Multi-Loop Hybrid Frequency Synthesis

Before discussing the $\Delta\Sigma$ fractional-N PLL in detail, let us take a look at the block diagram of a traditional signal source generator shown in Fig. 9.10, Hewlett Packard 8662A,[1] which was one of the lowest-noise signal source generators in the 1980s and early 1990s. To achieve a frequency resolution of 0.1 Hz and low noise performance, a multi-loop PLL with an auxiliary fractional-N loop is employed to increase the PD frequency. The fractional-N operation was done with the traditional DAC-based phase-compensation method. To achieve good spur performance, a S/H PD is employed. As discussed, the S/H PD is highly effective for the fractional-N PLL and the integer-N PLL to have low fractional spur and reference spur, respectively. A frequency-translation PLL that has a

---

1 Now, Keysight/Agilent 8662A.

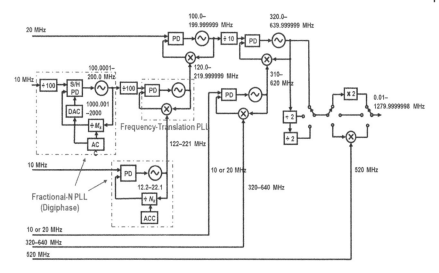

Figure 9.10 Simplified block diagram of an early-stage frequency synthesizer.

mixer instead of the frequency divider in the feedback path is also used to shift a frequency to other frequency for flexible frequency synthesis. Like other expensive testing equipment, it fully utilizes discrete components so that nonlinearity and mismatching effects could be minimized for high spectral purity. Numerous modern signal source generators adopt the $\Delta\Sigma$ fractional-N PLL not only for high performance but also for low cost. Now, let us move to the main part of this chapter.

9.2 Frequency Synthesis with Delta-Sigma Modulation

The $\Delta\Sigma$ modulation has become an important technique not only in the area of data conversion but also in the area of clock and frequency generation these days. The basic operation is to use an oversampling $\Delta\Sigma$ modulator to interpolate fractional frequency with a coarse integer divider as depicted in Fig. 9.8. Thanks to the noise shaping property of the $\Delta\Sigma$ modulator, the $\Delta\Sigma$ fractional-N PLL has good phase noise near the carrier. While the traditional finite-modulo fractional-N PLLs encounter more difficulties in spur reduction with higher VCO frequency, the $\Delta\Sigma$ fractional-N PLL does not have the same problem as the frequency resolution does not depend on the VCO frequency. By simply increasing the number of modulation bits, a very fine frequency resolution can be obtained. For example, if we use a 20-bit $\Delta\Sigma$ modulator with a 10-MHz reference clock, a frequency resolution less than 1 Hz can be achieved at the PLL output. With such a fine frequency resolution, frequency modulation can also be achieved at the output of the $\Delta\Sigma$

fractional-N PLL if the control input of the $\Delta\Sigma$ modulator is dynamically modulated. The key features of the $\Delta\Sigma$ fractional-N PLL are summarized as follows:

- Spur-free randomization with high-order modulation
- Low in-band phase noise contribution with noise shaping
- Fine frequency resolution with digital oversampling
- Optional direct-digital frequency modulation

After the development of the $\Delta\Sigma$ fractional-N PLL, the overall performance of the frequency synthesizer was significantly improved. Moreover, the $\Delta\Sigma$ modulation method enables the fractional-N PLL to perform direct-digital frequency modulation for low-cost transmitter design, making the fractional-N PLL a key building block in modern transceiver systems.

To gain a good understanding of the $\Delta\Sigma$ fractional-N PLL, it is good to know the basic operation principle of a $\Delta\Sigma$ modulator. Then, the different design aspects of the $\Delta\Sigma$ modulator for the design of the $\Delta\Sigma$ fractional-N PLL will be discussed. We will also learn how to reduce the out-of-band phase noise caused by the quantization noise of a $\Delta\Sigma$ modulator, which is important for the design of wideband PLLs.

9.2.1 $\Delta\Sigma$ Modulation

As the power and area of high-speed digital circuits become less significant with advanced CMOS technology, oversampling data converters with $\Delta\Sigma$ modulation are widely adopted for high-resolution data conversion. Like a channel coding technique in digital communications, redundant output bits make the system robust against possible bit errors caused by analog mismatches. The noise-shaping property improves the signal-to-noise ratio (SNR) by filtering high-frequency quantization noise with a decimation filter. The use of the $\Delta\Sigma$ modulation for fractional-N frequency synthesis also alleviates the analog design constraints of the PLL and offers several advantages over the standard approach. Let us discuss the basic operation principle and the effect of quantization noise based on the traditional theory of the $\Delta\Sigma$ ADC.[2]

9.2.1.1 $\Delta\Sigma$ ADC and Quantization Noise

Figure 9.11 shows the simplified block diagram of a $\Delta\Sigma$ ADC having a 1-bit quantizer and the transient waveforms of each building block. The output of the quantizer is compared with an input signal, and the difference information is accumulated by an integrator. The naming of either a $\Delta\Sigma$ ADC or a $\Sigma\Delta$ ADC is used, depending on the way of description, that is "difference (Δ) and then integration (Σ) function based on the block diagram" or "integration (Σ) of the difference (Δ) based on the operation." In this book, the term $\Delta\Sigma$ ADC will be used to follow the

---

2 The analytic part of this section is mostly based on Norsworthy's book.

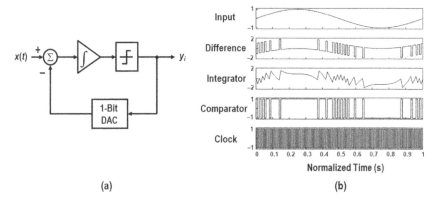

(a)

(b)

Figure 9.11 ΔΣ ADC with a 1-bit quantizer: (a) block diagram and (b) transient waveforms.

naming convention by those who analyzed the ΔΣ ADC in the early stage. The key properties of the ΔΣ ADC are oversampling and noise shaping, which makes it possible to achieve high SNR or high resolution without having a high-performance operational amplifier. However, the use of the 1-bit quantizer adds nonlinearity in the loop, requiring careful design especially for high-order modulators.

By having an integrator in the feedforward path, quantization noise will be high-pass filtered by the noise transfer function of a feedback system, exhibiting the noise-shaping property. This property is similar to the high-pass filtering characteristic to the VCO noise within the PLL. To understand the noise shaping property, we analyze the effect of quantization noise based on a sampled-data model where an integrator is replaced with an accumulator as shown in Fig. 9.12. For an input signal x_i, the quantized output y_i is given by

$$y_i = a_i + e_i = (x_{i-1} - y_{i-1} + a_{i-1}) + e_i = (x_{i-1} - e_{i-1}) + e_i = x_{i-1} + (e_i - e_{i-1})$$

(9.6)

where a_i is the output of the accumulator and e_i is the quantization error. Equation (9.6) implies that the ΔΣ modulator differentiates the quantization

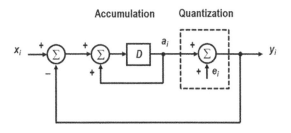

Figure 9.12 Equivalent sampled-data model.

error without affecting the input signal except a sampled delay. Assuming that the quantization error behaves as white noise with a busy input signal, we define modulation noise n_i by

$$n_i = e_i - e_{i-1} \tag{9.7}$$

We obtain the spectral density of the modulation noise $N(f)$ given by

$$N(f) = E(f)|1 - \exp^{-j\omega T}| = 2e_{rms}\sqrt{2T}\sin\left(\frac{\omega T}{2}\right) \tag{9.8}$$

where $E(f)$ is the spectral density of e_i, T is the sampling period or the inverse of the sampling frequency f_s, and e_{rms} is the root-mean-square value of the quantization error e_i.

Figure 9.13 shows the shaped noise spectrum of $N(f)$ with an oversampling ratio (OSR) of 16 in comparison with $E(f)$. The total noise power n_o^2 in the signal band is given by

$$n_o^2 = \int_0^{f_o} |N(f)|^2 df \approx e_{rms}^2 \frac{\pi^2}{3}(2f_o T)^3, \qquad f_s^2 \gg f_o^2 \tag{9.9}$$

Then, the rms value n_o is given by

$$n_o \approx e_{rms}\frac{\pi}{\sqrt{3}}(2f_o T)^{3/2} = e_{rms}\frac{\pi}{\sqrt{3}}(OSR)^{-3/2} \tag{9.10}$$

Equation (9.10) shows that the SNR improvement of 9 dB could be achieved simply by doubling the OSR.

9.2.1.2 High-Order Modulator

When a high-order $\Delta\Sigma$ modulator is considered, the SNR can be further enhanced. Figure 9.14 shows the sampled-data model of a second-order $\Delta\Sigma$ modulator. The quantized output y_i is given by

$$y_i = x_{i-1} + (e_i - 2e_{i-1} + e_{i-2}) \tag{9.11}$$

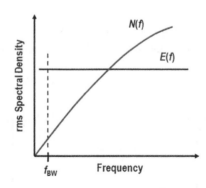

Figure 9.13 Spectral density of the quantization noise.

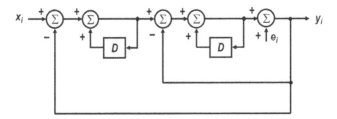

Figure 9.14 Second-order modulator.

where the modulation noise is expressed as the second difference of the quantization error. Then, the spectral density of the modulation noise is given by

$$N(f) = E(f)(1 - \exp^{-j\omega T})^2 \tag{9.12}$$

Assuming a busy input signal, the absolute value of $N(f)$ becomes

$$|N(f)| = 4e_{\text{rms}}\sqrt{2T}\sin^2\left(\frac{\omega T}{2}\right) \tag{9.13}$$

and the rms value n_o is given by

$$n_o \approx e_{\text{rms}}\frac{\pi^2}{\sqrt{5}}(2f_oT)^{\frac{5}{2}} = e_{\text{rms}}\frac{\pi^2}{\sqrt{5}}(\text{OSR})^{-\frac{5}{2}}, \quad f_s^2 \gg f_o^2 \tag{9.14}$$

The second-order modulator achieves the SNR improvement of 15 dB for every doubling of the sampling frequency, while the first-order modulator achieves 9 dB from (9.10). It implies that the SNR performance can be significantly improved by having a higher-order loop for the given OSR.

The modulation noise of a higher-order modulator can be obtained by adding more feedback loops in Fig. 9.14 and extending the difference equation from (9.11). For the L-th order modulator, the spectral density $N_L(f)$ of the modulation noise is given by

$$|N_L(f)| = e_{\text{rms}}\sqrt{2T}\left[2\sin\left(\frac{\omega T}{2}\right)\right]^L \tag{9.15}$$

and the rms value n_o is approximately given by

$$n_o = e_{\text{rms}}\frac{\pi^L}{\sqrt{2L+1}}(2f_oT)^{L+1/2} \tag{9.16}$$

From (9.16), the SNR is improved by $3(2L - 1)$ dB for every doubling of the sampling rate, providing $(L - 0.5)$ extra bits of resolution in the design of a $\Delta\Sigma$ ADC. Figure 9.15 shows the rms noise performance over different OSR values and loop orders. The high-order modulator, however, has some restrictions on input dynamic range and loop dynamics for stability, requiring careful design.

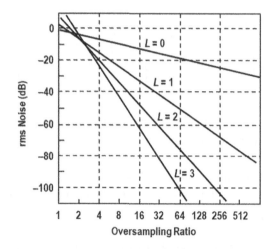

Figure 9.15 Modulation quantization noise with different OSRs.

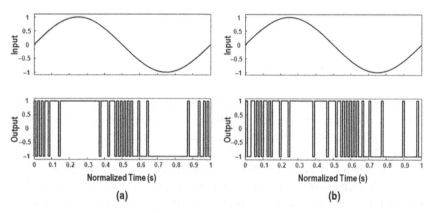

Figure 9.16 Transient waveforms: (a) the first-order modulator and (b) the second-order modulator.

Figure 9.16 shows the transient comparison of the first-order and the second-order $\Delta\Sigma$ modulators with a sinusoidal input. The output of the modulator behaves as a pulse-width modulated signal depending on the input signal. Compared with the output of the first-order modulator, the output of the second-order modulator shows more high-frequency fluctuation, while keeping the same level on average.

9.2.1.3 Cascaded Modulator
A high-order $\Delta\Sigma$ modulator can also be implemented by cascading first-order modulators as shown in Fig. 9.17. The cascaded modulator is also called a

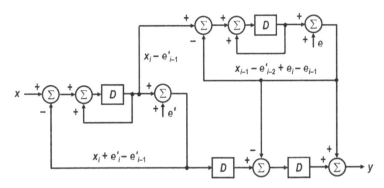

Figure 9.17 Cascaded modulator.

"Multi-stAge noise-SHaping" (MASH) modulator. A second-stage modulator takes the integrator output of a first-stage modulator as an input, and the output of the second-stage modulator is subtracted from the output of the first-stage modulator so that the quantization noise of the first-stage modulator is canceled. After adding the output of the second-stage modulator in the final stage, the remaining noise is the second difference of the quantization error from the second-stage modulator. By considering an imperfect cancellation with an error factor g, we have

$$y_i = x_{i-2} + (1-g)\left(e'_{i-2} - e'_{i-3}\right) + (e_i - 2e_{i-1} + e_{i-2}) \tag{9.17}$$

where e' denotes the quantization error in the first-stage modulator. With ideal matching and linearity, g becomes unity, making (9.17) the same form as the noise of the second-order $\Delta\Sigma$ modulator in (9.11).

Even though a high-order MASH modulator does not suffer from stability, it is difficult to make the error factor g close to unity in practice. As a result, noise-shaping performance is limited by the noise smeared from the first-stage modulator, and the use of a high-order MASH modulator has little advantage. For that reason, the MASH modulator is often combined with a high-order single-loop $\Delta\Sigma$ modulator (SLDSM). For example, a fourth-order $\Delta\Sigma$ modulator can be designed with a two-stage topology by having a second-order SLDSM followed by a second-order MASH modulator. In the case of the $\Delta\Sigma$ fractional-N PLL, an all-digital $\Delta\Sigma$ modulator is employed. Therefore, the digital MASH modulator can have g of unity. We will discuss different design aspects.

9.2.2 All-Digital $\Delta\Sigma$ Modulators for Fractional-N Frequency Synthesis

For fractional-N frequency synthesis, an all-digital $\Delta\Sigma$ modulator is designed since both the input and output of the modulator are digital. An integrator in the $\Delta\Sigma$

ADC is replaced with an accumulator, and 1-bit quantization is done simply by taking the carry output of the accumulator. An effective oversampling ratio OSR_{eff} can be defined by the ratio of the PD frequency f_{PD} to the PLL noise bandwidth B_n

$$\text{OSR}_{\text{eff}} = \frac{f_{\text{PD}}}{2B_n} \tag{9.18}$$

Therefore, either a high f_{PD} or a low B_n is good to reduce the quantization noise by having a high OSR_{eff}. Note that we use B_n instead of the 3-dB frequency of the system transfer function to analyze the dynamic range performance based on the analytic results of a $\Delta\Sigma$ ADC, which will be discussed later. In general, the order of a PLL should be higher than that of a $\Delta\Sigma$ modulator since the quantization noise is filtered by the low-pass filter (LPF) characteristic of the PLL.

The $\Delta\Sigma$ modulator design for the frequency synthesizer has some different aspects. Since it is an all-digital modulator, the coefficients of the modulator can be accurately set. Hence, the MASH modulator does not have the noise-smearing problem from the first stage due to imperfect gain mismatch. Accordingly, the digital MASH modulator becomes a dominant architecture for the $\Delta\Sigma$ fractional-N PLL for its simple implementation and guaranteed stability. However, the modulator design for the fractional-N PLL still faces matching and nonlinearity problems as the digital information is transformed into a phase error in the analog domain when combined with the PLL. We will learn the pros and cons of modulator architectures and discuss different design aspects from the $\Delta\Sigma$ ADC in the following sections.

9.2.2.1 MASH Modulator

The MASH modulator consisting of cascaded first-order modulators is inherently stable, having the input range of the MASH modulator fully utilized. To the contrary, the SLDSM faces the stability problem by saturating the output of an accumulator when the input value is close to the upper or low level of the input range of the modulator. Without having a feedback or feedforward path, the design of a high-order MASH modulator is relatively easy as there is no need of careful selection of loop parameters. Figure 9.18 shows a third-order all-digital MASH modulator where the quantization noise is expressed in the z domain as follows

$$Q(z) = (1 - z^{-1})^3 \tag{9.19}$$

A high-order MASH modulator has a high-order noise shaping by canceling the residual noise of a previous stage. For an n-th order modulator, the noise transfer function (NTF) is given by

$$H_n(z) = (1 - z^{-1})^n \tag{9.20}$$

The drawback of this architecture for the fractional-N PLL is that a high-order modulator generates a widespread output bit pattern since the nth order MASH

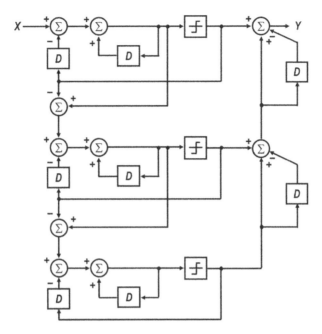

Figure 9.18 Third-order MASH modulator.

modulator has an n-bit output. For example, third- and fourth-order MASH modulators require 8-modulus and 16-modulus dividers, respectively. Such a widespread bit pattern adds design complexity not only for a multi-modulus divider but also for a PD, which will be discussed later in this chapter.

9.2.2.2 SLDSM with 1-Bit Quantizer

Figure 9.19 shows the block diagram of a third-order SLDSM with a single-bit quantizer. Since the single-bit quantizer generates only two-level outputs, a

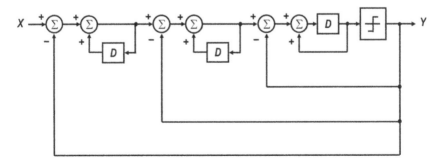

Figure 9.19 Third-order SLDSM with a 1-bit quantizer.

dual-modulus divider can be used in the design of a fractional-N PLL. The NTF of the third-order SLDSM in Fig. 9.19 is given by

$$H_n(z) = (1 - z^{-1})^3 \tag{9.21}$$

In general, the NTF of an nth order SLDSM is the same as that of the MASH modulator in (9.20). However, a different NTF can also be designed in the case of the SLDSM by adding feedforward and feedback coefficients. That is, a customized NTF can be designed based on the known types of high-pass filters to further suppress the quantization noise in high-frequency region. One example would be a third-order Butterworth high-pass filter that exhibits a flat passband gain and achieves more reduction of quantization noise in high frequencies than the NTF from (9.21).

The main drawback of the SLDSM for the fractional-N PLL is that the input range of the modulator is limited due to stability. The problem of the limited input range gets more serious with a higher-order SLDSM. As a result, the full-scale range of a dual-modulus divider cannot be used, causing a dead-band problem for fractional-N frequency synthesis. A possible solution is to use a high reference frequency or to expand the quantizer level by using an $N/(N+2)$ dual-modulus divider rather than an $N/(N+1)$ dual-modulus divider. By overlapping the integer boundary with the quantizer level set by an $(N+1)/(N+3)$ dual-modulus divider, all range of the channels can be covered at the cost of increased quantization noise.

9.2.2.3 SLDSM with Multi-Level Quantizer

If the SLDSM is designed with a multi-level quantizer, the dead-band problem of a $\Delta\Sigma$ fractional-N PLL due to the limited input range can be solved. Figure 9.20

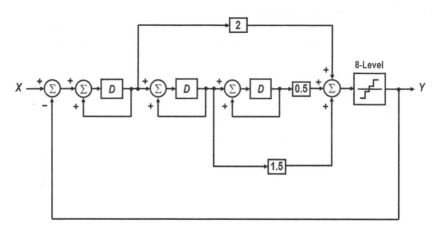

Figure 9.20 Third-order SLDSM with a 3-bit quantizer.

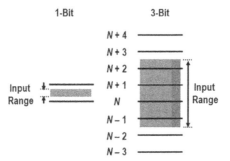

Figure 9.21 Input range comparison of the single-bit and the multi-bit quantizer.

shows a third-order SLDSM that employs an eight-level quantizer. The eight-level quantizer expands the active division range from $\{N, N+1\}$ to $\{N-3, N-2, \ldots, N+3, N+4\}$. Note that the total quantization noise power is almost the same since the minimum quantization error is still set by a single period of a VCO. As shown in Fig. 9.21, an actual range occupied by the $N/(N+1)$ dual-modulus divider for frequency synthesis is about 12% of the full range of the eight-level quantizer, which is low enough to ensure good stability of the third-order SLSDM.

The SLDSM has feedforward coefficients to realize a customized NTF such as the Butterworth high-pass filter that features a flat pass-band gain. Based on the Butterworth filter design, the NTF can be designed as

$$H_n(z) = \frac{(1 - z^{-1})^3}{1 - z^{-1} + 0.5z^{-2} - 0.1z^{-3}} \tag{9.22}$$

For stability, the poles of the NTF should be within the unit circle in the z-domain as shown in Fig. 9.22(a). Compared with the MASH modulator, low-Q Butterworth poles significantly reduce the high-frequency-shaped noise energy as

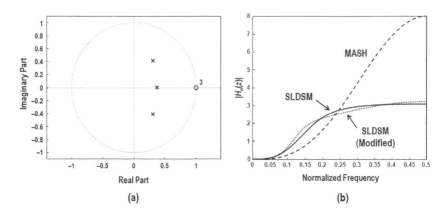

Figure 9.22 SLDSM characteristics: (a) pole-zero plot and (b) NTF comparison.

shown in Fig. 9.22(b). The reduced high-frequency noise in the frequency domain results in a low-spread output bit pattern in the time domain. In the design of a multi-level output SLDSM, using more number of quantizer levels allows a higher passband gain for the NTF. As a result, the corner frequency of the NTF can be further increased for the same noise power. In that way, the quantization noise in low frequencies can be suppressed more with the multi-level quantizer. For instance, the NTF shown in Fig. 9.22(b) has a passband gain of 3.1 and a corner frequency of 0.18 times the clock frequency.

In practice, the digital coefficients of {2, 0.5, 1.5} are implemented by using bit-shifting, addition, and subtraction to avoid digital multiplication. For example, a digital value of 1.5 is achieved by $(2 - 0.5)$ where the digital values of 2 and 0.5 are realized by the left bit-shifting and right bit-shifting of an input control word. As a result, the actual NTF of the digital SLDSM shown in Fig. 9.20 is given by

$$H_n(z) \approx \frac{(1 - z^{-1})^3}{1 - z^{-1} + 0.5z^{-2}} \tag{9.23}$$

Figure 9.22(b) shows the NTF from (9.23) in comparison with the original NTF from (9.22). The NTF of a third-order MASH modulator is also compared. The simplified implementation slightly modifies the original NTF, but the NTF variation is tolerable for $\Delta\Sigma$ fractional-N PLLs.

Based on what we have discussed so far, we could see that the NTF must be carefully designed for the stability and noise performance of a high-order SLDSM. To have a valid NTF, three conditions should be met for the NTF of the SLDSM. The first one is the causality condition that prevents a delay-free loop. Otherwise, the SLDSM circuit cannot be implemented since the delay-free loop does not exist in hardware. The causality requirement can be met by setting the leading coefficients of the numerator and the denominator polynomials of $H_n(z)$ to 1, or

$$H_n(\infty) = 1 \tag{9.24}$$

That is, the ratio of the constant in the numerator to the constant in the denominator must be exactly 1, which also holds for the NTF in (9.22) and (9.23). The second condition is the small-signal stability. The poles of the NTF need to be within the unit circle in the z-domain as depicted in Fig. 9.22(a). Even though the digital modulator accurately controls the coefficients, the location of the poles must be checked carefully after simplifying hardware implementation with the bit-shifting as done in (9.23). The third condition is the large-signal or nonlinear stability, which is related with an input dynamic range. An empirical study of the nonlinear stability shows that the passband gain should be limited. Therefore, the NTF design based on the Butterworth high-pass filter is a good choice to have a flat frequency response over the passband. It also reduces the high-frequency noise energy resulting in a low-spread output bit pattern. For a given filter, the passband

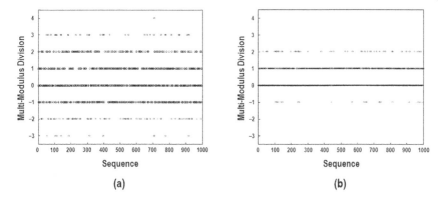

Figure 9.23 Comparison of the output-bit patterns: (a) MASH and (b) SLDSM.

gain is determined once the corner frequency is set. Hence, there is no degree of freedom for choosing the coefficients to satisfy both the causality and the passband gain requirements.

Figure 9.23 shows an example of the output patterns of digital $\Delta\Sigma$ modulators. The transient output bits of a third-order MASH modulator and a third-order SLDSM are shown in Fig. 9.23(a) and (b), respectively. It is clearly shown that the MASH modulator exhibits a more spread pattern than the SLDSM. The third-order MASH has a 3-bit output, having the output bits spread over up to eight levels. To the contrary, even though the third-order SLDSM also has the eight-level quantizer, the output-bit range is limited mostly to 4 levels over 1,000 samples. The simulation result verifies that reduced high-frequency noise power in the frequency domain results in less spread pattern in the time domain.

9.2.3 Phase Noise by Quantization Error

Suppose that the quantization error is uniformly distributed. Then, the noise power is given by the 1/12 of the minimum quantizer level. Having the noise power spread over a bandwidth of the PD frequency f_{PD}, frequency fluctuation in the z-domain $S_v(z)$ with the NTF $H_n(z)$ is given by

$$S_v(z) = |H_n(z)f_{PD}|^2 \left(\frac{1}{12 f_{PD}} \right) = |H_n(z)|^2 \frac{f_{PD}}{12} \tag{9.25}$$

Phase fluctuation $S_\Phi(z)$ is the integration of the frequency fluctuation, that is,

$$S_\Phi(z) = \left(\frac{2\pi}{|1 - z^{-1}|f_{PD}} \right)^2 \frac{|H_n(z)|^2 f_{PD}}{12} \tag{9.26}$$

If $S_\Phi(z)$ is two-sided power spectral density (PSD), then the single-sided PSD $L(z)$ is the same as $S_\Phi(z)$. Converting it to the frequency domain and generalizing

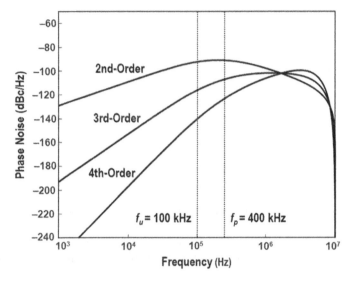

Figure 9.24 Phase noise due to quantization noise ($f_s = 10\,\text{MHz}$).

to any modulator order, we have the phase noise of the quantization error expressed as

$$L(f) = \frac{(2\pi)^2}{12 f_{\text{PD}}}\left[2\sin\left(\pi\frac{f}{f_{\text{PD}}}\right)\right]^{2(m-1)} \tag{9.27}$$

where m is the order of the modulator.

Figure 9.24 shows the phase noise plots of the second-, third-, and fourth-order MASH modulators in a fractional-N PLL based on (9.26). A clock frequency f_s of 10 MHz is used. The slopes of the phase noise are 20, 40, and 60 dB/dec for the second-, third-, and fourth-order modulators, respectively. If a PLL bandwidth of 100 kHz and a pole frequency of 400 kHz are assumed, we could see that phase noise suppression in high frequencies is not enough for the fourth-order modulator even though it gives much lower in-band noise than other modulators. It also implies that the order of the PLL should be higher than that of the ΔΣ modulator so that the quantization noise should gradually decrease in high frequencies. If out-of-band phase noise is important for some wireless applications, high-order poles need to be added at the cost of reduced phase margin unless a loop bandwidth is further reduced. It is good to note that the phase noise contribution at the PLL output does not depend on the division ratio. It is because the dual-modulus or multi-modulus divider swallows a pulse whose period is the same as that of a VCO. In other words, the minimum quantization level in the phase domain is set by the period of the VCO, making the phase noise contribution of the ΔΣ modulator independent of the division ratio. Another effective way of mitigating the quantization noise effect is to use a high clock frequency for the $\Delta\Sigma$ modulator. To do that, a frequency doubler circuit could be designed in the reference clock path to double the PD frequency.

Example 9.2   *Loop dynamics of $\Delta\Sigma$ fractional-N PLL*
In the design of a $\Delta\Sigma$ fractional-N PLL design, identifying the phase noise contribution of a $\Delta\Sigma$ modulator is important for optimum phase noise. Noise contributions from various sources for a type-2 fourth-order fractional-N PLL having a third-order MASH modulator are plotted along with an open-loop gain in Fig. 9.25. Since the quantization noise is generated by divider modulation, the PLL has a low-pass filter transfer function to the quantization noise. As discussed previously, the quantization noise contribution does not depend on the division ratio since the divider modulation has the resolution of one VCO clock period. If the PLL does not suffer from PD nonlinearity, the in-band phase noise due to the $\Delta\Sigma$ modulator is much lower than the noise of other sources. On the other hand, the out-of-band phase noise is possibly determined by the $\Delta\Sigma$ modulator if a loop bandwidth is not narrow enough. In Fig. 9.25, an example of how the quantization noise affects the out-of-band phase noise of the PLL by exhibiting a noise bump is shown. The phase noise of the third-order modulator exhibits a noise slope of 40 dB/dec and has a flat noise response after the third pole of the PLL before decreasing from the fourth pole of the PLL. This kind of noise plateau in high frequencies is the unique feature of a wideband $\Delta\Sigma$ fractional-N PLL. Like the spur reduction methods in the

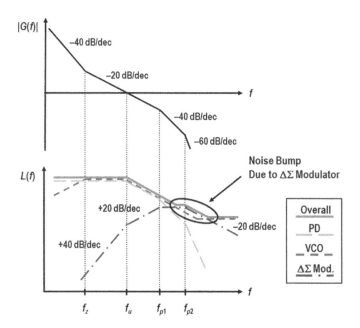

Figure 9.25   Phase noise contribution in the type-2 fourth-order fractional-N PLL.

traditional fractional-N PLL, a quantization noise reduction method should be considered for the design of a wideband $\Delta\Sigma$ fractional-N PLL.

Example 9.3 *Quantization noise effect*

Let us consider an interesting example of the quantization noise effect on phase noise with three similar $\Delta\Sigma$ fractional-N PLL topologies. We consider a fourth-order PLL with a third-order $\Delta\Sigma$ modulator and assume that loop parameters, the reference frequency, and the output frequency are kept the same. The only difference is the placement of $\Delta\Sigma$ modulation within the PLL. The first one has a divider-by-2 circuit followed by an 8/9 dual-modulus divider with $\Delta\Sigma$ modulation in the feedback path. The second one has the 8/9 dual-modulus divider followed by the divider-by-2 circuit. The third one has a 16/17 dual-modulus divider only. Let us suppose that the same fractional division ratio is set for each PLL. Then, what would be the quantization noise effect on the phase noise of each PLL?

Let us compare the first two cases shown in Fig. 9.26(a) and (b). As discussed previously, the quantization noise effect of the $\Delta\Sigma$ modulator will not be amplified

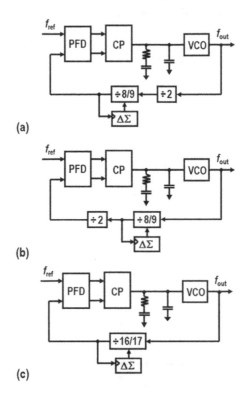

Figure 9.26 Comparison of quantization noise effect with three PLLs: (a) with 2 and 8/9 dividers; (b) with 8/9 and 2 dividers; and (c) with 16/17 divider.

by the dual-modulus divider itself since the minimum quantization error in the phase domain is still one VCO period. However, if the 8/9 dual-modulus divider follows the divide-by-2 circuit in the feedback path, the 8/9 dual-modulus divider performs $\Delta\Sigma$ modulation with two VCO periods, resulting in increased phase noise by 6 dB at the VCO output. Therefore, the PLL in Fig. 9.26(a) has worse noise performance than the PLL in Fig. 9.26(b). What about two PLLs in Fig. 9.26(b) and (c)? Even if the 16/17 dual-modulus divider is used, it is modulated with the minimum quantization error that equals to one VCO period. Hence, there is no 6-dB increment of the phase noise due to the quantization noise at the VCO output. On the other hand, the clock frequency of the $\Delta\Sigma$ modulator in Fig. 9.26(b) is twice the reference frequency since the $\Delta\Sigma$ modulation is performed before the divide-by-2 circuit. From (9.27), the quantization noise of the modulator within the PLL bandwidth can be reduced with a higher PD frequency. Accordingly, the quantization noise of the PLL in Fig. 9.26(c) should be worse than that in Fig. 9.26(b). However, the following divide-by-2 circuit after the $\Delta\Sigma$ modulator will induce an aliasing effect at the divider output, significantly increasing the in-band phase noise.[3] Therefore, the PLL in Fig. 9.26(c) has the best noise performance.

9.2.4 Dynamic Range and Bandwidth

The quantization error of a $\Delta\Sigma$ modulator appears as phase noise in fractional-N frequency synthesizers, but it can be analyzed as the voltage noise of $\Delta\Sigma$ data converters. By using well-established theories of the $\Delta\Sigma$ ADC, we can define the dynamic range of the $\Delta\Sigma$ modulation in the frequency domain and derive a closed-loop equation of the upper boundary of the PLL bandwidth for a given integrated phase error. Figure 9.27 shows a conceptual diagram of how to define the dynamic range of the $\Delta\Sigma$ modulation and the requirement of the bandwidth and in-band noise for the design of a fractional-N PLL. If in-band phase noise A_n in units of rad^2/Hz and noise bandwidth B_n are given, the integrated frequency noise Δf_n within B_n is calculated as

$$\Delta f_n = \sqrt{2 \int_{f_1}^{B_n} (A_n f^2) df} \approx \sqrt{\frac{2}{3} A_n B_n^{\frac{3}{2}}} \tag{9.28}$$

where f_1 is the minimum noise-integration frequency with $f_1 \ll B_n$ assumed. Quantizer levels are represented by $+f_{\text{PD}}/2$ and $-f_{\text{PD}}/2$ in the frequency domain as depicted in Fig. 9.27. Based on the dynamic range requirement of the Lth-order $\Delta\Sigma$ modulator in the ADC, a similar derivation is obtained for the Lth-order $\Delta\Sigma$ modulator in the fractional-N PLL where the dynamic range is set by the ratio of f_{PD} to Δf_n. Then, we get

$$\frac{2}{3} \left(\frac{2L+1}{\pi^{2L}} \right) (\text{OSR}_{\text{eff}})^{2L+1} > \left(\frac{f_{\text{PD}}}{\Delta f_n} \right)^2 \tag{9.29}$$

---

3 Detailed analysis can be found in [37].

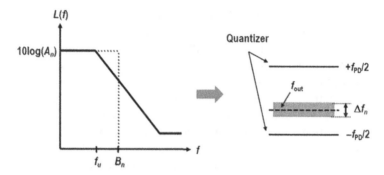

Figure 9.27 Dynamic range consideration in $\Delta\Sigma$ fractional-N division.

where OSR_{eff} is defined in (9.18). Using (9.18), (9.28), and (9.29), we have

$$B_n < \left[A_n \frac{L+0.5}{(2\pi)^{2L}}\right]^{\frac{1}{2L-2}} f_{PD}^{\frac{2L-1}{2L-2}} \tag{9.30}$$

An integrated phase error θ_{rms} is an important parameter for transceivers in digital communications, and it is given by

$$\theta_{\text{rms}} = \sqrt{2A_n B_n} \tag{9.31}$$

From (9.30) and (9.31), an approximate upper bound for the noise bandwidth is obtained

$$B_n < \left[\left(\frac{\theta_{\text{rms}}}{\sqrt{2}}\right)^2 \frac{L+0.5}{(2\pi)^{2L}}\right]^{\frac{1}{2L-1}} f_{PD} \tag{9.32}$$

For an overdamped loop, we approximate B_n as the unity-gain frequency f_u multiplied by $\pi/2$ from (2.31). Then, the upper boundary of the PLL bandwidth is given by

$$f_u < \frac{2}{\pi}\left[\left(\frac{\theta_{\text{rms}}}{\sqrt{2}}\right)^2 \frac{L+0.5}{(2\pi)^{2L}}\right]^{\frac{1}{2L-1}} f_{PD} \tag{9.33}$$

In practice, an actual loop bandwidth in the PLL design should be narrower than the bandwidth given by (9.33). For instance, if a third-order $\Delta\Sigma$ modulator is used, the quantization noise is tapered off after the fourth pole of the PLL rather than B_n. As addressed in Example 9.2, the role of high-order poles is important not only for spur suppression but also for quantization noise reduction in the design of the $\Delta\Sigma$ fractional-N PLL.

9.2.5 Nonideal Effects

Unlike the integer-N PLL, the $\Delta\Sigma$ fractional-N PLL suffers from PD nonlinearity and coupling. The PD nonlinearity can degrade both the in-band phase noise and fractional spur performance, while the coupling can cause severe spur generation when an output frequency is close to the integer multiples of a reference frequency. If those facts are not well considered in the design of a $\Delta\Sigma$ fractional-N PLL, we will see a large performance discrepancy between circuit-level simulation and hardware performance.

9.2.5.1 Nonlinearity

Before discussing the nonlinearity problem, let us consider whether the $\Delta\Sigma$ fractional-N PLL is analogous to a $\Delta\Sigma$ DAC or to a $\Delta\Sigma$ ADC. The former case is reasonable since the output of a digital $\Delta\Sigma$ modulator controls the analog phase of the PLL. However, in that case, it is difficult to understand the nonlinearity problem of the $\Delta\Sigma$ fractional-N PLL. Figure 9.28 shows the comparison of a $\Delta\Sigma$ ADC and a $\Delta\Sigma$ fractional-N PLL, both having an eight-level quantizer and a third-order $\Delta\Sigma$ modulator. The eight-level quantizer is analogous to an 8-modulus divider as the third-order $\Delta\Sigma$ modulator generates up to eight quantized phases. When a CP-PLL is designed, the charge pump is the block that converts multiple phases into multiple voltages at a loop filter. Hence, the role of the charge pump is similar to that of a multi-bit DAC in the $\Delta\Sigma$ ADC. In the $\Delta\Sigma$ ADC with the multi-level quantizer, the linearity of the multi-bit DAC is critical to determine the overall performance of the $\Delta\Sigma$ ADC. Hence, we can draw a similar conclusion that the nonlinearity of the charge pump in the fractional-N PLL significantly degrades the performance of the $\Delta\Sigma$ fractional-N PLL.

Figure 9.29 shows the simulated nonlinearity effect on the phase noise and spur performance. In the simulation, a fourth-order fractional-N PLL with a

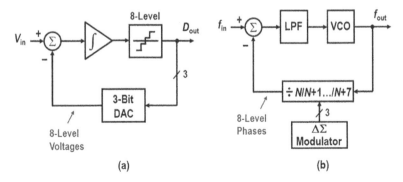

Figure 9.28 Comparison of multi-bit $\Delta\Sigma$ ADC and fractional-N PLL: (a) $\Delta\Sigma$ ADC with eight-level quantizer and (b) $\Delta\Sigma$ fractional-N PLL with the third-order MASH.

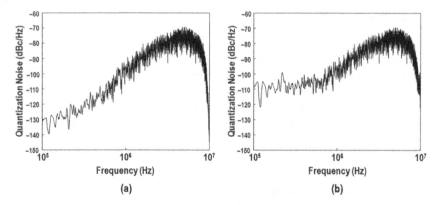

Figure 9.29 Output phase noise of the $\Delta\Sigma$ fractional-N PLL with a third-order modulator: (a) without nonlinearity and (b) with 10% nonlinearity.

third-order MASH modulator is used with a fractional division ratio of 1/32, a reference frequency of 10 MHz, and a PLL bandwidth of 1 MHz. To focus on the quantization noise only, a phase noise plot at the output of a multi-modulus divider instead of the output of a VCO is shown. When an ideal charge pump is used, the phase noise at the divider output shows clear noise shaping without any fractional spur as shown in Fig. 9.29(a). However, when a charge pump nonlinearity of 10% is added, the in-band phase noise is severely degraded as shown in Fig. 9.29(b). Moreover, a fractional spur is observed at an offset frequency of 312.5 kHz ($= 1/32 \times 10$ MHz). The simulation results also verify that the nonlinearity of a charge pump or a PD is critical in the design of a $\Delta\Sigma$ fractional-N PLL. Note that the use of a high-order $\Delta\Sigma$ modulator does not necessarily improve the noise and spur performance when the nonlinearity problem is encountered.

9.2.5.2 Integer-Boundary Spur

In the integer-N PLL, the generation of a reference spur can easily be understood and quantitatively analyzed as learned in Chapter 4. The spur generation of a $\Delta\Sigma$ fractional-N PLL, however, is quite complex and difficult to predict in the circuit-level simulation. The appearance of fractional spurs is well observed when the PLL output is tuned near the integer multiple of the reference frequency. For that reason, the fractional spur is often referred as *integer-boundary spur*. The integer-boundary spur is caused mainly by three mechanisms; nonlinearity of the PLL as discussed, idle tones of the $\Delta\Sigma$ modulator, and coupling.

In theory, high-order $\Delta\Sigma$ modulators do not generate idle tones. In the design of digital modulators, a limited sequence length from the digital modulator can generate the idle tone especially when the control word is close to a rational number. For instance, a fractional division ratio of 1/8 only excites three bits from the most significant bit (MSB), resulting in a fractional spur located at one-eighth of the reference frequency. A common way to mitigate the idle tone is to apply dithering to a control word, which expands the sequence length. A simple dithering method is to set the least significant bit (LSB) to high all the time. That is, an offset frequency equal to the minimum resolution frequency is added to the desired frequency to decorrelate the quantization error. For example, with the use of a 24-bit MASH modulator and a clock frequency of 10 MHz, one LSB corresponds to less than 1-Hz frequency error, or less than 0.001 ppm for 1-GHz output. Then, the effective fractional division ratio of $(1/8 + 1/2^{24})$ is used, which does not affect the accuracy of an output frequency. To further expand the sequence length, a dynamic dithering method can also be considered. When the fractional division value is set to a large rational number such as 1/2 or 1/4, it may be difficult to suppress the spur even with dithering. For those fractional division values, the dithering needs to be automatically disabled for the given channel. Since the fractional spur due to the fractional division value of 1/2 or 1/4 is higher than the PLL bandwidth, those spurs can be suppressed by the loop filter of a PLL.

In the $\Delta\Sigma$ ADC, the high-order SLDSM provides better randomization than the same-order MASH modulator as discussed previously. When a digital modulator is considered for the $\Delta\Sigma$ fractional-N PLL, the high-order MASH modulator does not have the smearing effect of the first-stage modulator due to an imperfect gain. On the other hand, the SLDSM suffers from an internal truncation error if digital coefficients are realized by bit-shifting, addition, and subtraction as seen in (9.23). Contrary to the case of the $\Delta\Sigma$ ADC, the digital MASH modulator generates better uncorrelated output with smaller idle tones than the SLDSM does. Therefore, the MASH modulator is a popular choice for most commercial applications unless the PLL nonlinearity becomes a dominant factor for the in-band phase noise.

The fractional spur generation due to coupling is quite a complex behavior and difficult to predict from the circuit design of a $\Delta\Sigma$ fractional-N PLL. As illustrated in Fig. 9.30, there are three mechanisms to cause the generation of integer-boundary spurs. The first mechanism is the direct coupling between a VCO frequency and the harmonics of a reference frequency. When the VCO frequency is the non-integer multiple of the reference frequency, it induces cross-coupling with the harmonic frequency of the reference, resulting in the integer-boundary spur. The closer the VCO frequency is to the harmonic frequency of the reference, the higher the integer-boundary spur. The second mechanism is the coupling between a charge pump and a VCO. Since the charge pump operates at a reference clock period, any coupling path through supply and substrate noise causes integer-boundary spurs with the harmonics of the charge pump frequency. The third one is a beat tone caused by intermodulation between the reference frequency and the divider output frequency. Since a fractional division ratio N_{frac} is achieved by a periodic operation of an $N/(N + 1)$ dual-modulus

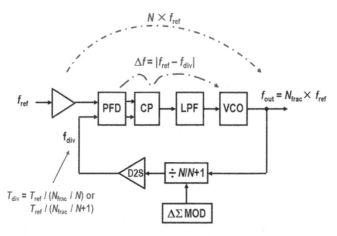

Figure 9.30 Integer-boundary spur generation by coupling.

divider, noise coupling between the reference clock path and the feedback clock path can generate the beat tone that modulates the VCO as illustrated in Fig. 9.30.

When the PLL is locked, the intermodulation frequency products at the PFD input f_d consist of multiples of the reference frequency f_{ref} and the divider output frequency f_{div}. The frequency difference Δf at the PFD input is given by

$$\Delta f = f_{ref}\left(\frac{f_d}{f_{ref}} - \text{round}\left(\frac{f_d}{f_{ref}}\right)\right)$$

$$= f_{ref}\left[\frac{mf_{ref} \pm nN_{frac}f_{ref}}{f_{ref}} - \text{round}(mf_{ref} \pm nN_{frac}f_{ref})\right] \quad (9.34)$$

Then,

$$\Delta f = f_{ref}[m \pm n(N + \alpha) - \text{round}(m \pm n(N + \alpha))] \quad (9.35)$$

where the function round( ) is for getting the nearest integer value, and N and α are the integer and the fractional part of the division ratio N_{frac}, respectively. The effective frequency difference of $f_{ref}(n\alpha - \text{round}(n\alpha))$ is a set of discrete frequency components, becoming the modulation frequency of the VCO.

9.2.6 Practical Design Aspects for the $\Delta\Sigma$ Fractional-N PLL

Below is the summary of different design aspects for the $\Delta\Sigma$ fractional-N PLL in comparison with the integer-N PLL.

- Charge pump nonlinearity is critical for in-band noise and fractional spur performance, while charge pump matching matters for the integer-N PLL. Note that achieving good matching of the up and down currents does not necessarily guarantee high linearity. The phase-to-voltage transfer function over small phase errors should be considered.

- Even with a linear charge pump, integer-boundary spurs would appear due to coupling by intermodulation between the harmonics of the reference frequency and the fractional frequency at the output of the VCO or between the reference frequency and the output frequency of the frequency divider. This kind of coupling due to the intermodulation mechanism is somewhat similar to the fractional spur generation in a digital PLL even with a linear time-to-digital converter.

- It is good to know some unique features of the $\Delta\Sigma$ fractional-N PLL in hardware. With wide bandwidth, phase noise plateau in high frequencies could be observed. Also, noise shaping near the reference spur can be observed because of the digital modulation with the reference clock.

- High-order poles are important not only for the reference spur but also for out-of-band phase noise. Due to an integration factor for frequency-to-phase conversion, the phase noise contributed by an nth-order $\Delta\Sigma$ modulator exhibits $(n-1)$th-order noise shaping. To have the quantization noise gradually decrease in high frequencies, an $(n+1)$th-order $\Delta\Sigma$ fractional-N PLL should be designed to have the nth-order $\Delta\Sigma$ modulator as illustrated in Fig. 9.25.

- As a rule of thumb, the order of the $\Delta\Sigma$ modulator needs to be higher or equal to three. Even though the second-order $\Delta\Sigma$ modulator can satisfy the phase noise requirement and do not generate idle tone in simulation, it is possible that time-varying spur or phase noise fluctuation could be observed in hardware.

9.3 Quantization Noise Reduction Methods

With the noise-shaping property, quantization noise from a $\Delta\Sigma$ modulator has a negligible effect on the phase noise of a fractional-N PLL. However, when a wideband PLL is designed, the quantization noise could increase out-of-band phase noise due to insufficient low-pass filtering by the PLL. High out-of-band phase noise can potentially violate the spectrum mask in some wireless systems or substantially increase the short-term jitter of a clock for wireline systems. A simple way of suppressing the quantization noise would be adding more high-order poles for the given bandwidth, but it will significantly reduce a phase margin especially in the design of a wideband PLL. Therefore, the quantization noise must be suppressed by additional circuits to have the full benefit of designing a wideband fractional-N PLL. There are several methods proposed in the literature, and most of them are classified into two categories. One is based on phase compensation in the analog domain, and the other is based on noise filtering in the digital or mixed-signal domain.

9.3.1 Phase Compensation

Even though a high-order $\Delta\Sigma$ modulator generates nearly uncorrelated outputs, the output sequence is predictable for a given control word like a pseudo-random number generator. Like the conventional spur reduction method using a DAC or DTC as introduced in Section 9.1.2, direct phase compensation at the output of a PD could be considered. Here, we only consider a CP-PLL. A straightforward way of cancelling instantaneous phase errors generated by $\Delta\Sigma$ modulation is to compensate for the residual charge at the output of a charge pump by using a current DAC. Figure 9.31(a) shows the block diagram of a charge compensation method. The amount of the charge Q_n to be compensated by a current DAC is given by

$$Q_n[k] = I_{CP}\Delta t[k] = I_{CP}T_{ref}\sum_{m=0}^{k-1}e[m] \tag{9.36}$$

where Δt is the phase error in time and e is the quantization error. Unlike the traditional fractional-N PLL that employs a few-bit accumulator, the $\Delta\Sigma$ fractional-N PLL requires a large dynamic range of the DAC. To improve the compensation performance, several techniques including the dynamic element matching (DEM) and the adaptive calibration based on a least mean square (LMS) algorithm have been proposed in the literature.

The charge compensation by the current DAC cannot completely remove a voltage ripple since the PFD delivers the phase error information by pulse width modulation in the time domain. Accordingly, a phase compensation method in the time domain based on a DTC shown in Fig. 9.31(b) would be more effective. Thanks to advanced CMOS technology, the design of a high-resolution DTC with background digital calibration is available these days, achieving decent

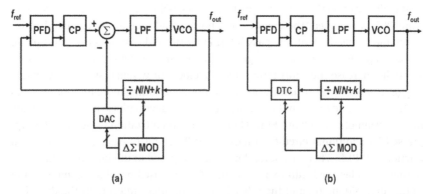

Figure 9.31 Quantization noise reduction: (a) by charge compensation and (b) by DTC.

noise reduction. With the advent of a digital-intensive PLL (DPLL) architecture, various noise reduction methods by using the TDC, the DTC, or both have been proposed.

The use of a multi-phase ring VCO also reduces the quantization noise. The quantizing step will be $1/k$ of the original size by using k phases, leading to $20 \log k$ dB quantization noise reduction. Having the multi-phase operation in high frequencies and maintaining linearity with good matching are critical for the $\Delta\Sigma$ fractional-N PLL performance. In general, the quantization noise reduction based on the phase compensation in the analog domain suffers from large performance variation over PVT corners. Now let us consider a digital-intensive filtering method.

9.3.2 Noise Filtering

In integrated circuits, having a better worst-case performance with a worse best-case performance over PVT variations is considered more valuable than having a better best-case performance. We consider a robust way of reducing the out-of-band quantization noise when moderate noise reduction is enough to meet the system requirement. Figure 9.32 shows a hybrid finite-impulse response (FIR) filtering method that effectively reduces high-frequency quantization noise at

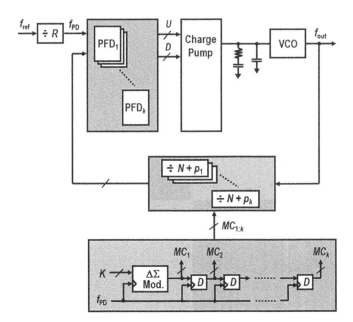

Figure 9.32 Fractional-N PLL with the hybrid FIR filtering technique.

the output of a charge pump, while a noise transfer function for the quantization noise is formed in the digital domain. Since the FIR filtering is performed in parallel paths without creating any latency in the loop, it does not affect the loop dynamics of the PLL. To realize the hybrid FIR filter, multiple PFDs, a multi-input charge pump, and multi-modulus dividers (MMDs) are used. The number of the PFDs and the MMDs sets the number of the FIR filter taps. The output of a ΔΣ modulator is loaded into a D-type flip-flop (DFF) and then shifted in sequence by following DFFs. Then, the multiple MMDs in parallel are controlled by the sequential control bits from the DFFs. Each PFD in the parallel path generates a phase error which is fed to the multi-input charge pump. Therefore, the phase errors from the parallel paths are summed at the output of the multi-input charge pump, which realizes the FIR filter with respect to the quantization noise of the ΔΣ modulator.

Figure 9.33 shows the combined *s*-domain and *z*-domain model of the hybrid FIR filter. The charge pump output at frequency offset Δ*f* can be derived as

$$Y(\Delta f) = [\phi_{\text{ref}}(s) - \phi_{\text{div}}(s) + \phi_{\text{qn}}(z)H_{\text{FIR}}(z)]\frac{\sum_{i=0}^{k-1}I_i}{2\pi}, \quad s = j2\pi\Delta f, \quad z = e^{j\frac{2\pi\Delta f}{f_{\text{ref}}}}$$

(9.37)

where ϕ_{ref} is the phase of the reference signal, ϕ_{div} is the nominal phase of the divider output, ϕ_{qn} is the quantization noise, n_i is the delay depth of the modulator output, I_i is the current of each branch for the multi-input charge pump, f_{ref} is the

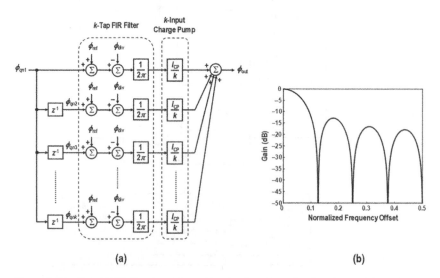

(a)   (b)

Figure 9.33 Hybrid FIR filtering: (a) combined *s*-domain and *z*-domain models and (b) transfer function of the 8-tap FIR filter.

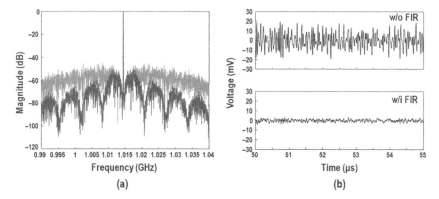

Figure 9.34 Performance comparison: (a) output spectra and (b) control voltage.

reference frequency, and $H_{\mathrm{FIR}}(z)$ is the transfer function of the embedded FIR filter given by

$$H_{\mathrm{FIR}}(z) = \frac{I_0 + I_1 z^{-n_1} + I_2 z^{-n_2} + \ldots + I_{k-1} z^{-n_{k-1}}}{\sum_{i=0}^{k-1} I_i}, \quad z = e^{j\frac{2\pi\Delta f}{f_{ref}}} \qquad (9.38)$$

From (9.38), it can be seen that the embedded FIR filter does filtering with respect to the quantization noise only.

Figure 9.34 shows the effect of the hybrid FIR filter on the output spectrum and the instantaneous phase error of the PLL in the time domain. To clearly demonstrate the FIR-filtering effect on quantization noise, high-order poles of the PLL are put at higher than optimum location. Figure 9.34(a) shows the output spectra of two $\Delta\Sigma$ fractional-N PLLs; upper one without the FIR filter, and bottom one with the FIR filter. As seen clearly, the phase noise in high frequencies is substantially reduced with the FIR filter. The peak-to-peak phase error is also significantly reduced after all phase errors are summed at the output of the charge pump as shown in Fig. 9.34(b), which is good for the linearity of the charge pump circuit. In addition, the time-interleaving operation of the multi-input charge pump can further enhance the linearity, thus improving in-band noise and fractional spur performance.

Even though the DAC cancellation method achieves better performance with perfect matching and high linearity assumed, the hybrid FIR filtering method provides a straightforward way of noise reduction by simply implementing multiple dividers and PDs. Since noise reduction is done by the semi-digital filtering method, the quantization noise suppression can be well predicted with less dependency on PVT variations. In addition, the power and area of the FIR filtering circuitry can be scaled with the advanced CMOS technology when the phase-selection multi-modulus divider is employed. Several improved FIR filtering methods have been proposed by using a multi-phase VCO or a high-attenuation FIR filter driven by a VCO clock.

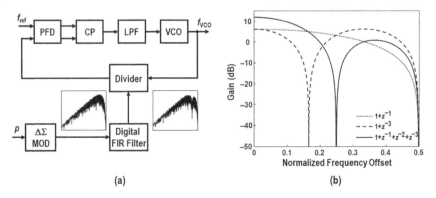

(a)   (b)

Figure 9.35 Digital FIR filter: (a) block diagram and (b) transfer function showing gain amplification in low frequencies.

Example 9.4   *Quantization noise reduction with a digital FIR filter?*

Since a digital $\Delta\Sigma$ modulator is implemented, you may consider having a digital FIR filter at the output of the digital modulator instead of the hybrid FIR filter at the output of a charge pump. As shown in Fig. 9.35(a), the digital FIR filter is put at the output of the $\Delta\Sigma$ modulator so that the quantization noise at high frequencies is suppressed by the filter. With all-digital implementation, this approach does not suffer from any analog mismatch. However, a fractional coefficient such as 1/2 can be achieved by bit-shifting only. The LSB of the data will be lost during such operation, resulting in a truncation error. Therefore, the transfer function of the digital FIR filter can employ integral coefficients only, which causes a DC gain. For example, let us consider a single-stage digital FIR filter whose transfer function is given by $(1 + z^{-1})$. Unlike the hybrid FIR filter, the digital FIR filter has a 6-dB DC gain with $z = 0$. Figure 9.35(b) shows the noise transfer functions of digital FIR filters with different orders. Even though increasing the order of the FIR filter gives a steeper roll-off in high frequencies, the increased gain makes the high-order digital filter not so effective for quantization noise suppression in the design of $\Delta\Sigma$ fractional-N PLLs.

Example 9.5   *Finite-modulo spur reduction with S/H PDs*

The FIR filtering method can also be useful for the spur reduction in a finite-modulo fractional-N PLL. Figure 9.36 shows the example of an 8-modulo fractional-N PLL. A S/H PD is used since it achieves good phase error cancellation in the voltage domain. A multi-input S/H PD circuit is designed in a compact way as shown in Fig. 9.36 and does not generate high-frequency voltage ripple when perfect matching is assumed. The 8-modulo fractional-N PLL with an 3-bit accumulator generates a periodic tone at one-eighth of the reference frequency. Since the 8-tap FIR filter creates a null at integer multiples of one-eighth of the reference frequency, all the fractional spurs could be suppressed. Figure 9.37

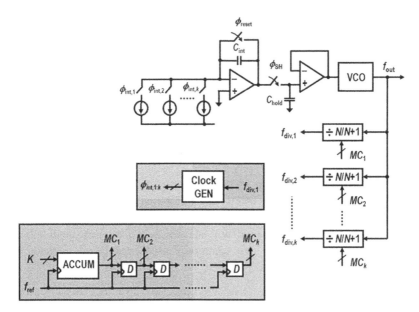

Figure 9.36 Finite-modulo fractional-N PLL with the S/H PD and hybrid FIR filter.

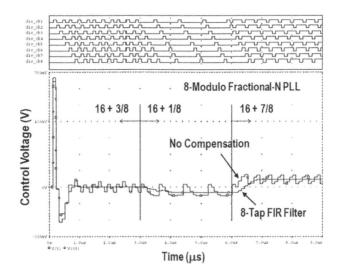

Figure 9.37 Simulated control voltages of 8-modulo fractional-N PLL with and without the 8-tap FIR filter: Division ratio changes as $(16 + 3/8)$, $(16 + 1/8)$, and $(16 + 7/8)$.

shows the behavioral simulation results of the fractional-N PLL with a division ratio of $(16 + 7/8)$. Control voltages with and without the 8-tap FIR filter are compared. It is shown that the periodic tone is completely cancelled when the FIR filter is enabled. In the case of using a PFD, complete spur reduction cannot be achieved due to pulse width modulation, but the FIR filtering technique can still be useful to reduce the fractional spur significantly.

9.4 Frequency Modulation by Fractional-N PLL

Compared with the open-loop method that directly modulates a VCO or a digitally-controlled oscillator (DCO), the PLL-based modulation achieves an accurate frequency control. Nowadays, the $\Delta\Sigma$ fractional-N PLL plays an important role not only as a local oscillator but also as a frequency modulator in wireless transceiver systems. Two commonly used modulation methods are briefly discussed in this section.

9.4.1 One-Point Modulation

The $\Delta\Sigma$ fractional-N frequency synthesizer generates an arbitrary frequency with fine resolution by sending a fixed control word to a $\Delta\Sigma$ modulator. If the control word of the $\Delta\Sigma$ modulator is dynamically changed, the output frequency of the $\Delta\Sigma$ fractional-N PLL will be modulated, enabling the PLL to achieve direct-digital frequency modulation. Suppose that a third-order MASH modulator is implemented. Then, the z-domain transfer function of the $\Delta\Sigma$ modulator from an input signal $X(z)$ to an output signal $Y(z)$ including quantization noise $Q(z)$ is given by

$$Y(z) = X(z)z^{-3} + Q(z)(1 - z^{-1})^3 \tag{9.39}$$

The phase noise caused by divider modulation with $Y(z)$ will be low-pass filtered by the loop dynamics of the PLL, and the information of the digital modulation signal $X(z)$ appears at the PLL output if the PLL bandwidth is wider than the signal bandwidth. The main disadvantage of this method is that the maximum data rate is constrained by the PLL bandwidth. To have an optimum performance of the phase noise and the spur, it is difficult to increase the PLL bandwidth for a higher data rate.

To further increase the data rate for a given PLL bandwidth, a pre-distortion or pre-emphasis digital filter is used before the $\Delta\Sigma$ modulator as shown in Fig. 9.38. Knowing that the modulated signal $X(z)$ is attenuated outside the PLL bandwidth, we employ the digital pre-emphasis filter with a boosted high-frequency gain to compensate for the open-loop gain of the PLL in high frequencies. Since the transfer function of the pre-emphasis filter can be controlled well in the digital domain,

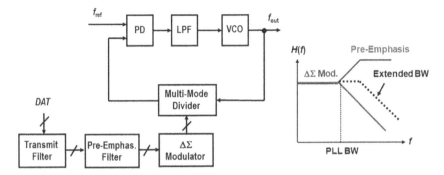

Figure 9.38 Frequency modulation by the $\Delta\Sigma$ fractional-N PLL.

an effective modulation bandwidth is extended after combining both the transfer functions of the pre-emphasis filter and the PLL as depicted in Fig. 9.38. However, the modulation linearity will be degraded if there is a mismatch between the transfer functions of the digital filter and the analog PLL. In fact, the loop dynamics of the PLL varies a lot over process and temperature due to the variation of circuit parameters such as a VCO gain, a charge pump current, and loop filter values. To overcome the limited bandwidth of the PLL, we will discuss an elegant modulation method in the next section.

9.4.2 Two-Point Modulation

A PLL has a low-pass filter characteristic to an input phase, while showing a high-pass filter characteristic to a VCO phase. Both the low-pass filter and the high-pass filter have the same 3-dB corner frequency. Therefore, an all-pass filter transfer function can be obtained if the same modulation data is injected to both the VCO and the $\Delta\Sigma$ modulator as shown in Fig. 9.39. The modulation path to the VCO is called a high-pass modulation path, and the modulation path to the $\Delta\Sigma$ modulator is called a low-pass modulation path. Since this technique utilizes both modulation paths, we call it two-point modulation.

Figure 9.40 shows the simplified linear model of the two-point modulator. For the given open-loop gain $G(s)$ of a PLL, the system transfer functions of the low-pass and the high-pass modulation paths denoted by $H_{\mathrm{LP}}(s)$ and $H_{\mathrm{HP}}(s)$ are given by

$$H_{\mathrm{LP}}(s) = \frac{k_1 e^{-s\tau_1} G(s)}{1 + G(s)} \quad \text{and} \quad H_{\mathrm{HP}}(s) = \frac{k_2 e^{-s\tau_2}}{1 + G(s)} \tag{9.40}$$

where k_1 and k_2 are the gain factors, and τ_1 and τ_2 are the delay times of the low-pass and the high-pass modulation paths, respectively. When both

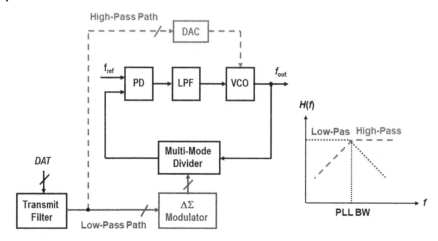

Figure 9.39 Two-point modulation.

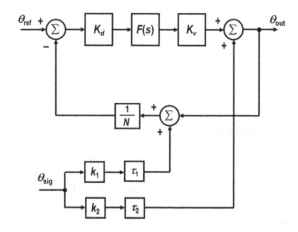

Figure 9.40 Simplified linear model of the two-point modulator.

modulation paths are combined, the transfer function $H_{\text{TPM}}(s)$ of the two-point modulation is given by

$$H_{\text{TPM}}(s) = \frac{k_1 e^{-s\tau_1} G(s)}{1 + G(s)} + \frac{k_2 e^{-s\tau_2}}{1 + G(s)} \tag{9.41}$$

If both modulation paths have the same gain and the time delay, we set $k_1 = k_2 = k$ and $\tau_1 = \tau_2 = \tau$. Then, we have

$$H_{\text{TPM}}(s) = \frac{k e^{-s\tau} G(s)}{1 + G(s)} + \frac{k e^{-s\tau}}{1 + G(s)} = k e^{-s\tau} \tag{9.42}$$

Therefore, the magnitude of $H_{\text{TPM}}(s)$ exhibits an all-pass filter characteristic with a fixed gain k. Equation (9.42) implies that the all-pass modulation transfer function

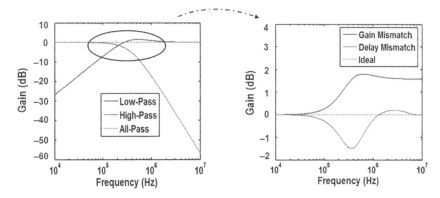

Figure 9.41 Nonideal effects of gain and delay mismatches.

can be achieved regardless of the PLL bandwidth in the two-point modulation. In other words, the PLL bandwidth can be determined mainly to optimize the PLL performance such as phase noise and spur without considering the modulation bandwidth.

Even though the two-point modulation method does not suffer from the PLL bandwidth variation, it is highly sensitive to the gain and the time mismatches of two modulation paths as implied in (9.39). Figure 9.41 shows the effects of gain and delay mismatches on the transfer function. The all-pass transfer function is severely distorted due to those mismatches. The gain mismatch between two paths could be calibrated, but the gain variation in the high-pass modulation path due to VCO nonlinearity is critical and difficult to be calibrated. The gain variation and nonlinearity of the VCO can degrade the out-of-band spectral emission of a transmitter and modulation linearity. For example, the VCO nonlinearity to meet the modulation linearity[4] of Global System for Mobile communications (GSM) wireless system is typically less than 2%. When the data rate becomes high, say >10 Mb/s, the delay mismatch should be a few nsec depending on the requirement of the modulation linearity, which is also challenging. After the development of the DPLL having a DCO, the two-point modulation becomes a viable method as the DCO nonlinearity can be calibrated well within the DPLL.

References

1 W. Egan, *Frequency Synthesis by Phase Lock*, 2nd ed., Wiley, New York, 2000.
2 U. L. Rohde, *Microwave and Wireless Frequency Synthesizers: Theory and Design*, Wiley, New York, 1997.

---

4 In general, the modulation linearity is quantified by error vector magnitude (EVM).

3 V. Manassewitsch, *Frequency Synthesizers, Theory and Design*, Wiley, New York, 1987.

4 K. Shu and E. Sanchez-Sinencio, *CMOS PLL Synthesizers; Analysis and Design*, Springer, 2005.

5 W. Rhee (ed.), *Phase-Locked Frequency Generation and Clocking: Architectures and Circuits for Modern Wireless and Wireline Systems*, The Institution of Engineering and Technology, United Kingdom, 2020.

6 S. R. Norsworthy, R. Schreier and G. C. Temes, *Delta-Sigma Data Converters, Theory, Design, and Simulation*, IEEE Press, New York, 1997.

7 J. C. Candy and G. C. Temes, *Oversampling Delta-Sigma Data Converters*, Wiley-IEEE Press, New York, 1992.

8 W. Egan, *Advanced Frequency Synthesis by Phase Lock*, Wiley-IEEE Press, Hoboken, NJ, 2011.

9 G.C. Gillette, "Digiphase synthesizer," in *Proc. IEEE Frequency Control Symposium*, 1969, pp. 201–210.

10 J. Gibbs and R. Temple, "Frequency domain yields its data to phase-locked synthesizer," *Electronics*, pp. 107–113, Apr. 1978.

11 *Dana Series 7000 Digiphase Frequency Synthesizers*, Publication 980428, Dana Laboratories, Inc., 1973.

12 W. Rhee and A. Ali, "An on-chip phase-compensation technique in fractional-N frequency synthesis," in *Proc. IEEE International Symposium on Circuits and Systems*, May 1999, pp. 363–366.

13 T. Weigandt and S. Mehta, "A phase interpolation technique for fractional-N frequency synthesis," in *Website of RF Integrated Circuit Design Group* (http://kabuki.eecs.berkeley.edu/rf/), University of California, Berkeley, 1997.

14 W. Rhee, "Design of low jitter 1-GHz phase-locked loops for digital clock generation," in *Proc. IEEE International Symposium on Circuits and Systems*, May, 1999, pp. 520–523.

15 V. Reinhardt, "Spur reduction techniques in direct digital synthesizers," in *Proc. 47th Frequency Control Symposium*, Oct. 1993, pp. 230–241.

16 B. Miller and R.J. Conley, "A multiple modulator fractional divider," in *Proc. IEEE Frequency Control Symposium*, Mar. 1990, pp. 559–568.

17 T. A. Riley, M. A. Copeland and T. A. Kwasniewski, "Delta-sigma modulation in fractional-N frequency synthesis," *IEEE Journal of Solid-State Circuits*, vol. 28, pp. 553–559, May 1993.

18 I. Galton, "Delta-sigma data conversion in wireless transceivers," *IEEE Transactions on Microwave Theory and Techniques*, vol. 50, pp. 3092–3315, Jan. 2002.

19 W.L. Lee, *A Novel Higher-Order Interpolative Modulator Topology for High Resolution Oversampling A/D Converter*, M.S. Thesis, Massachusetts Institute of Technology, 1987.

20 K. Chao, S. Nadeem, W. Lee *et al.*, "A higher-order topology for interpolative modulation for oversampling A/D converters," *IEEE Transactions on Circuits and Systems I*, vol. 37, pp. 309–318, Mar. 1990.

21 M. H. Perrott, T. L. Tewksbury III, and C. G. Sodini, "A 27-mW CMOS fractional-N synthesizer using digital compensation for 2.5-Mb/s GFSK modulation," *IEEE Journal of Solid-State Circuits*, vol. 32, pp. 2048–2060, Dec. 1997.

22 N. Filiol, T. Riley, C. Plett *et al.*, "An agile ISM band frequency synthesizer with built-in GMSK data modulation," *IEEE Journal of Solid-State Circuits*, vol. 33, pp. 998–1008, July 1998.

23 S. Willingham, M. Perrott, B. Setterberg *et al.*, "An integrated 2.5 GHz $\Sigma\Delta$ frequency synthesizer with 5 μs settling and 2 Mb/s closed loop modulation," in *Proc. IEEE International Solid-State Circuits Conference*, Feb. 2000, pp. 200–201.

24 W. Rhee, *Multi-Bit Modulation Technique for Fractional-N Frequency Synthesizers*, Ph.D. Thesis, University of Illinois, Urbana-Champaign, Aug. 2000.

25 W. Rhee, B. Song and A. Ali, "A 1.1-GHz CMOS fractional-N frequency synthesizer with a 3-b third-order Δ-Σ modulator," *IEEE Journal of Solid-State Circuits*, pp. 1453–1460, Oct. 2000.

26 W. Rhee, B. Bisanti and A. Ali, "An 18-mW 2.5-GHz/900-MHz BiCMOS dual frequency synthesizer with <10-Hz RF carrier resolution," *IEEE Journal of Solid-State Circuits*, vol. 37, pp. 515–520, Apr. 2002.

27 B. De Muller and M. Steyaert, "A CMOS monolithic $\Delta\Sigma$-controlled fractional-N frequency synthesizer for DCS-1800," *IEEE Journal of Solid-State Circuits*, vol. 37, pp. 835–844, July 2002.

28 B. De Muller and M. Steyaert, "On the analysis of $\Delta\Sigma$ fractional-N frequency synthesizers for high-spectral purity," *IEEE Transactions on Circuits and Systems II*, vol. 50, pp. 784–793, Nov. 2003.

29 P. V. Brennan, P. M. Radmore and D. Jiang, "Intermodulation-borne fractional-N frequency synthesizer spurious components," *IEE Circuits and Systems*, vol. 151, pp. 536–542, Dec. 2004.

30 P. V. Brennan, H. Wang, D. Jiang *et al.*, "A new mechanism producing discrete spurious components in fractional-N frequency synthesizers," *IEEE Transactions on Circuits and Systems I*, vol. 55, pp. 1279–1288, June 2008.

31 W. Rhee, K. Jenkins, J. Liobe *et al.*, "Experimental analysis of substrate noise effect on PLL performance," *IEEE Transactions on Circuits and Systems II*, vol. 55, pp. 638–642, July 2008.

32 K. Wahee, R. B. Staszewski, F. Dulger *et al.*, "Spurious-free time-to-digital conversion in an ADPLL using short dithering sequences," *IEEE Transactions on Circuits and Systems I*, vol. 58, pp. 2051–2060, Sept. 2011.

33 C. Heng and B. Song, "A 1.8-GHz fractional-N frequency synthesizer with randomized multi-phase VCO," *IEEE Journal of Solid-State Circuits*, vol. 38, pp. 848–854, June 2003.

34 E. Temporiti, G. Albasini, I. Bietti *et al.*, "A 700-kHz bandwidth ΣΔ fractional synthesizer with spurs compensation and linearization techniques for WCDMA applications," *IEEE Journal of Solid-State Circuits*, vol. 39, pp. 1446–1454, Sept. 2004.

35 M. Gupta and B. Song, "A 1.8-GHz spur cancelled fractional-N frequency synthesizer with LMS-based DAC gain calibration," *IEEE Journal of Solid-State Circuits*, vol. 41, pp. 2842–2851, Dec. 2006.

36 S. E. Meninger and M. H. Perrott, "A 1-MHz bandwidth 3.6-GHz 0.18-μm CMOS fractional-N synthesizer," *IEEE Journal of Solid-State Circuits*, vol. 41, pp. 966–980, Apr. 2006.

37 P. Park, D. Park and S. Cho, "A 2.4 GHz fractional-N frequency synthesizer with high-OSR ΔΣ modulator and nested PLL," *IEEE Journal of Solid-State Circuits*, vol. 47, no. 10, pp. 2433–2443, Oct. 2012.

38 D. Park and S. Cho, "A 14.2 mW 2.55-to-3 GHz cascaded PLL with reference injection and 800 MHz delta-sigma modulator in 0.13 μm CMOS," *IEEE Journal of Solid-State Circuits*, vol. 47, pp. 2989–2998, Dec. 2012.

39 X. Yu, Y. Sun, W. Rhee *et al.*, "An FIR-embedded noise filtering method for fractional-N PLL clock generators," *IEEE Journal of Solid-State Circuits*, vol. 44, pp. 2426–2436, Sept. 2009.

40 X. Yu, Y. Sun, W. Rhee *et al.*, "A ΔΣ fractional-N frequency synthesizer with customized noise shaping for WCDMA/HSDPA applications," *IEEE Journal of Solid-State Circuits*, vol. 44, pp. 2193–2201, Aug. 2009.

41 D.-W. Jee, Y. Suh, B. Kim *et al.*, "A 0.1-fref BW 1GHz fractional-N PLL with FIR-embedded phase-interpolator-based noise filtering," in *Proc. IEEE International Solid-State Circuits Conference*, Feb. 2011, pp. 94–95.

42 X. Yu, Y. Sun, W. Rhee *et al.*, "A 65 nm CMOS 3.6 GHz fractional-N PLL with 5th-order delta-sigma modulation and weighted FIR filtering," in *Proc. IEEE Asian Solid-State Circuits Conference*, Nov. 2009, pp. 77–80.

43 L. Kong and B. Razavi, "A 2.4-GHz RF fractional-N synthesizer with BW = $0.25 f_{REF}$," *IEEE Journal of Solid-State Circuits*, vol. 53, pp. 1707–1718, June 2018.

44 A. M. Fahim and M. I. Elmasry, "A wideband sigma–delta phase-locked-loop modulator for wireless applications," *IEEE Journal of Solid-State Circuits*, vol. 50, pp. 53–62, Feb. 2003.

45 B. Chi, X. Yu, W. Rhee *et al.*, "A fractional-N PLL for digital clock generation with an FIR-embedded frequency divider," in *Proc. IEEE International Symposium on Circuits and Systems*, May 2007, pp. 3051–3054.

46 D. R. McMahill and C. G. Sodini, "A 2.5-Mb/s GFSK 5.0-Mb/s 4-FSK automatically calibrated Σ-Δ frequency synthesizer," *IEEE Journal of Solid-State Circuits*, vol. 37, pp. 18–26, Jan. 2002.

47 S. Pamarti, L. Jansson and I. Galton, "A wide-band 2.4-GHz delta-sigma fractional-N PLL with 1-Mb/s in-loop modulation," *IEEE Journal of Solid-State Circuits*, pp. 49–62, Jan. 2004.

48 H. Shanan, G. Retz, K. Mulvaney *et al.*, "A 2.4 GHz 2 Mb/s versatile PLL-based transmitter using digital pre-emphasis and auto calibration in 0.18 μm CMOS for WPAN," in *Proc. IEEE International Solid-State Circuits Conference*, Feb. 2009, pp. 420–421.

49 M. J. Underhill and R. I. H. Scott, "Wideband frequency modulation of frequency synthesisers," *Electronics Letters*, vol. 15, no. 13, pp. 541–542, June 1979.

50 C. Durdodt, M. Friedrich, C. Grewing *et al.*, "A low-IF RX two-point ΣΔ-modulation TX CMOS single-chip Bluetooth solution," *IEEE Transactions on Microwave Theory and Techniques*, vol. 49, no. 9, pp. 1531–1537, Sept. 2001.

51 D. Theil, C. Durdodt, and A. Hanke *et al.*, "A fully integrated CMOS frequency synthesizer for Bluetooth," in *Proc. IEEE RFIC Symposium*, May 2001, pp. 103–106.

52 S. A. Yu and P. Kinget, "A 0.65-V 2.5-GHz fractional-N synthesizer with two-point 2-Mb/s GFSK data modulation," *IEEE Journal of Solid-State Circuits*, vol. 44, pp. 2411–2425, Sept. 2009.

53 S. Lee, J. Lee, H. Park *et al.*, "Self-calibrated two-point Delta-Sigma modulation technique for RF transmitters," *IEEE Transactions on Microwave Theory and Techniques*, vol. 58, no. 7, pp. 1748–1757, Jul. 2010.

54 G. Marzin, S. Levantino, C. Samori *et al.*, "A 20 Mb/s phase modulator based on a 3.6 GHz digital PLL with −36 dB EVM at 5 mW power," *IEEE Journal of Solid-State Circuits*, vol. 47, no. 12, pp. 2974–2988, Dec. 2012.

55 N. Xu, W. Rhee and Z. Wang, "A hybrid loop two-point modulator without DCO nonlinearity calibration by utilizing 1-bit high-pass modulation," *IEEE Journal of Solid-State Circuits*, vol. 49, pp. 2172–2186, Oct. 2014.

56 N. Markulic, K. Raczkowski, E. Martens *et al.*, "A self-calibrated 10 Mb/s phase modulator with −37.4 dB EVM based on a 10.1-to-12.4 GHz, −246.6 dB FOM, fractional-N subsampling PLL," in *Proc. IEEE International Solid-State Circuits Conference*, Feb. 2016, pp. 176–177.

10

Digital-Intensive PLL

The fractional-N PLL was developed to improve the performance of integer-N
PLLs. To the contrary, high attention towards a digital-intensive PLL, namely
digital PLL (DPLL) in the early days was mainly associated with technology
issues such as the gate leakage current and poor scalability problems of a passive
loop filter in the design of traditional CP-PLLs with advanced CMOS technology. Figure 10.1 shows the simplified block diagram of a conventional DPLL.
Consisting of a time-to-digital converter (TDC), a digital loop filter (DLF), and a
digitally-controlled oscillator (DCO), the DPLL is free from the leakage current
problem of a loop filter and offers good technology scalability by replacing a large
capacitor with digital logic circuits. As learned, an accurate frequency control is
one of the most important features of the PLL. Therefore, the DCO must generate
output frequencies with a very fine frequency resolution, while covering a wide
tuning range. The dynamic range requirement of the DCO makes it difficult to
design an LC DCO having a capacitor array with a large number of MOS switches.
To relax the resolution requirement, frequency interpolation by oversampling is
typically done by having a delta-sigma ($\Delta\Sigma$) modulator in front of the DCO. Note
that the $\Delta\Sigma$ modulator after the DLF in Fig. 10.1 should not be considered a part
of the DLF but a part of the DCO.

The DPLL also offers reconfigurable system parameters and background digital
calibration for robust performance. A digital coefficient α in the DLF sets the
gain of a proportional-gain path, while a digital coefficient β sets the gain of an
integral path. The basic concept of two control paths in a type 2 PLL was learned
in Chapter 2 and analogy with a DPLL was described in Chapter 3. The DPLL
achieves fast frequency acquisition by having the automatic frequency calibration
of a DCO with fine resolution. High-performance two-point modulation by
having the adaptive calibration of DCO nonlinearity is another good example of
DPLL merits. Since the passive loop filter is replaced with the DLF and there
is no building block with current biasing, easy design migration can be done
for the design with different CMOS technologies. If a TDC has fine resolution

Phase-Locked Loops: System Perspectives and Circuit Design Aspects, First Edition.
Woogeun Rhee and Zhiping Yu.
© 2024 The Institute of Electrical and Electronics Engineers, Inc. Published 2024 by John Wiley & Sons, Inc.

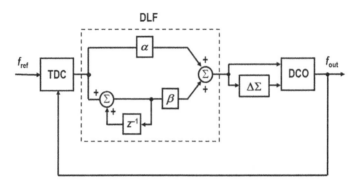

Figure 10.1 Digital-intensive PLL.

and good linearity, the DPLL can achieve better in-band phase noise than the traditional CP-PLL since there is no noise contribution of bias currents. Below is the summary of the merits of the DPLL over the traditional CP-PLL.

- Free from the leakage current problem of the loop filter
- High integration and good area scalability with advanced CMOS technology
- Robust performance with reconfigurable design parameters
- Easy design migration with bias-current-free digitally controlled building blocks
- Fast frequency acquisition and linear modulation with DCO calibration
- Low in-band phase noise without the charge pump and analog bias current

Adopting the DPLL architecture, however, does not necessarily guarantee superior performance over conventional PLLs in spite of the fancy term "digital-intensive" or "all-digital." In practice, the main bottleneck for the design of high-performance DPLLs lies in the implementation of a high-resolution TDC in addition to the design of a high-performance DCO. Poor resolution or nonlinearity of the TDC results in degraded in-band phase noise or fractional spur performance. Accordingly, not only various TDC circuits but also hybrid architectures have been proposed for the design of high-performance DPLLs even with the state-of-the-art CMOS technology. In this chapter, three digital-intensive architectures based on phase detection methods will be mainly discussed.

10.1 DPLL with Linear TDC

A TDC generates quantized phase boundaries with fine time resolution and outputs a digital value proportional to an input phase variation referenced to a reference clock. Figure 10.2 depicts the functional diagram of the TDC in which the quantized phases are formed by delay cells and DFFs. If the time resolution and linearity of the TDC are good enough to control a DCO with a negligible

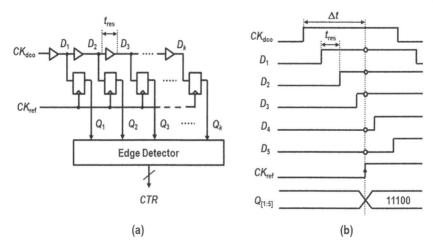

Figure 10.2 Time-to-digital converter: (a) a block diagram and (b) timing diagram example.

frequency error, the behavior of a DPLL with the linear TDC is not much different from that of a traditional PLL. Rigorous analyses based on the discrete-time model are found in the literature, but we focus on intuitive analyses by leveraging what we have learned from traditional type 2 PLLs. Two approximation methods will be discussed in the following section.

10.1.1 Loop Dynamics

10.1.1.1 Time-Continuous Approximation

As discussed in Chapter 2, the loop dynamics based on the time-continuous approximation instead of the discrete-time analysis can be used to analyze the loop dynamics of the PLL when the loop time constant is much longer than the reference clock period. That is, the loop bandwidth is lower than the reference frequency by at least one decade. Such a narrowband approximation can still be applied to analyze the DPLL. We first model the DPLL based on a z-domain model and convert it to an s-domain model with the time-continuous approximation.

To build the z-domain model of a DPLL, let us define loop parameters based on the discrete-time domain at a reference clock rate instead of the phase domain. A TDC converts the time difference to digital values with a time resolution t_{res} at every reference period T_{ref}. That is, the TDC generates one bit for a time difference of t_{res}. Then, the TDC gain K_{td} is simply given by

$$K_{td} = \frac{1}{t_{res}} \tag{10.1}$$

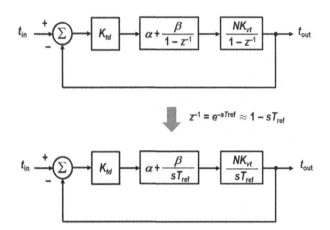

Figure 10.3 DPLL modeling with time-continuous approximation.

A DCO generates quantized frequencies with a frequency resolution f_{res} per one bit. To define the DCO gain K_{vt} in the time domain, we have

$$K_{vt} = \frac{f_{res}}{f_v} T_v = \frac{f_{res}}{f_v^2} \tag{10.2}$$

where f_v and T_v are the frequency and period of the DCO, respectively. When a frequency divider is included in the feedback path of the DPLL, the output period of the N divider becomes NT_v in the discrete-time model for the time interval of T_{ref}. The upper plot in Fig. 10.3 shows the linear model of the DPLL in the z-domain.[1] Note that the division ratio N is included in the DCO model to have the z-domain model at the T_{ref} rate. The open-loop gain $G(z)$ of the DPLL is given by

$$G(z) = K_{td} F(z) \frac{NK_{vt}}{1 - z^{-1}} \tag{10.3}$$

where

$$F(z) = \alpha + \frac{\beta}{1 - z^{-1}} \tag{10.4}$$

Now we transform the z-domain model into the s-domain model. If the loop time constant of the DPLL is much longer than the sampling period, that is T_{ref}, we can convert z-domain equations to s-domain equations with following approximation:

$$z^{-1} = e^{-sT_{ref}} \approx 1 - sT_{ref} \tag{10.5}$$

---

1 For consistency, we will keep using α for the proportional-gain path and β for the integral path.

Then, we have an open-loop gain $G(s)$ with a time-continuous approximation given by

$$G(s) = \frac{NK_{td}K_{vt}}{sT_{ref}} \left(\alpha + \frac{\beta}{sT_{ref}} \right) \tag{10.6}$$

From (10.6), we get the unity-gain frequency ω_u and the zero frequency ω_z given by

$$\omega_u \approx \frac{\alpha NK_{td}K_{vt}}{T_{ref}} = \frac{\alpha N}{t_{res}T_{ref}} \frac{f_{res}}{f_v^2} = \frac{\alpha}{f_v} \frac{f_{res}}{t_{res}} \tag{10.7}$$

We also have the zero frequency ω_z given by

$$\omega_z = \frac{\beta}{\alpha} \frac{1}{T_{ref}} \tag{10.8}$$

The bottom plot in Fig. 10.3 shows the equivalent s-domain model of the DPLL. Like the traditional type 2 PLL, the value of α determines ω_u, while the ratio of β to α determines the stability of the DPLL. Unlike the traditional PLL, having a large value of α is not a good way to increase the loop bandwidth since it will increase the truncation error of the DLF as well as the DCO granularity, resulting in increased quantization noise. Therefore, increasing the TDC gain K_{td} is the best way to have a wide bandwidth, but it is limited by the time resolution t_{res} whose value depends on CMOS technology. For that reason, various TDC architectures have been studied to further improve the time resolution.

In (10.7), the role of N is confusing since a large N will increase the loop bandwidth, which is different from what we learned in conventional PLLs. For better understanding, we express the loop parameters in the phase domain instead of the time domain based on the s-domain model in Fig. 10.3. The TDC gain T_{tdc} in the phase domain can be expressed with the phase resolution θ_{res}

$$K_{tdc} = \frac{1}{\theta_{res}} = \frac{1}{2\pi \left(\frac{t_{res}}{T_{ref}} \right)} = \frac{T_{ref}}{2\pi t_{res}} \tag{10.9}$$

Similarly, the DCO gain K_{dco} in units of angular frequency per bit is given by

$$K_{dco} = 2\pi f_{res} \tag{10.10}$$

Then, we will get the same unity-gain frequency ω_u as

$$\omega_u \approx \frac{\alpha K_{tdc}K_{dco}}{N} = \frac{\alpha}{N} \frac{T_{ref}(2\pi f_{res})}{2\pi t_{res}} = \frac{\alpha}{f_v} \frac{f_{res}}{t_{res}} \tag{10.11}$$

Figure 10.4 shows the equivalent s-domain model of the DPLL in the phase domain, which is more straightforward from what we learned in Chapter 2. Let us take a numerical example to learn how to calculate the DLF values based on the linear model of the DPLL.

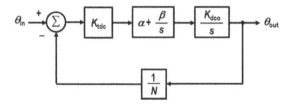

Figure 10.4 Phase-domain model of the DPLL.

Example 10.1 *Loop filter calculation in the DPLL*

Suppose that we generate an output frequency of 2.4 GHz from a reference frequency of 50 MHz and that a TDC with a time resolution of 10 ps and a DCO with a frequency resolution of 10 kHz are designed. What are the values of α and β to have a unity frequency of 500 kHz and a zero frequency of 100 kHz for the DPLL?

From (10.11), we obtain α by

$$\alpha = 2\pi f_u f_v \frac{t_{res}}{f_{res}} = 2\pi \times (500 \times 10^3) \times (2.4 \times 10^9) \times \frac{10 \times 10^{-12}}{10 \times 10^3} = 7.54$$

From (10.8), we get β by

$$\beta = 2\pi \alpha \frac{f_z}{f_{ref}} = 2\pi \times 7.54 \times \frac{10 \times 10^3}{50 \times 10^6} = 0.0947$$

Like the conventional CP-PLL, α in the proportional-gain path mainly determines the value of the loop bandwidth, while β in the integral path determines the phase margin. As those values are digitally programmable, the loop dynamics of the DPLL can easily be controlled by a system, which is one of the strong advantages of the DPLL architecture.

10.1.1.2 CP-PLL Analogy

One way of designing a digital filter is to begin with a prototype analog filter and transform the analog filter into the digital filter by the commonly used bilinear transform. We can do a similar work for the DPLL design. That is, we begin with the desired loop dynamics of an analog PLL, say, a type 2 CP-PLL. After obtaining loop parameters, we convert them into the loop parameters of the DPLL. As briefly discussed in Chapter 3, a digital coefficient β in the integral path is analogous to $1/C_1$ in the loop filter of a type 2 CP-PLL, while a digital coefficient α in the proportional-gain path is analogous to R_1. We can find an accurate expression by using the bilinear transform. The bilinear transform converts an s-domain transfer function to a z-domain transfer function with

$$s = \frac{2}{T_s} \frac{1 - z^{-1}}{1 + z^{-1}} \tag{10.12}$$

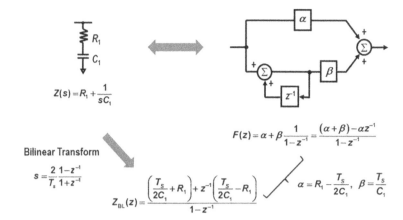

Figure 10.5 CP-PLL analogy.

where T_s is the sampling period. For the sake of simplicity, we consider the impedance transfer function of the loop filter as the loop filter transfer function of a type 2 CP-PLL, that is,

$$F(s) = R_1 + \frac{1}{sC_1} \tag{10.13}$$

With the bilinear transform by having $T_s = T_{ref}$, we have an equivalent z-domain transfer function $F_{BL}(z)$ given by

$$F_{BL}(z) = \frac{\left(\frac{T_{ref}}{2C_1} + R_1\right) + z^{-1}\left(\frac{T_{ref}}{2C_1} - R_1\right)}{1 - z^{-1}} \tag{10.14}$$

As illustrated in Fig. 10.5, we compare $F_{BL}(z)$ with $F(z)$ in (10.4). Then, we have

$$\alpha = R_1 - \frac{T_{ref}}{2C_1}, \quad \beta = \frac{T_{ref}}{C_1} \tag{10.15}$$

Now, let us consider the loop dynamics of a type 2 CP-PLL with the target values of the loop bandwidth ω_u and the phase margin ϕ_M as illustrated in Fig. 10.6(a). From Chapter 2, ω_u and ω_z of a second-order type 2 CP-PLL are approximated by

$$\omega_u = \frac{I_{CP}R_1K_v}{2\pi N}, \quad \omega_z = \frac{\omega_u}{\tan\phi_M} = \frac{1}{R_1C_1} \tag{10.16}$$

Then, R_1 and C_1 are given by

$$R_1 = \frac{2\pi N\omega_u}{I_{CP}K_v}, \quad C_1 = \frac{1}{R_1\omega_z} \tag{10.17}$$

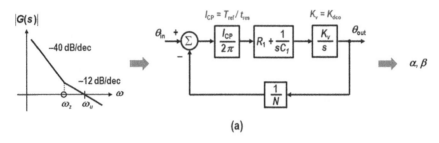

(a)

	DPLL	CP-PLL
	K_{tdc} ($= T_{ref}/(2\pi\, t_{res})$)	$\dfrac{I_{CP}}{2\pi}$
	K_{dco}	K_v
α		$\dfrac{2\pi N\omega_u}{I_{CP}K_v} - \dfrac{T_{ref}R_1\omega_z}{2}$
β		$T_{ref}R_1\omega_z$

(b)

Figure 10.6 Loop dynamics with charge pump analogy: (a) equivalent linear model and (b) corresponding loop parameters.

Therefore, the values of α and β can be obtained from (10.15). That is,

$$\alpha = \frac{2\pi N\omega_u}{I_{CP}K_v} - \frac{T_{ref}R_1\omega_z}{2}, \quad \beta = T_{ref}R_1\omega_z \tag{10.18}$$

In (10.17), I_{CP} and K_v need to be set. Unlike the CP-PLL, those values cannot be determined by a circuit designer since they are related with K_{tdc} and K_{dco} and should be minimized for a given CMOS technology. By comparing the phase-domain model of the DPLL with the linear model of the CP-PLL, we get

$$I_{CP} = 2\pi K_{tdc} = \frac{T_{ref}}{t_{res}} \tag{10.19}$$

and

$$K_v = K_{dco} = 2\pi f_{res} \tag{10.20}$$

Using (10.18)–(10.20), we can express α by

$$\alpha = \frac{2\pi N\omega_u}{I_{CP}K_v} - \frac{T_{ref}R_1\omega_z}{2} = \frac{N\omega_u}{T_{ref}}\left(\frac{t_{res}}{f_{res}}\right) - \frac{T_{ref}R_1\omega_z}{2} \approx \frac{N\omega_u}{T_{ref}}\left(\frac{t_{res}}{f_{res}}\right) = 2\pi f_u f_v \frac{t_{res}}{f_{res}} \tag{10.21}$$

where approximation is done for an overdamped loop. Note that the α value is matched with that from (10.11) if the overdamped loop is assumed. Figure 10.6(b) shows the corresponding loop parameters of the CP-PLL in comparison with those of the DPLL.

Example 10.2 *Calculation of loop parameters based on the CP-PLL analogy*
Let us use the same conditions of the PLL in Example 10.1 and obtain the loop parameters of an equivalent CP-PLL. From (10.19), we have the CP current by

$$I_{CP} = \frac{T_{ref}}{t_{res}} = \frac{1}{(50 \times 10^6) \times (10 \times 10^{-12})} = 2 \times 10^3$$

From (10.20), we obtain the VCO gain by

$$K_v = 2\pi f_{res} = 2\pi(10 \times 10^3)$$

Note that the CP current is unusually high due to a very low value of K_v in the equivalent model. Based on the linear model of the CP-PLL shown in Fig. 10.6(a), the loop filter values can be obtained from (10.17), that is,

$$R_1 = \frac{2\pi N \omega_u}{I_{CP} K_v} = \frac{2\pi \times 48 \times 2\pi(500 \times 10^3)}{(2 \times 10^3) \times 2\pi(10 \times 10^3)} = 7.54$$

and

$$C_1 = \frac{1}{R_1 \omega_z} = \frac{1}{(2 \times 10^3) \times 2\pi(100 \times 10^3)} = 2 \times 10^3$$

For an overdamped loop, we get the values of α and β can be approximated as

$$\alpha \approx R_1 = \frac{2\pi \times 48 \times 2\pi(500 \times 10^3)}{(2 \times 10^3) \times 2\pi(10 \times 10^3)} = 7.54$$

and

$$\beta = T_{ref} R_1 \omega_z = \frac{7.54 \times 2\pi(100 \times 10^3)}{50 \times 10^6} = 0.0947$$

Note that the values of α and β are the same as the results in Example 10.1 with the assumption of an overdamped loop. The results of two examples show that both the DPLL models based on the time-continuous approximation and the CP-PLL analogy are valid in analyzing the loop dynamics of the DPLL under the condition that the loop bandwidth is substantially lower than the reference frequency.

10.1.2 TDC

10.1.2.1 Time Resolution and Phase Noise

The conventional TDC shown in Fig. 10.2 consists of inverter-based delay cells and DFFs. To achieve the finest resolution offered by CMOS technology, the delay cell is typically implemented by one or two inverters. The time resolution of the

TDC is an important parameter for the in-band phase noise and open-loop gain of a DPLL. To intuitively understand how the phase noise is related with the TDC resolution, let us take a heuristic example.

Suppose that a TDC having poor resolution is designed in a DPLL. We consider two cases of initial conditions. The first case is that the initial phase of a reference clock sits in the middle of two phase boundaries. For instance, the rising edge of a reference clock is in the middle of D_3 and D_4 in Fig. 10.2. Since the time interval between two quantized phases is long for the poor-resolution TDC, it takes time for the phase of the reference clock to touch the phase boundary unless the jitter of the DCO or the reference clock is substantial. In other words, the TDC does not generate any output until the reference phase crosses any phase boundary set by the delay cells. If there is no output from the TDC, the DCO operates like an open-loop DCO without getting any control bit from the DPLL. As a result, the phase noise of the DCO increases since the DPLL cannot suppress the phase noise of the DCO based on the noise transfer function of the PLL. This phenomenon is quite similar to the PFD dead-zone problem in a CP-PLL. Hence, the bandwidth of the DPLL becomes narrow due to the reduced TDC gain. Now, we consider the second case that the initial phase of the reference clock is near the phase boundary. In that case, it is likely that the TDC output is toggled around the phase boundary, making the DPLL operate like the one with a 1-bit TDC, namely bang-bang phase detector (BBPD). The TDC has a large PD gain like a phase-domain comparator, resulting in a wide loop bandwidth. Two cases show that achieving a fine TDC resolution is critical to obtain the linear loop dynamics of the DPLL with good noise performance.

Assuming that the TDC resolution is fine enough to be considered a linear TDC for the given clock jitter from a DCO and a reference source, let us formulate a phase noise equation. With the uniform distribution of TDC quantization, we have the variance σ_{tdc}^2 of quantization noise is given by

$$\sigma_{\text{tdc}}^2 = \frac{(t_{\text{res}})^2}{12} \tag{10.22}$$

Phase noise power normalized to T_{out} is expressed as

$$\sigma_{n,\text{out}}^2 = \left(2\pi \frac{\sigma_{\text{tdc}}}{T_{\text{out}}} \right)^2 \tag{10.23}$$

Over the frequency span from DC to Nyquist frequency, that is $0.5 f_{\text{ref}}$, the single-sideband spectral density of the TDC noise at the DCO phase noise is given by

$$L = \frac{\sigma_{n,\text{out}}^2}{f_{\text{ref}}} = \frac{(2\pi)^2}{12} \left(\frac{t_{\text{res}}}{T_{\text{out}}} \right)^2 \frac{1}{f_{\text{ref}}} \tag{10.24}$$

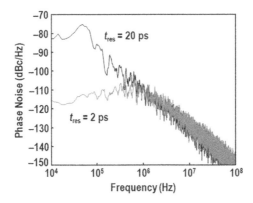

Figure 10.7 TDC resolution effect on the phase noise and bandwidth.

The above equation implies two things. For a given output frequency, poor TDC resolution degrades phase noise at the DPLL output. For a given TDC resolution, phase noise at the DPLL output becomes worse with higher output frequency like the PD noise in a conventional PLL. Figure 10.7 shows the simulated phase noise plots with different TDC resolutions. When a TDC resolution of 2 ps is used, in-band phase noise below −110 dBc/Hz could be achieved. If the TDC resolution is increased by ten times, the in-band phase noise is increased by more than 30 dB. Therefore, designing a fine-resolution TDC is one of the most important tasks for the design of a low-noise DPLL.

10.1.2.2 Enhanced Architectures and Variations
In addition to the phase noise, the TDC must provide much finer phase resolution than a DCO to obtain the linear loop dynamics of a DPLL. When the required TDC resolution is limited by CMOS technology, circuit-level solutions should be considered. There are numerous TDC architectures to overcome the limited time resolution set by technology. Figure 10.8 shows two typical TDC circuits that enhance the time resolution of the conventional TDC. The first one in Fig. 10.8(a) employs a pseudo-differential DFF so that generating quantized phases is insensitive to PMOS and NMOS mismatches. In addition, the time resolution of the TDC is reduced to the gate-delay time of a single inverter instead of two inverters. The second one is a Vernier TDC shown in Fig. 10.8(b). If we use two kinds of inverters having different gate-delay times for the upper and bottom delay lines, a TDC resolution less than the gate-delay time of a single inverter can be achieved even without using the pseudo-differential logic. As depicted in Fig. 10.8(b), the TDC resolution is set by the delay difference of two inverters, $\tau_1 - \tau_2$, where τ_1 and τ_2 are the gate-delay times of the upper and bottom inverters, respectively. However,

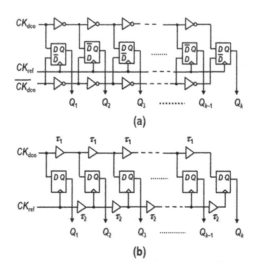

Figure 10.8 TDC with enhanced resolution: (a) pseudo-differential-delay TDC and (b) Vernier TDC.

maintaining good matching among inverters for such a fine resolution requires careful circuit design over process, voltage, and temperature (PVT) variations.

Figure 10.9 shows some nonconventional TDC architectures. A TDC shown in Fig. 10.9(a) is based on a gated ring oscillator. Instead of a simple time-gating function, the residual phase at the time of gating is stored when a ring oscillator is disabled and used as an initial phase at the next enable period. Since the residual phase (Δ) from quantization is accumulated (Σ) for next cycles like the $\Delta\Sigma$ modulator, the gated ring oscillator exhibits a noise-shaping property with respect to the quantization noise of the TDC. As the high-frequency quantization noise is suppressed by a digital loop filter, a high signal-to-noise ratio (SNR) is achieved, thus improving the effective dynamic range of the TDC within the DPLL. The second TDC shown in Fig. 10.9(b) employs an injection-locked ring oscillator instead of delay cells for the generation of quantized phases. Thanks to the property of the injection-locked oscillator, the output period of a DPLL is always an integer multiple of the minimum time step of the TDC regardless of the output frequencies of the DPLL. Therefore, the gain normalization of a TDC is not needed, and a stable TDC gain is obtained over PVT variations. The third one shown in Fig. 10.9(c) performs a two-step conversion where a time amplifier is employed after a first-stage TDC to make a second-stage TDC achieve fine resolution with relaxed condition. The basic operation of the two-step TDC architecture is analogous to that of a two-step flash analog-to-digital converter (ADC) or sub-ranging ADC where a digital-to-analog converter (DAC) amplifies the residual error of a first-stage ADC. Recently, more complex TDC architectures

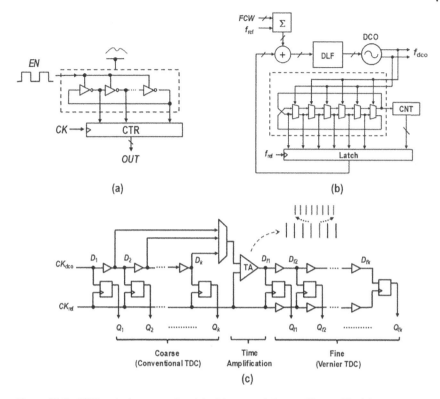

Figure 10.9 TDC variation examples: (a) with a gated ring oscillator; (b) with an injection-locked oscillator; and (c) with a time amplifier and two-step conversion.

such as a two-dimensional Vernier TDC, an ADC-assisted TDC, a high-order MASH-modulated TDC, and a TDC assisted by a digital-to-time converter (DTC) could be found in the literature. Having so many proposed TDC architecture simply indicates how the TDC design is critical for the overall performance of the DPLL.

10.1.3 DCO

10.1.3.1 Basic Architectures

Figure 10.10 shows two typical design approaches for the design of a DCO. The first one shown in Fig. 10.10(a) realizes the digital control of an output frequency by having a digitally-controlled capacitor array in the case of an LC DCO or a digitally-controlled resistor or capacitor array for a ring DCO. To achieve a fine frequency resolution as well as a wide tuning range, the digitally-controlled array consists of many passive elements and switches, which increases parasitic

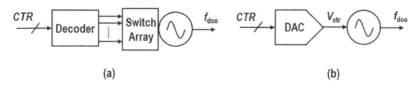

Figure 10.10 DCO topologies: (a) switched-array based and (b) DAC based.

capacitance. If advanced CMOS technology is not used, the portion of the parasitic capacitance could be substantial, making it difficult to design a high-performance DCO. The second architecture shown in Fig. 10.10(b) employs a DAC to control a VCO with a digital input. Unlike the first one, the DAC-based DCO does not need a passive-element array, and the DCO design is similar to the VCO design if the DAC does not limit the DCO performance. Unfortunately, the DAC in front of the VCO adds significant phase noise at the VCO output since the DAC noise is amplified by a VCO gain. To mitigate the DAC noise, a low-pass filter can be put between the DAC and the VCO. The 3-dB corner frequency of the low-pass filter should be much lower than the bandwidth of the DPLL so that an extra pole does not affect stability. If a simple *RC* filter is used as the low-pass filter, the noise contribution of *R* will be low-pass filtered by the *RC* network and band-pass filtered by the loop dynamics of the DPLL at the same time. In general, the challenge of design complexity and the problem of noise contribution of the DAC make the array-based DCO architecture shown in Fig. 10.10(a) more popular when an advanced CMOS technology is available.

Figure 10.11 shows the schematic of a typical LC DCO core having capacitor arrays. Depending on the number of unit capacitors set by an input control word, different output frequencies are generated with fine resolution. If we represent all parasitic capacitance and main capacitance by a single capacitor, the DCO frequency f_{dco} is simply given by

$$f_{dco} = \frac{1}{2\pi\sqrt{LC}} \tag{10.25}$$

With the unit capacitance change ΔC in the capacitor array, the output frequency variation Δf_{dco} is approximated as

$$\Delta f_{dco} = -\frac{\Delta C}{2C} f_{dco} \tag{10.26}$$

For example, we consider an LC DCO with $f_{dco} = 2.4\,\text{GHz}$, $L = 1.2\,\text{nH}$, and $C = 3.6\,\text{pF}$. Then, to have a frequency resolution Δf_{dco} of 10 kHz, a capacitance of about 30 aF is required. As discussed previously, the frequency-interpolation method with $\Delta\Sigma$ modulation could relax the required value of unit capacitance. To enhance the dynamic range of the DCO frequency, two kinds of capacitor arrays can be used as shown in Fig. 10.11. The capacitor array consisting of

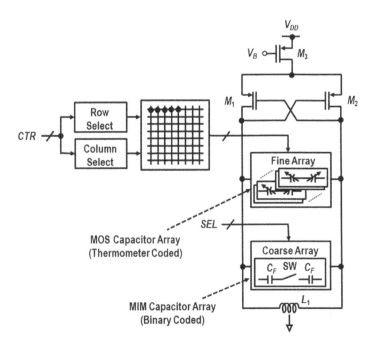

Figure 10.11 LC DCO schematic.

MOS capacitors is the normal one that is to be controlled by a loop filter and determines a DCO gain. To achieve good matching and linearity, it is important to use a thermometer-coded configuration with the same unit capacitors. The other capacitor array is used for coarse-tuning. This capacitor array is mainly to provide the initial calibration of center frequencies over different frequency channels or process variation. Accordingly, the array of metal-to-metal capacitors that have much higher minimum capacitance than the MOS capacitor can be used with a binary-coded control to save the capacitor area.

Figure 10.12 shows two ring DCO topologies. The first one shown in Fig. 10.12(a) employs an inverter array. By having a matrix configuration with serial and parallel connections, the oscillation frequency of the ring DCO is reduced with serial connection or increased with parallel connection. It is because the serial connection increases the number of stages in the ring DCO, while the parallel connection reduces the effective gate delay time of each delay cell. The second topology shown in Fig. 10.12(b) uses capacitor and resistor arrays to control the delay time of a pseudo-differential delay cell. The digitally controlled capacitor array consisting of MOS capacitors and logic gates provides variable load capacitance at the output of each delay cell. The digitally-controlled resistor array changes the effective supply voltage of the delay cell. To save the area, linear-region PMOS

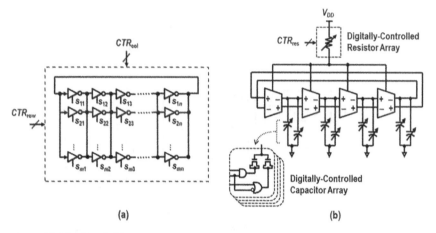

Figure 10.12 Ring DCO schematic: (a) using an inverter matrix and (b) using the programmable resistor and capacitor arrays.

transistors can be used instead of resistors. Note that this topology can also be used with singed-ended delay cells.

10.1.3.2 Phase Noise Due to Quantization

In the DPLL, a quantized frequency error Δf is generated by a DCO. If a nonzero frequency error is inevitable due to quantization, the quantized frequency error should be minimized not to affect the performance of coherent communication systems. At the same time, the DCO needs to provide a wide tuning range in many applications. Therefore, the required dynamic range of the DCO is typically in the range of 14–16 bits, which is quite challenging with current CMOS technology. To alleviate the resolution requirement of a DCO circuit, an oversampled modulation method is employed. For that, a $\Delta\Sigma$ modulator is commonly used to modulate the least-significant bit (LSB) of a control word to generate an interpolated frequency with noise-shaped dithering, that is, high-frequency jitter only.

Before considering the $\Delta\Sigma$ modulation effect on phase noise, let us begin with the generic quantization noise of a DCO. We model the quantization noise at the DCO output in the frequency domain and have it converted to phase noise in the phase domain as depicted in Fig. 10.13. Similar to what we did for the noise analysis of the TDC, the variance $\sigma_{n,f}^2$ of DCO quantization noise for a given frequency resolution f_{res} can be obtained by assuming the uniform distribution of DCO quantization, that is,

$$\sigma_{n,f}^2 = \frac{(f_{\text{res}})^2}{12} \tag{10.27}$$

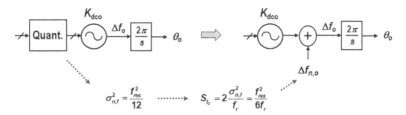

Figure 10.13 DCO resolution effect on phase noise.

Over the frequency span from DC to Nyquist frequency, that is $0.5f_{\text{ref}}$, the single-sideband spectral density L_f of the frequency quantization noise is given by

$$L_f = \frac{\sigma_{n,f}^2}{f_{\text{ref}}} = \frac{f_{\text{res}}^2}{12f_{\text{ref}}} \tag{10.28}$$

To obtain phase noise, $2\pi/s$ term should be considered as illustrated in Fig. 10.13. At an offset frequency f_m outside the loop bandwidth, the phase noise $L(f_m)$ is given by

$$L(f_m) = \frac{\sigma_{n,f}^2}{f_{\text{ref}}}\frac{1}{f_m^2} = \frac{1}{12}\left(\frac{f_{\text{res}}^2}{f_m}\right)^2\frac{1}{f_{\text{ref}}} \tag{10.29}$$

Note that 2π is removed in (10.29) as $L(f_m)$ is used instead of $L(\omega_m)$. The above equation is based on an impulse signal at the DCO input to have white-noise assumption. In fact, the DCO input is held constant until the TDC updates control bits. Considering such a zero-order hold pattern, we add the sinc function in (10.29). That is,

$$L(f_m) = \frac{1}{12}\left(\frac{f_{\text{res}}^2}{f_m}\right)^2\frac{1}{f_{\text{ref}}}\left(\text{sinc}\frac{f_m}{f_{\text{ref}}}\right)^2 \tag{10.30}$$

Now, we consider DCO noise with the $\Delta\Sigma$ modulation. Since the $\Delta\Sigma$ modulation is directly applied to the input of the DCO, the quantization noise of the $\Delta\Sigma$ modulator cannot be filtered by the loop filter of the DPLL. Therefore, the clock frequency f_{clk} of the $\Delta\Sigma$ modulator should be much higher than the reference frequency so that the shaped quantization noise is pushed to a very high frequency and does not affect the phase noise performance of the DCO. Typically, a quarter or an eighth of the DCO frequency is used for the clock frequency. By including an Lth-order $\Delta\Sigma$ modulator operating at f_{clk}, the phase noise becomes

$$L(f_m) = \frac{1}{12}\left(\frac{f_{\text{res}}^2}{f_m}\right)^2\frac{1}{f_{\text{clk}}}\left(2\sin\frac{\pi f_m}{f_{\text{clk}}}\right)^{2L} \tag{10.31}$$

In practice, a simple first-order MASH modulator is employed to modulate the DCO because of the high-speed operation.

10.2 DPLL with 1-Bit TDC

The time resolution of the TDC is important to the DPLL for achieving a low in-band phase noise and the concern of not inducing a narrower-than-expected bandwidth, which is somewhat analogous to the dead-zone problem of the PFD in the CP-PLL. For that reason, the TDC design is quite complex, resulting in various TDC architectures as discussed. To the contrary, a 1-bit TDC, so called a BBPD, greatly simplifies the DPLL architecture as shown in Fig. 10.14 without requiring high linearity and fine resolution for phase detection. The loop dynamics of a BBPD-based DPLL (BB-DPLL), however, is nonlinear since the BBPD produces only early-late information without concerning the amount of an actual phase error. The theoretical analysis of the BB-DPLL is quite complex yet well studied with good explanation in the literature including papers by Da Dalt and Levantino. We will discuss the key properties of the BB-DPLL based on established results and also consider practical design aspects for frequency synthesis and modulation.

10.2.1 Loop Behavior of BB-DPLL

10.2.1.1 Limit-Cycle Regime

Since the BBPD performs binary phase detection, the gain of the BBPD should be very high like a time-domain comparator. However, the gain of the BBPD will decrease if random jitter is added at the BBPD input. Let us begin with the loop behavior of a BB-DPLL without the random noise. In that case, the BBPD behaves as an ideal hard limiter in the time domain, and the noise performance of the BB-DPLL is dominated by quantization noise. Figure 10.15 shows a BB-DPLL model with quantization-induced jitter only. Suppose that there is no latency in a DLF due to extra clock cycles, that is $D = 0$. Then, peak-to-peak jitter Δt_{pp} is dominated by a proportional-gain path and given by

$$\Delta t_{pp} = 2\alpha N K_{vt} \tag{10.32}$$

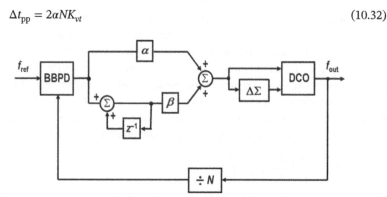

Figure 10.14 DPLL with a 1-bit TDC.

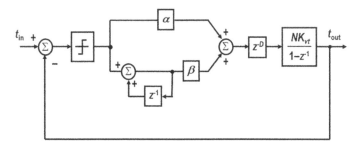

Figure 10.15 BB-DPLL model with quantization-induced jitter only.

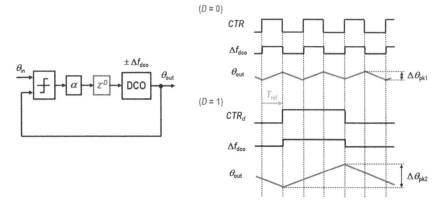

Figure 10.16 Limit-cycle jitter with and without loop latency (first-order loop assumed).

where K_{vt} is expressed in units of seconds per bits. Note that the loop quantization of the BB-DPLL is not determined by TDC resolution but by DCO granularity.

Figure 10.16 illustrates how with the loop latency. For the sake of simplicity, a first-order BB-DPLL without a frequency divider is assumed. In the first-order loop, the output of the BBPD directly toggles the output frequency of the DCO with a step frequency Δf_{dco}. As depicted in Fig. 10.16, the output phase θ_{out} exhibits a periodic tone as it integrates the toggling frequency of the DCO. When there is an extra clock delay in the loop filter, the amount of jitter generation will increase since the delayed transition from an up output to a down output, or vice versa, increases the toggling period as illustrated in Fig. 10.16. With such a fixed pattern and dependency on the loop latency, the BB-DPLL operates in a so-called *limit-cycle regime*. With the additional number of clock cycles, the peak-to-peak time error Δt_{pp} is given by

$$\Delta t_{pp} = 2(1 + D)\alpha N K_{vt} \tag{10.33}$$

where D is the number of clock cycles in a feedforward loop. It is shown that a time error at the steady state has a uniform distribution around the threshold at the BBPD input in the first- or second-order BB-DPLL. With the uniform distribution of the time error, the variance $\sigma_{t,lc}^2$ of the time error at the BBPD input is given by

$$\sigma_{t,lc}^2 = \frac{(\Delta t_{pp})^2}{12} \tag{10.34}$$

From (10.33) and (10.34), the quantization-induced RMS jitter is given by

$$\sigma_{t,lc} = \frac{(1+D)\alpha N K_{vt}}{\sqrt{3}} \tag{10.35}$$

In addition to the quantization-induced jitter, a large spur or noise peaking could be observed in the spectrum of the BB-DPLL in the limit-cycle regime. The limit-cycle frequency f_{lc} is shown to be related with the loop latency and the reference frequency and bounded by

$$f_{lc} \leq \frac{f_{ref}}{2(D+2)} \tag{10.36}$$

It is also known that a heavily overdamped loop by minimizing the integral-path gain β is desirable to achieve low jitter for the given α and D in the design of the BB-DPLL.

10.2.1.2 Random-Noise Regime

Let us consider the case with the random noise, namely *random-noise regime*. In the random-noise regime, it is shown that the average output of a BBPD follows the average time error and that the PD characteristic of the BBPD gets smoothened near the threshold. Figure 10.17 illustrates how the BBPD characteristic is linearized with input jitter. The output of the BBPD E_p is replaced with the averaged output $\overline{E_p}$. If the jitter distribution gets larger, the effective gain of the BBPD becomes lower, showing that the BBPD gain is inversely proportional to the amount of the input jitter.

For quantitative analysis, let us obtain the BBPD characteristic by assuming that a reference clock CK_{ref} is sampled by a divider clock CK_{div} with the random jitter that has Gaussian distribution with the probability density function $p(x)$ given by

$$p(x) = \frac{1}{\sqrt{2\pi}\sigma_t} e^{-\frac{x^2}{2\sigma_t^2}} \tag{10.37}$$

where σ_t is the RMS jitter. As shown in Fig. 10.17, the rising edge of CK_{div} has a chance of generating an opposite output with the probability given by the error function that is defined for Gaussian distribution. Therefore, by using the

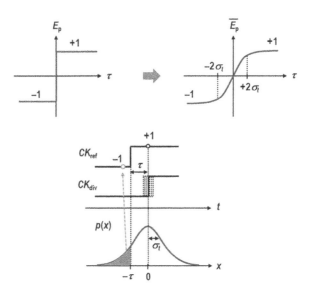

Figure 10.17 BBPD characteristic with input jitter.

even-symmetric property of $p(x)$, the mean value of the BBPD output can be obtained by

$$\overline{E_p}(\tau) = -P(E_p = -1) + P(E_p = +1) = -\int_{-\infty}^{-\tau} p(x)dx + \int_{-\tau}^{+\infty} p(x)dx$$

$$= 2\int_0^\tau p(x)dx = \frac{2}{\sqrt{2\pi}\sigma_t}\int_0^\tau e^{-\frac{x^2}{2\sigma_t^2}}dx \qquad (10.38)$$

As depicted in Fig. 10.17, $\overline{E_p}(\tau)$ shows nearly linear characteristic for $|\tau| < 2\sigma_t$. The BBPD gain K_{td} is obtained by the derivative of at $\tau = 0$, that is,

$$K_{td} = \frac{d\overline{E_p}(\tau)}{d\tau}\Bigg|_{\tau=0} = \frac{d}{d\tau}\left(\frac{2}{\sqrt{2\pi}\sigma_t}\int_0^\tau e^{-\frac{x^2}{2\sigma_t^2}}dx\right)\Bigg|_{\tau=0} = \frac{\sqrt{2/\pi}}{\sigma_t}e^{-\frac{\tau^2}{2\sigma_t^2}}\Bigg|_{\tau=0} = \frac{\eta}{\sigma_t}$$
$$(10.39)$$

where $\eta = \sqrt{2/\pi}$.

Once the small-signal gain of the BBPD is defined, the BB-DPLL can be modeled as done in the conventional DPLL. Figure 10.18 shows the linear model of the BB-DPLL with the TDC gain defined by (10.39). Assuming an overdamped loop, the unity-gain frequency is expressed by

$$\omega_u \approx \sqrt{\frac{2}{\pi}\frac{\alpha N K_{vt}}{T_{\text{ref}}\sigma_t}} \qquad (10.40)$$

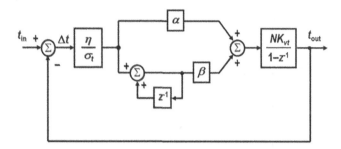

Figure 10.18 BB-DPLL model with random jitter included.

Compared with the conventional DPLL, the BB-DPLL can achieve a wide loop bandwidth without TDC resolution limitation. However, achieving a uniform bandwidth control is difficult as the BBPD gain depends on the input jitter. Accordingly, a background calibration method is employed to control the loop bandwidth over PVT variations by using the $\Delta\Sigma$ quantization error as a training sequence.

Since the BB-DPLL generates large spurs or noise peaking in the limit-cycle regime, it is good to have it operate in the random-noise regime by having the quantization-induced jitter less than the random-noise-induced jitter. That is,

$$\sigma_{t,lc} = \frac{(1 + D)\alpha NK_{vt}}{\sqrt{3}} \leq \sigma_t \tag{10.41}$$

For the given K_{vt} of a DCO, the loop latency should be minimized. Once the loop latency is fixed, the loop bandwidth should be optimized by considering both the quantization-induced jitter and the random-noise-induced jitter. From (10.41), it is good to note that the required DCO granularity for the given random jitter and loop latency is set by

$$\alpha K_{vt} \leq \frac{\sqrt{3}\sigma_t}{N(1 + D)} \tag{10.42}$$

The random-noise-induced jitter could be dominated by DCO noise in the frequency synthesizer, while it is dominated by data noise in the CDR PLL.

10.2.2 Fractional-N BB-DPLL

Even though the function of a BB-DPLL is simple with binary phase detection, designing a fractional-N BB-DPLL is not straightforward. In fact, the strong advantages of the BB-DPLL over the conventional DPLL get lost when the BB-DPLL is designed for fractional-N frequency synthesis. It is because a multi-modulus divider controlled by a $\Delta\Sigma$ modulator generates multiple phases with a fixed time interval of a DCO period. The instantaneous phase variation at the output of the multi-modulus divider caused by the $\Delta\Sigma$ modulation is considered the deterministic jitter (DJ) that does not contain multiple of the DCO period, the DJ contribution to the BBPD output is much bigger than the RJ contribution, which makes the BB-DPLL operate in the limit-cycle regime. The BBPD output is toggled mostly by the DJ rather than the RJ, making the high-pass noise transfer characteristic of the PLL not effective on the DCO noise. As a result, the in-band phase noise is significantly increased with a narrow loop bandwidth, which is analogous to the dead-zone problem of the PFD in the conventional CP-PLL. Figure 10.19 illustrates how a BB-DPLL operates dominantly in the limit-cycle regime due to the $\Delta\Sigma$-modulated multi-modulus divider. It also shows the phase noise comparison of the integer-N and fractional-N BB-DPLLs. The simulation result verifies that the loop bandwidth and the in-band phase noise of the fractional-N BB-DPLL are more than 20 times narrower and more than 30-dB worse than those of the integer-N BB-DPLL, respectively.

To have the BB-DPLL operate in the random-noise regime, the DJ component should be removed before the BBPD. Like the DTC-based cancellation in the conventional $\Delta\Sigma$ fractional-N PLL, the corresponding phase error can be cancelled in the time domain by using the DTC since the output pattern of the $\Delta\Sigma$ modulator is predictable for a given control word. Figure 10.20 shows the functional block diagram of the $\Delta\Sigma$ fractional-N BB-DPLL that operates in the random-noise regime. As the fixed pattern of the phase errors is cancelled by the DTC, the output of the BBPD is solely determined by the random noise at the input. Accordingly, the BB-DPLL does not suffer from the degraded in-band phase noise and reduced loop bandwidth. However, to properly cancel the phase pattern generated

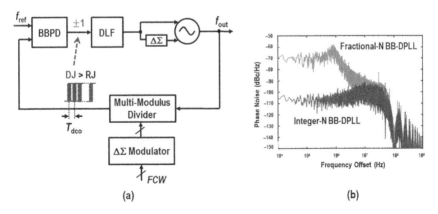

Figure 10.19 $\Delta\Sigma$ fractional-N BBPLL: (a) peak phase error due to modulation (DJ > RJ) and (b) in-band phase noise comparison of the integer-N and fractional-N BB-DPLLs.

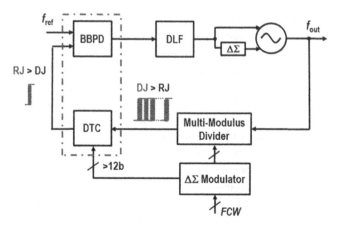

Figure 10.20 BB-DPLL with the DTC for linearized PD gain.

by the $\Delta\Sigma$ modulator, the dynamic range of the DTC should be high, typically requiring 12-bit or higher linearity. It would be very challenging to design such a high-linearity DTC with inverters and digitally-controlled capacitors over PVT variations. Therefore, the DTC design is critical for the fractional-N BB-DPLL, and digital calibration is a must for a high-performance DTC. The digital calibration can significantly enhance the DTC linearity, but a long calibration time could be problematic for fast-settling applications.

10.2.3 Different Design Aspects of BB-DPLL

To leverage the feature of the 1-bit TDC for low-power robust frequency generation systems, a $\Delta\Sigma$ fractional-*N* BB-DPLL that does not rely on the linearity and fine resolution of a DTC would be desirable. So, let us discuss what else we can do to improve the in-band noise performance of the $\Delta\Sigma$ fractional-N BB-DPLL without the high-resolution DTC or at least to relax the design requirement of the DTC. For that, we need to discuss some different design aspects of the BB-DPLL in comparison with conventional linear PLLs including the linear DPLL for fractional-N frequency synthesis and modulation.

10.2.3.1 SLDSM versus MASH Modulation

In the conventional fractional-*N* PLL, both the single-loop $\Delta\Sigma$ modulator (SLDSM) and the MASH modulator have a negligible effect on the in-band phase noise since the quantization noise is much lower than other circuit noise in low frequencies. In the $\Delta\Sigma$ fractional-N BB-DPLL, the in-band noise degradation is mainly due to DJ contribution from the $\Delta\Sigma$ modulator as discussed. Compared with the SLDSM, the MASH modulator generates a wide-spread bit pattern, resulting in

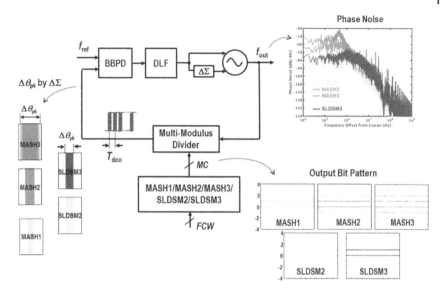

Figure 10.21 In-band noise reduction with the 1-bit output SLDSM.

a large DJ at the output of a multi-modulus divider. Therefore, the fractional-N BB-DPLL with the SLDSM can achieve lower in-band phase noise than that with the MASH modulator. Especially when the SLDSM is designed with a 1-bit quantizer for narrow frequency-range applications, the benefit of the SLDSM could be substantial for either improving the in-band noise or relaxing the dynamic range of the DTC.

Figure 10.21 illustrates how the in-band phase noise of a fractional-N BB-DPLL could be affected by different $\Delta\Sigma$ modulators. The first-, second-, third-order MASH modulators, the second-order, and third-order SLDSM are denoted by MASH1, MASH2, MASH3, SLDSM2, and SLDSM3, respectively. It is shown that in-band phase noise gets worse with wide-spread phase errors as the order of the MASH modulator increases. The MASH1 achieves good in-band phase noise with a 1-bit output but suffers from a large idle tone due to a lack of randomization. To the contrary, the SLDSM2 and the SLDSM3 with the 1-bit quantizer exhibit nearly the same in-band noise performance. When the in-band noise of the SLDSM3 is compared with that of the MASH3, the in-band noise is improved by more than 20 dB. Since the peak phase error is also reduced in the case of the SLDSM3, the dynamic range requirement of a DTC would be relaxed if the DTC is employed for the fractional-N BB-DPLL. The main drawback of the SLDMS3 is a limited input range for stability. Hence, it is good to have a high reference frequency to cover a wide frequency range. We will discuss a two-stage cascaded topology for the fractional-N BB-DPLL to operate with high reference frequency.

10.2.3.2 Two-Stage versus Single-Stage Topology

When cascaded integer-N PLLs are used to synthesize an output frequency with two-step up conversion as shown in Fig. 10.22(a), there is no definite advantage unless the first-stage PLL has a narrow bandwidth with a low-noise VCO and work as a clean-up PLL against a noisy reference source. It is because the noise and spur contributions of each block is nearly the same for output-referred performance. When a $\Delta\Sigma$ fractional-N PLL is used as a second-stage PLL as shown in Fig. 10.22(b), the two-stage architecture has some benefit with respect to quantization noise. Having the $\Delta\Sigma$ modulator operate at much higher frequency, the two-stage PLL achieves less quantization noise effect on out-of-band phase noise than a single-stage PLL does.

Now, we consider the case of having a $\Delta\Sigma$ fractional-N BB-DPLL in the second stage as shown in Fig. 10.22(c). The first-stage integer-N PLL can be either a linear DPLL or a BB-DPLL. In the second stage, the $\Delta\Sigma$ fractional-N BB-DPLL with a high reference frequency reduces not only the out-of-band phase noise but also the in-band phase noise. The classical $20\log N$ scaling rule of the phase noise with the division ratio N no longer holds for the fractional-N BB-DPLL. The BBPD itself generates quantization noise Δq_{BB} because of bi-level phase detection. If the BBPD operates at higher frequency, Δq_{BB} will be pushed to high frequencies and further

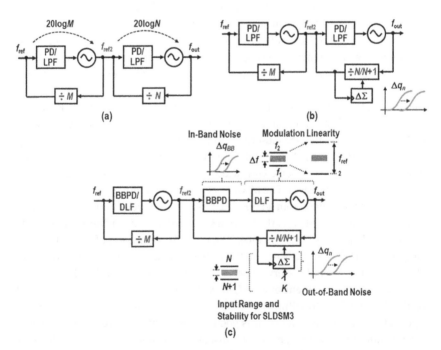

Figure 10.22 Two-stage architecture comparison: (a) cascaded integer-N linear PLLs; (b) integer-N and fractional-N linear PLLs and (c) integer-N and fractional-N BB-DPLLs.

filtered by a DLF. Therefore, for the same total frequency multiplication ratio, the cascaded topology consisting of the integer-N and the $\Delta\Sigma$ fractional-N BB-DPLLs can improve the overall phase noise compared with a single-stage BB-DPLL. In addition, having a high reference frequency is highly useful for the SLDSM3 to improve the input dynamic range as well as stability. If the BB-DPLL is used for frequency modulation, having the high reference frequency is good to enhance modulation linearity since increasing the reference frequency for a given modulation frequency deviation is equivalent to having a small input dynamic range for a given quantizer level as depicted in Fig. 10.22(c).

10.2.3.3 Effect of Phase-Domain Low-Pass Filter

We learned that the PLL behaves like a high-Q auto-tracking band-pass filter. If we put a nested PLL in the feedback path of a $\Delta\Sigma$ fractional-N PLL as shown in Fig. 10.23(a), the nested PLL can work as a phase-domain low-pass filter (PDLPF) to filter the quantization noise of a $\Delta\Sigma$ modulator. The PDLPF mitigates the effect of the quantization noise on out-of-band phase noise. If the PDLPF is used in the BB-DPLL, there is an additional benefit. A peak phase error at the BBPD input can be significantly reduced since high-frequency components of the quantization noise are suppressed. Accordingly, the PDLPF improves not only the out-of-band phase noise but also the in-band phase noise of the $\Delta\Sigma$ fractional-N BB-DPLL.

However, putting the nested PLL within a loop requires complex design and additional area. To achieve the PDLPF with compact area and low power, an injection-locked oscillator (ILO) can be used instead of the nested PLL as shown in Fig. 10.23(b). The ILO that makes the phase of a free-running oscillator locked to the phase of an incoming signal actually behaves like a first-order PLL whose bandwidth is nearly the same as the locking range of the ILO. If cascaded ILOs are used, they provide additional filtering. If an ILO-based divide-by-2 circuit is designed, the divide-by-2 circuit works as a frequency divider as well as the PDLPF.

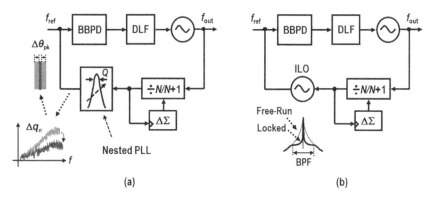

(a) (b)

Figure 10.23 Phase-domain low-pass filtering method: (a) with the nested PLL and (b) with the injection-locked oscillator.

10.2.3.4 Two-Point versus One-Point Modulation

When a modulation rate is much lower than the loop bandwidth, there is no difference between the one-point and two-point modulation methods in the conventional $\Delta\Sigma$ fractional-N PLL. If a $\Delta\Sigma$ fractional-N BB-DPLL with an overdamped loop is designed for frequency modulation, the two-point modulation method can achieve better in-band phase noise than the one-point modulation method. Let us consider an example of a low-rate wideband frequency modulation with a triangular modulation profile. In the case of the one-point modulation, a large phase deviation could be observed at the input of the BBPD since multiple phases are directly generated by the multi-modulus divider, which degrades the in-band phase noise. Unlike the one-point modulation, the role of the $\Delta\Sigma$-modulated multi-modulus divider in the two-point modulation is not to inject modulation but to cancel the low-frequency term of the DCO modulation so that the low-frequency information of the DCO modulation is not suppressed by the high-pass filter characteristic of the PLL. Since a low-rate triangular modulation is cancelled by the $\Delta\Sigma$-modulated multi-modulus divider, no large phase deviation is observed at the output of the frequency divider or at the input of the BBPD as depicted in Fig. 10.24(a). Accordingly, the use of the two-point modulation enables the BB-DPLL to improve the in-band phase noise since the large frequency deviation due to the triangular modulation profile is not seen at the input of the BBPD.

Figure 10.24(b) shows the simulated phase noise performance of a $\Delta\Sigma$ fractional-N BB-DPLL performing 300-kHz triangular modulation with a loop bandwidth of 800 kHz and a zero frequency of 80 kHz. To compare the noise performance of the one-point and two-point modulations, three cases are run; no modulation, one-point modulation, and two-point modulation. Even though the loop bandwidth is sufficiently higher than the modulation rate, the one-point modulator has worse in-band noise than the two-point modulator by about 10 dB, while the two-point modulator maintains almost the same in-band phase noise as the non-modulated BB-DPLL. The simulation result shows that the two-point modulation should be considered even for low-frequency modulation in the case of the $\Delta\Sigma$ fractional-N BB-DPLL.

Below is the summary of different design aspects of the BB-DPLL and other traditional bang-bang PLLs (BBPLLs).

- SLDSM versus MASH modulation
  - Linear PLL: MASH achieves better fractional spur performance
  - BBPLL: SLDSM with 1-bit quantizer achieves better in-band noise
- Two-stage topology versus single-stage topology
  - Linear PLL: Integer-N + $\Delta\Sigma$ fractional-N → Useful to reduce out-of-band noise
  - BBPLL: Integer-N + $\Delta\Sigma$ fractional-N → Useful to reduce in-band noise as well

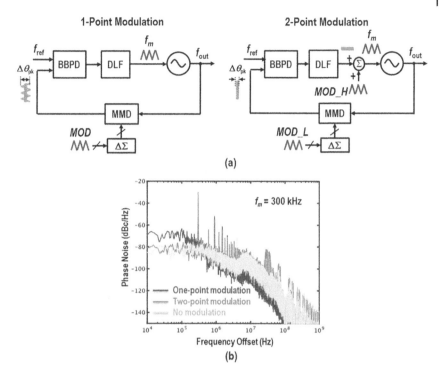

Figure 10.24 Comparison of the 1-point and 2-point modulations in the BB-DPLL: (a) a low-frequency modulation example; (b) phase noise performance.

- PDLPF effect
  - Linear PLL: Useful to reduce out-of-band noise
  - BBPLL: Useful to reduce in-band noise as well
- Two-point modulation versus one-point modulation
  - Linear PLL: No difference with low data rate
  - BBPLL: Two-point modulation achieves better in-band noise

10.3 Hybrid PLL

To design a low-noise DPLL, a high-bit TDC or DTC is a must to control a DCO with fine time resolution over a wide range. Accordingly, design complexity for phase detection is substantial for the design of a low-noise DPLL. Another concern in practice is the poor supply noise sensitivity of the TDC or DTC as their time resolution is determined by the delay time of a single-ended inverter. Figure 10.25 shows an example of how sensitive the TDC resolution is to supply noise variation especially for low-voltage design. In the plot, nearly 90% variation could be

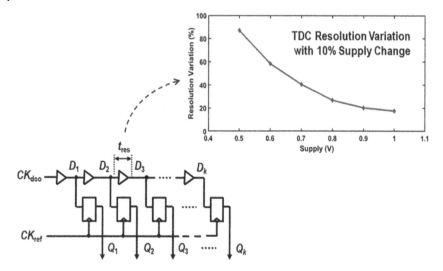

Figure 10.25 TDC resolution variation over supply voltage.

observed with only 50 mV change when 0.5-V supply voltage is used. The poor supply sensitivity could be more problematic in the system-on-chip (SoC) design suffering from the large digital switching noise that contains high-frequency components. For that reason, a huge decoupling capacitor is used for the TDC or DTC in practice, which could nullify the merit of the compact area of the DPLL.

10.3.1 Hybrid Loop Control

In the design of the CP-PLL with advanced CMOS technology, the critical issue lies in the presence of a large capacitor in the integral path, which causes poor technology scalability, large area consumption, and substantial leakage current. Knowing that the main role of the integral path is to achieve frequency acquisition like a large-signal DC-tracking path in a type 2 PLL, we consider replacing the integral path with a digital one, while keeping a proportional-gain path as it is. Since the integral path does not require fast frequency tracking and has less impact on the loop dynamics of the PLL for an overdamped loop, the digital integral path can be implemented with a BBPD and an accumulator.

Figure 10.26 shows a conceptual diagram with single-ended configuration to illustrate how the analog loop filter is converted to a hybrid filter. For the proportional-gain path, the conventional analog control with the PFD and the charge pump followed by a passive loop filter is designed. Since the capacitance values for high-order poles are not high, the area of the loop filter is negligible compared with that a traditional loop filter. To accommodate the hybrid loop

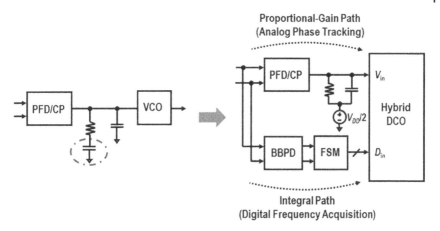

Figure 10.26 Semi-digital PLL with a hybrid loop control.

control, a digital/voltage-controlled oscillator (D/VCO) with analog and digital inputs is employed to form a hybrid PLL (HPLL). Similar to the way of controlling the DCO in the DPLL, the $\Delta\Sigma$ modulator operating at high clock frequency is placed at the digital input of the D/VCO to have a negligible frequency quantization error from the digital integral path. Like the DPLL, the HPLL offers good technology scalability and has negligible leakage current problem with the minimal implementation of passive filters. At the same time, the analog phase tracking offers linear phase tracking, resulting in linear loop dynamics. Simply speaking, the small-signal behavior of the HPLL is mainly determined by the analog control path, while the large-signal frequency acquisition is done by the digital control path. Therefore, the loop bandwidth of the HPLL is similar to that of the conventional CP-PLL when an overdamped loop is designed. More importantly, the in-band phase noise does not vary with the supply fluctuation such as a few tens of mV.

Figure 10.27 shows the linear model of the HPLL. If we assume an overdamped loop, βK_{dco} becomes small, which could make the input jitter σ_t dominant at the BBPD input as deduced from (10.42). Therefore, the BBPD can be linearized with the gain given by (10.39). By combining the transfer functions of the analog proportional-gain and digital integral paths, the open-loop gain of the HPLL $G(z,s)$ is given by

$$G(z,s) = \frac{1}{N}\left(\frac{I_{CP}R_1}{2\pi}\frac{K_v}{s} + \frac{\eta}{\sigma_t}\frac{\beta z^{-1}}{1-z^{-1}}\frac{K_{dco}}{1-z^{-1}} \right) \tag{10.43}$$

Since the loop gain by the digital path is sufficiently lower than the reference frequency, we use the time-continuous approximation based on the phase domain

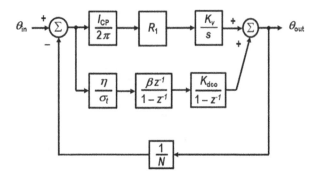

Figure 10.27 Linear model of the HPLL.

model in Fig. 10.4. Then, we obtain an approximated equation as

$$G(s) \approx \frac{1}{Ns} \left(\frac{I_{CP}R_1 K_v}{2\pi} + \frac{\eta}{\sigma_t} \frac{\beta K_{dco}}{s} \right) \tag{10.44}$$

Note that the BBPD is already linearized in Fig. 10.27. Since the digital coefficient β can be set to a very small value, the overall loop behavior of the HPLL can be analyzed based on the analog proportional-gain path only, that is, a type 1 CP-PLL in the initial design.

10.3.2 Design Aspects of the HPLL

Figure 10.28(a) shows the practical implementation of the hybrid control circuits where a fully differential analog control path is designed. If the digital integral path is disabled, the HPLL becomes a third-order type 1 CP-PLL. This is a good example of how to realize a type 1 CP-PLL. When the digital integral path is enabled, it controls the D/VCO with fine frequency resolution with $\Delta\Sigma$ modulation like the DPLL. As a result, a differential control voltage of the analog path becomes very small, which is good for the linearity of the charge pump as well as the varactor in the analog control path of the D/VCO. As shown in the simulation plot in Fig. 10.28(b), the HPLL can achieve frequency acquisition with a fast-settling analog path at the beginning. At this time, the HPLL behaves like a type 1 PLL, exhibiting a large static phase error. Once the digital integral path slowly acquires the frequency and settles, a differential control voltage in the analog control path is nearly 0 V. If another small frequency jump is taken, the digital path tracks the frequency promptly, and a small perturbation in the analog control voltage could be observed. With proper design, the tuning voltage range in the analog path can be very small over entire tuning range if the digital control path sets the desired output frequency of the D/VCO with fine resolution. The settling time of the PLL can

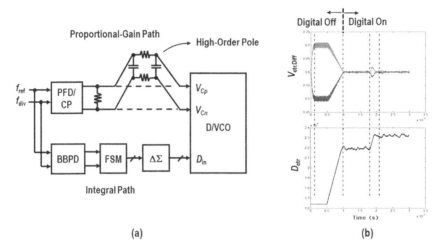

Figure 10.28 Practical implementation of the hybrid loop control: (a) with differential analog path; (b) transient response of analog and digital control paths.

be improved by pre-tuning the D/VCO with an automatic frequency-calibration block in the digital path.

It is worth mentioning that the HPLL architecture can still have the digital calibration feature for the DCO gain and nonlinearity when a fractional-N HPLL is designed for two-point modulation. You may wonder if the hybrid loop control aggravates the gain and delay mismatches of the low-pass and high-pass modulation paths in the two-point modulation loop. In fact, the DCO or VCO non-linearity is important in the high-pass modulation path only. Since the high-pass modulation can be performed by the digital integral path, the DCO nonlinearity calibration feature is still valid in the HPLL-based two-point modulation. The gain and delay mismatches of the low-pass and high-pass modulation paths can also be calibrated by the digital high-pass modulation path. By having two control paths for the D/VCO, improved high-pass modulation or low-pass modulation techniques are proposed in the literature.

Table 10.1 shows the architecture comparison of the CP-PLL, DPLL, and HPLL. The HPLL has good immunity against the leakage current of a loop filter, compact area, and moderate digital calibration feature, while offering linear loop dynamics and robust in-band phase noise performance over supply voltage. Since the TDC or DTC has serious performance degradation under low-supply voltage, the HPLL could be an alternative architecture for digital-intensive design of low-voltage clock generation and modulation circuits if a low-voltage charge pump is designed. For the given fact that a large decoupling capacitor is needed for phase detection in digital-intensive PLL architectures in practice, a

Table 10.1 Architecture comparison.

	CP-PLL	DPLL	HPLL
I_{leak} in LPF	Yes	No	Negligible
Loop Filter Area	Large	Small	Small
Reconfigurable Design	Poor	Good	Fair
Loop Linearity	Good	Fair	Good
In-Band Noise Source	I_{CP}	t_{res}	I_{CP}
In-Band Noise Variation over Supply	Low	High	Low
Technology Dependency	Low	High	Fair

voltage-mode PD can also be considered in the HPLL so that a bias-current-free PLL architecture is implemented. Main difference would be that the in-band noise variation over the PD supply in the HPLL is not as sensitive as that in the DPLL. In addition, the use of advanced CMOS technology is not a must for the design of HPLLs. Therefore, the bias-current-free HPLL with the voltage-mode PD could be a useful architecture for low-voltage or low-cost clock generation and modulation systems.

References

1 R. B. Staszewski and P. T. Balsara, *All-Digital Frequency Synthesizer in Deep-Submicron CMOS*, Wiley-Interscience, Hoboken, NJ, 2006.

2 R. B. Staszewski, C.-M. Hung, D. Leipold *et al.*, "A first multigigahertz digitally controlled oscillator for wireless applications," *IEEE Transactions on Microwave Theory and Techniques*, vol. 51, no. 11, pp. 2154–2164, Nov. 2003.

3 R. B. Staszewski and P. T. Balsara, "Phase-domain all-digital phase-locked loop," *IEEE Transactions on Circuits and Systems II*, vol. 52, no. 3, pp. 159–163, Mar. 2005.

4 R. B. Staszewski, J. Wallberg, S. Rezeq *et al.*, "All-digital PLL and transmitter for mobile phones," *IEEE Journal of Solid-State Circuits*, vol. 40, no. 12, pp. 2469–2482, Dec. 2005.

5 R. B. Staszewski, K. Waheed, F. Dulger *et al.*, "Spur-free multirate all-digital PLL for mobile phones in 65 nm CMOS," *IEEE Journal of Solid-State Circuits*, vol. 46, pp. 2904–2919, Dec. 2011.

6 S. Levantino, "Bang-bang digital PLLs," in *Proc. European Solid-State Circuits Conference*, Sept. 2016, pp. 329-334.

7 V. Kratyuk, P. K. Hanumolu, U.-K. Moon *et al.*, "A design procedure for all-digital phase-locked loops based on a charge pump phase-locked-loop analogy," *IEEE Transactions on Circuits and Systems II*, vol. 54, no. 3, pp. 247–251, Mar. 2007.

8 W. Rhee (ed.), *Phase-Locked Frequency Generation and Clocking: Architectures and Circuits for Modern Wireless and Wireline Systems*, The Institution of Engineering and Technology, United Kingdom, 2020.

9 P. Dudek, S. Szczepanski and J. V. Hatfield, "A high-resolution CMOS time-to-digital converter utilizing a Vernier delay line," *IEEE Journal of Solid-State Circuits*, vol. 35, no. 2, pp. 240–247, Feb. 2000.

10 V. Ramakrishnan and P.T. Balsara, "A wide-range, high-resolution, compact, CMOS time to digital converter," in *Proc. VLSI Design*, Jan. 2006, pp. 197–202.

11 C.-M. Hsu, M. Z. Straayer and M. H. Perrott, "A low-noise wide-BW 3.6-GHz digital fractional-N frequency synthesizer with a noise-shaping time-to-digital converter and quantization noise cancellation," *IEEE Journal of Solid-State Circuits*, vol. 43, pp. 2776–2786, Dec. 2008.

12 A. Rylyakov, J. Tierno, H. Ainspan *et al.*, "Bang-bang digital PLLs at 11 and 20GHz with sub-200 fs integrated jitter for high-speed serial communication applications," in *Proc. IEEE International Solid-State Circuits Conference*, Feb. 2009, pp. 94–96.

13 M. Z. Straayer and M. H. Perrott, "A multi-path gated ring oscillator TDC with first-order noise shaping," *IEEE Journal of Solid-State Circuits*, vol. 44, pp. 1089–1098, Apr. 2009.

14 M. Lee, M. E. Heidari and A. A. Abidi, "A low-noise wideband digital phase-locked loop based on a coarse-fine time-to-digital converter with subpicosecond resolution," *IEEE Journal of Solid-State Circuits*, vol. 44, pp. 2808–2816, Oct. 2009.

15 J. Yu, F. Dai and R. C. Jaeger, "A 12-Bit Vernier ring time-to-digital converter in 0.13 μm CMOS technology," *IEEE Journal of Solid-State Circuits*, vol. 45, pp. 830–842, Apr. 2010.

16 T. Tokairin, M. Okada, M. Kitsunezuka *et al.*, "A 2.1-to-2.8-GHz low-phase-noise all-digital frequency synthesizer with a time-windowed time-to-digital converter," *IEEE Journal of Solid-State Circuits*, vol. 45, no. 12, pp. 2582–2590, Dec. 2010.

17 E. Temporiti, C. Weltin-Wu, D. Baldi *et al.*, "A 3.5 GHz wideband ADPLL with fractional spur suppression through TDC dithering and feedforward compensation," *IEEE Journal of Solid-State Circuits*, vol. 45, no. 12, pp. 2723–2736, Dec. 2010.

18 S.-K. Lee, Y.-H. Seo, H.-J. Park *et al.*, "A 1 GHz ADPLL with a 1.25 ps minimum-resolution sub-exponent TDC in 0.18 CMOS," *IEEE Journal of Solid-State Circuits*, vol. 45, no. 12, pp. 2874–2881, Dec. 2010.

19 M. Zanuso, S. Levatino, C. samori *et al.*, "A wideband 3.6 GHz digital fractional-N PLL with phase interpolation divider and digital spur cancellation," *IEEE Journal of Solid-State Circuits*, vol. 46, no. 3, pp. 627–638, Mar. 2011.

20 A. Elshazly, R. Inti, W. Yin *et al.*, "A 0.4-to-3 GHz digital PLL with PVT insensitive supply noise cancellation using deterministic background calibration," *IEEE Journal of Solid-State Circuits*, vol. 46, pp. 1626–1635, Dec. 2011.

21 F. Opteynde, "A 40nm CMOS all-digital fractional-N synthesizer without requiring calibration," in *Proc. IEEE International Solid-State Circuits Conference*, Feb. 2012, pp. 346–347.

22 G. Marzin, S. Levantino, C. Samori *et al.*, "A 20Mb/s phase modulator based on a 3.6GHz digital PLL with -36dB EVM at 5mW power," *IEEE Journal of Solid-State Circuits*, vol. 47, no. 12, pp. 2974–2988, Dec. 2012.

23 G. Marzin, S. Levantino, C. Samori et al., "A background calibration technique to control bandwidth in digital PLLs," in *Proc. IEEE International Solid-State Circuits Conference*, Feb. 2014, pp. 54–55.

24 V.K. Chillara, Y.-H. Liu, B. Wang et al., "An 860µW 2.1-to-2.7GHz all-digital PLL-based frequency modulator with a DTC-assisted snapshot TDC for WPAN (Bluetooth Smart and ZigBee) applications," in *Proc. IEEE International Solid-State Circuits Conference*, Feb. 2014, pp. 172–173.

25 J. Lee, K. S. Kundert and B. Razavi, "Analysis and modeling of bang–bang clock and data recovery circuits," *IEEE Journal of Solid-State Circuits*, vol. 39, no. 9, pp. 1571–1580, Sept. 2004.

26 N. Da Dalt, "A design-oriented study of the nonlinear dynamics of digital bang-bang PLLs," *IEEE Transactions on Circuits and Systems I*, vol. 52, no. 1, pp. 21–31, Jan. 2005.

27 N. Da Dalt, "Linearized analysis of a digital bang-bang PLL and its validity limits applied to jitter transfer and jitter generation," *IEEE Transactions on Circuits and Systems I*, vol. 55, no. 11, pp. 3663–3675, Dec. 2008.

28 M. Zanuso, D. Tasca, S. Levantino *et al.*, "Noise analysis and minimization in bang-bang digital PLLs," *IEEE Transactions on Circuits and Systems II*, vol. 56, no. 11, pp. 835–839, Nov. 2009.

29 P. Hanumolu, "Clock and data recovery architectures and circuits," *IEEE International Solid-State Circuits Conference*, Tutorial, Feb. 2015.

30 D. Tasca, M. Zanuso, G. Marzin *et al.*, "A 2.9-to-4.0GHz fractional-N digital PLL with bang-bang phase detector and 560fsrms integrated jitter at 4.5mW power," *IEEE Journal of Solid-State Circuits*, vol. 46, no. 12, pp. 2745–2758, Dec. 2011.

31 S. Levantino, G. Marzin and C. Samori, "An adaptive pre-distortion technique to mitigate the DTC non-linearity in digital PLLs," *IEEE Journal of Solid-State Circuits*, vol. 49, no. 8, pp. 1762–1772, Aug. 2014.

32 L. Avallone, M. Mercandelli, A. Santiccioli *et al.*, "A comprehensive phase noise analysis of bangbang digital PLLs," *IEEE Transactions on Circuits and Systems I*, vol. 68, no. 7, pp. 2775–2786, July 2021.

33 N. Xu, Y. Shen, S. Lv et al., "A spread-spectrum clock generator with FIR-embedded binary phase detection and 1-bit high-order ΔΣ modulation," in *Proc. IEEE Asian Solid-State Circuits Conference*, Nov. 2015, pp. 1-4.

34 P. Park, D. Park and S. Cho, "A 2.4 GHz fractional-N frequency synthesizer with high-OSR ΔΣ modulator and nested PLL," *IEEE Journal of Solid-State Circuits*, vol. 47, no. 10, pp. 2433–2443, Oct. 2012.

35 D. Park and S. Cho, "A 14.2 mW 2.55-to-3 GHz cascaded PLL with reference injection and 800 MHz delta-sigma modulator in 0.13 μm CMOS," *IEEE Journal of Solid-State Circuits*, vol. 47, pp. 2989–2998, Dec. 2012.

36 X. Huang, K. Zeng, W. Rhee *et al.*, "A noise and spur reduction technique for ΔΣ fractional-N bang-bang PLLs with embedded phase domain filtering," in *Proc. IEEE International Symposium on Circuits and Systems*, May 2019, pp. 1-4.

37 Y. Liu, W. Rhee, and Z. Wang, "A 1Mb/s 2.86% EVM GFSK modulator based on delta-sigma BB-DPLL without background digital calibration," in *Proc. IEEE Radio Frequency Integrated Circuits Symposium*, Aug. 2020, pp. 7-10.

38 Z. Wan, W. Rhee, and Z. Wang, "Design and analysis of DTC-free ΔΣ bang-bang phase-locked loops," in *Proc. IEEE International Symposium on Circuits and Systems*, May 2021, pp. 1-4.

39 L. Feng, W. Rhee and Z. Wang, "A Quantization noise reduction method for delta-sigma fractional-N PLLs using cascaded injection-locked oscillators," *IEEE Trans. Circuits and Systems II*, vol. 69, pp. 2448–2452, May 2022.

40 M. H. Perrott, Y. Huang, R. Baird *et al.*, "A 2.5-Gb/s multi-rate 0.25-μm CMOS clock and data recovery circuit utilizing a hybrid analog/digital loop filter and all-digital referenceless frequency acquisition," *IEEE Journal of Solid-State Circuits*, vol. 41, no. 12, pp. 2930–2944, Dec. 2006.

41 P.-Y. Wang, J.-H. Zhan, H.-H. Chang *et al.*, "A digital intensive fractional-N PLL and all-digital self-calibration schemes," *IEEE Journal of Solid-State Circuits*, vol. 44, pp. 2182–2192, Aug. 2009.

42 W. Yin, R. Inti, and P. K. Hanumolu, "A 1.6 mW 1.6 ps-rms-Jitter 2.5 GHz digital PLL with 0.7-to-3.5 GHz frequency range in 90 nm CMOS," in *Proc. IEEE Custom Integrated Circuits Conference*, Sept. 2010, pp. 1-4.

43 R. He, C. Liu, X. Yu et al., "A low-cost, leakage-insensitive semi-digital PLL with linear phase detection and FIR-embedded digital frequency acquisition," in *Proc. IEEE Asian Solid-State Circuits Conference*, Nov. 2010, pp. 197-200.

44 W. Yin, R. Inti, A. Elshazly *et al.*, "A TDC-less 7mW 2.5Gb/s digital CDR with linear loop dynamics and offset-free data recovery," *IEEE Journal of Solid-State Circuits*, vol. 46, no. 12, pp. 3163–3173, Dec. 2011.

45 A. Sai, T. Yamaji, and T. Itakura, "A 570fsrms integrated-jitter ring-VCO-based 1.21GHz PLL with hybrid loop," in *Proc. IEEE International Solid-State Circuits Conference*, Feb. 2011, pp. 98-99.

46 Y. Sun, Z. Zhang, N. Xu *et al.*, "A 1.75 mW 1.1 GHz semi-digital fractional-N PLL with TDC-less hybrid loop control," *IEEE Microwave and Wireless Components Letters*, vol. 22, pp. 654–656, Dec. 2012.

47 M. Ferriss, A. Rylyakov, J. A. Tierno *et al.*, "A 28 GHz hybrid PLL in 32 nm SOI CMOS," *IEEE Journal of Solid-State Circuits*, vol. 49, pp. 1027–1035, Apr. 2014.

48 N. Xu, W. Rhee and Z. Wang, "A hybrid loop two-point modulator without DCO nonlinearity calibration by utilizing 1-bit high-pass modulation," *IEEE Journal of Solid-State Circuits*, vol. 49, pp. 2172–2186, Oct. 2014.

49 T.-H. Tsai, R-B. Sheen, C.-H. Chang et al., "A 0.2 GHz to 4 GHz hybrid PLL (ADPLL/charge-pump-PLL) in 7 nm FinFET CMOS featuring 0.619 ps integrated jitter and 0.6 μs settling time at 2.3 mW," in *Proc. Symposium on VLSI Circuits*, June 2018, pp. 183-184.

50 X. Xu, Z. Wang, W. Rhee *et al.*, "A bias-current-free fractional-N hybrid PLL for low-voltage clock generation," *IEEE Transactions on Circuits and Systems I*, vol. 68, pp. 3611–3620, Sept. 2021.

11

Clock-and-Data Recovery PLL

The phase-locked loop (PLL) is the only device that can perform high-Q band-pass filtering with a wide auto-tracking range. As learned from Chapter 5, such a property is useful for clock-and-data recovery (CDR) systems since the PLL retimes a noisy input data with a clean clock extracted from the input data itself. Being a feedback system, the CDR PLL has fundamental trade-offs among jitter generation (JGEN), jitter transfer (JTRAN), and jitter tolerance (JTOL). Figure 11.1 reviews the basic operation and design trade-offs of the conventional CDR PLL. Since those three system parameters depend on the loop bandwidth of the CDR PLL, defining an optimum bandwidth by considering system requirements is important in the design of the CDR PLL. We will also discuss some CDR architectures that overcome the trade-offs among those parameters.

As to the circuit design of a CDR PLL, designing a high-speed phase detector (PD) to work with a non-periodic NRZ data is challenging. For that reason, the architectures of CDR PLLs could be categorized based on phase detection methods. We will discuss three CDR PLL architectures; linear CDR, binary CDR, and baud-rate CDR PLLs. Since the NRZ PDs suffer from limited frequency acquisition, frequency acquisition aid circuits should be considered for the CDR PLLs. In addition, there are some CDR architectures not based on the conventional PLL structure. In the following sections, we will study the jitter characteristics of the conventional CDR PLL associated with the loop dynamics of the PLL. Then, basic CDR architectures as well as advanced structures will be discussed.

11.1 Loop Dynamics Considerations for CDR

11.1.1 JGEN and Noise Sources

The JGEN refers to the amount of jitter generated by the PLL itself with a clean input data, so it depends on how to optimize the phase noise performance of the

Phase-Locked Loops: System Perspectives and Circuit Design Aspects, First Edition.
Woogeun Rhee and Zhiping Yu.

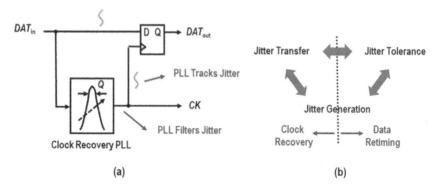

Figure 11.1 PLL-based CDR: (a) conceptual block diagram and (b) design trade-offs in clock recovery and data retiming.

PLL by considering the noise transfer function of each circuit. In the CDR PLL, the noise contribution of a PD is less significant than that of a voltage-controlled oscillator (VCO) since there is no frequency divider. Accordingly, it is good to have a wide bandwidth to suppress the VCO noise by the loop gain of a PLL. However, the PD that takes an NRZ data can generate a pattern-dependent jitter. In the case of a binary PD, there is an additional source of a systematic jitter caused by a limit cycle. For those deterministic jitters, a narrow loop bandwidth is desirable. Therefore, we still need to consider an optimum bandwidth for the JGEN performance of CDR PLLs. From Chapter 5, we learned how to estimate peak-to-peak jitter in terms of a clock period or unit interval (UI) from random jitter for a given bit-error rate (BER), which could set the minimum loop bandwidth to suppress the VCO noise.

11.1.2 JTRAN and Jitter Peaking

The JTRAN determines the jitter filtering performance of a CDR PLL and is characterized by the low-pass filter property of a PLL to an input phase. The JTRAN function from an input phase to an output phase is simply driven by the system transfer function of a PLL. From (2.23) and (2.43), we have the JTRAN function $H_{\mathrm{JTRAN}}(s)$ of a type 2 CP-PLL as

$$H_{\mathrm{JTRAN}}(s) = \omega_n^2 \frac{1 + s/\omega_z}{s^2 + 2\zeta\omega_n s + \omega_n^2} = \frac{K_d' K_v (1 + sR_1 C_1)}{s^2 C_1 + s K_d' K_v R_1 C_1 + K_d' K_v} \tag{11.1}$$

where the 3-dB bandwidth $\omega_{3\mathrm{dB}}$ from (2.29) is

$$\omega_{3\mathrm{dB}} = K_d' K_v R_1 \left(\frac{1}{2} + \frac{1}{4\zeta^2} + \frac{1}{2}\sqrt{1 + \frac{1}{\zeta^2} + \frac{1}{2\zeta^4}} \right)^{\frac{1}{2}} \approx \omega_u \tag{11.2}$$

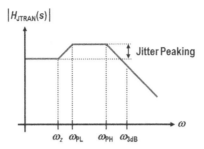

Figure 11.2 Jitter transfer and jitter peaking characteristics of the CDR PLL.

for an overdamped loop, that is, $\zeta > 1$. Compared with clock generators or frequency synthesizers, the CDR PLL should consider another factor, that is, gain peaking near ω_{3dB} in $H_{JTRAN}(s)$, which results in jitter peaking. As discussed in Chapter 5, jitter peaking causes jitter accumulation in the repeaters of optical networks, requiring a heavily overdamped PLL to have a tight control on the jitter peaking. Figure 11.2 depicts the Bode plot of $H_{JTRAN}(s)$ to address the jitter peaking behavior near a loop bandwidth by defining two poles between ω_z and ω_{3dB}. Those poles in low and high bands are denoted by ω_{PL} and ω_{PH}, respectively. For an overdamped loop, they are approximated as

$$\omega_{PL} = \omega_z + \frac{\omega_n}{8\zeta^3} \approx \omega_z \tag{11.3}$$

and

$$\omega_{PH} = \omega_u - \frac{\omega_n}{2\zeta} - \frac{\omega_n}{8\zeta^3} \approx \omega_u \approx \omega_{3dB} \tag{11.4}$$

Then, the amount of jitter peaking JP can be approximated as

$$JP \approx 20\log\left(1 + \frac{1}{4\zeta^2}\right) \approx \frac{8.686}{4\zeta^2} = \frac{8.686}{K'_d K_v R_1^2 C_1} = \frac{8.686}{\omega_u R_1 C_1} \text{ [dB]} \tag{11.5}$$

For example, to meet the synchronous optical network (SONET) requirement, the jitter peaking should be less than 0.1 dB. With a typical loop bandwidth range of 1–5 MHz, the value of C_1 should be several nF based on (11.5).

11.1.3 JTOR and Jitter Tracking

The JTOL characteristic indicates how agile the CDR PLL is to track the phase variation of an input data for data retiming with a tolerable BER. Ideally, the clock phase should track all frequency components of the data phase. In practice, the highest frequency component of the data jitter that the PLL can track is

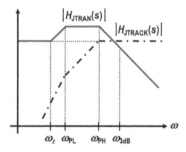

Figure 11.3 Jitter-tracking characteristic of the PLL in comparison with JTRAN.

determined by the loop bandwidth of a PLL. The jitter-tracking transfer function $H_{\mathrm{JTRACK}}(s)$ is the same as the error transfer function of the PLL and given by

$$H_{\mathrm{JTRACK}}(s) = \frac{\theta_e(s)}{\theta_i(s)} = 1 - H_{\mathrm{JTRAN}}(s) \tag{11.6}$$

Therefore, the 3-dB frequency of the jitter-tracking bandwidth ω_{JTRACK} is the same as the 3-dB frequency of $H_{\mathrm{JTRAN}}(s)$

$$\omega_{\mathrm{JTRACK}} = \omega_{3\mathrm{dB}} \approx \omega_{\mathrm{PH}} \approx \omega_{\mathrm{u}} \tag{11.7}$$

where an overdamped loop is assumed. Figure 11.3 shows the Bode plot of $H_{\mathrm{JTRACK}}(s)$ in comparison with $H_{\mathrm{JTRAN}}(s)$ for a type 2 PLL. For the overdamped loop, ω_{PL} and ω_{PH} are nearly determined by ω_z and ω_u, respectively. Therefore, defining an optimum performance between $H_{\mathrm{JTRAN}}(s)$ and $H_{\mathrm{JTRACK}}(s)$ is governed by ω_u, which is similar to the case of the phase noise optimization by considering the noise transfer functions of a reference source and a VCO at the PLL output.

To avoid one-bit-error, a phase error θ_e between the data and clock edges should be approximately less than half the clock period, that is <0.5 UI since the minimum pulse width of the NRZ data is the same as one UI. Therefore, we have

$$\theta_e(s) = H_{\mathrm{JTRACK}}(s)\theta_i(s) < 0.5 \tag{11.8}$$

From (11.6) and (11.8), we express the JTOL transfer function $H_{\mathrm{JTOL}}(s)$ of a type-2 second-order PLL by

$$H_{\mathrm{JTOL}}(s) = \frac{0.5}{H_{\mathrm{JTRACK}}(s)} = \frac{0.5}{1 - H_{\mathrm{JTRAN}}(s)} = \frac{s^2 + 2\zeta\omega_n s + \omega_n^2}{2s^2} \tag{11.9}$$

Figure 11.4(a) shows the Bode plots of $H_{\mathrm{JTRACK}}(s)$ and $H_{\mathrm{JTOL}}(s)$ for a type 2 PLL. The tolerable amount of jitter becomes large as the jitter frequency decreases. Therefore, the critical JTOL performance occurs around $\omega_{3\mathrm{dB}}$ where the PLL loses the jitter-tracking capability. In high frequencies beyond $\omega_{3\mathrm{dB}}$, the noise power of the data jitter itself will be diminished. Figure 11.4(b) shows an example of the

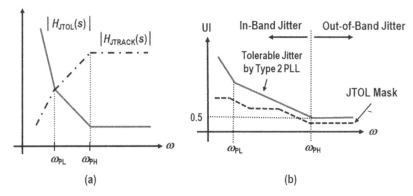

Figure 11.4 Jitter tolerance: (a) JTOL and jitter-tracking characteristics of the type 2 second-order PLL and (b) JTOL mask example.

JTOL mask redrawn from Fig. 5.17(b). If $H_{\mathrm{JTOL}}(s)$ is above the JTOL mask for all frequencies, the CDR PLL satisfies the JTOL requirement. Note that an accurate JTOL measurement below ω_z is not so meaningful as the loop gain of the type 2 PLL is very high, making the measurement of a maximum tolerable phase error limited by the testing equipment.

11.2 CDR PLL Architectures Based on Phase Detection

11.2.1 CDR with Linear Phase Detection

11.2.1.1 Phase Detection with NRZ Data

Designing a PD to detect a phase difference between an NRZ data and a clock is not straightforward since the NRZ data does not contain clock information. A common way is to convert the NRZ signal to an RZ-like signal before phase comparison. If the NRZ signal is differentiated and squared, the RZ-like signal can be generated as illustrated in Fig. 11.5(a). After the conversion, a mixer is used as a PD to generate error information in the voltage domain. If input signals are large enough, we can also generate the RZ-like signal by using digital logic gates, typically designed with current-mode-logic (CML) circuits for high-speed operation. As learned in Chapter 6, the XOR gate works as a PD like a digital mixer. If a delayed NRZ data is compared with the original NRZ data by the XOR gate as shown in Fig. 11.5(b), the overall function is analogous to the combination of a differentiator (edge detection) and a rectifier (unipolar output). Since the second configuration shown in Fig. 11.5(b) would be more robust than the first one in Fig. 11.5(a), it is worth amplifying input signals at the cost of additional power. In

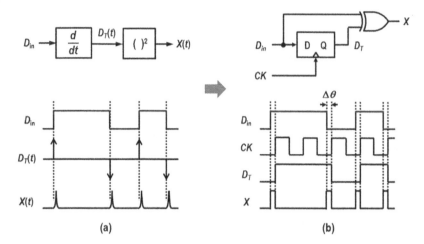

Figure 11.5 Converting NRZ to an RZ-like signal: (a) differentiator and squarer and (b) delay cell and XOR gate.

view of phase detection, the XOR gate in Fig. 11.5(b) generates a U pulse whose rising and falling edges are determined by a data transition edge and a clock rising edge, respectively. Therefore, the pulse width of the U signal is proportional to a phase error between the data transition edge and the clock rising edge. A charge pump can be added after the XOR gate to realize a type 2 PLL with a passive filter. The potential problem of the level-sensitive PDs like the XOR PD is to produce a large pattern-dependent jitter when there is a long sequence of identical bits in the NRZ data. It is because the bi-level output, U and D, of the XOR PD exhibits a long duration of U or D outputs, making the VCO phase drift for the sequence of identical bits.

11.2.1.2 Hogge PD
To mitigate the pattern dependency of the control voltage on the NRZ data, a time window for active phase detection can be defined by having additional logic gates. Figure 11.6(a) shows such a PD consisting of two D-type flip-flops (DFFs) and two XOR gates, named as a Hogge PD. The first DFF generates a retimed data D_T by the rising edge of clock, and the second DFF generates another retimed data D_R by the falling edge of clock. The upper XOR gate generates the U pulse that contains a phase error information as depicted in Fig. 6(b). On the other hand, the bottom XOR gate generates a D pulse whose pulse width simply defines one clock period. If the U and D signals are connected to the charge pump as shown in Fig. 11.6(a), the output of the charge pump represents the difference of two pulse areas, thus conveying the phase offset information between the NRZ data and the clock. Since only the rising edge of the U signal indicates the data transition, the net charge

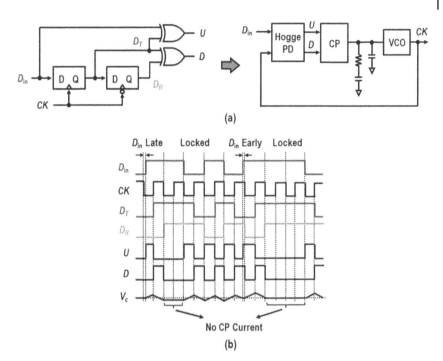

Figure 11.6 Hogge PD: (a) block schematic and its use in the CP-PLL and (b) operation example with different phase conditions.

after the subtraction of two pulses at the charge pump is linearly proportional to a phase error. When the falling edge of the clock is aligned to the data transition edge, the pulse widths of U and D signals are equal, resulting in balanced control voltage V_c as depicted in Fig. 11.6(b).

For a long absence of data transitions, the U and D outputs of the Hogge PD remain low and do not activate the charge pump, thus mitigating the data-pattern effect on the control voltage of a VCO. When the PLL is locked, the transition edge of the NRZ data is phase locked to the falling edge of the clock, and the rising edge of the clock is used to have the data sampled at the middle of transition times. Therefore, an optimum data retiming function is embedded in the Hogge PD. The basic operation of the Hogge PD is similar to that of the flip-flop phase detector (FF-PD), having the linear range of $\pm\pi$ and the PD gain of $V_{DD}/2\pi$. Since an effective PD gain is dependent on the density of data transitions, the loop gain of a PLL also varies. For that reason, many systems specify the maximum number of consecutive identical bits (CID).

The Hogge PD has an inherent phase offset problem for data retiming. As illustrated in Fig. 11.7, a delay from the clock input to the Q output in the first DFF

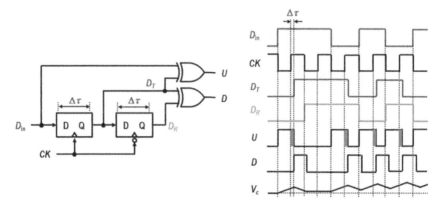

Figure 11.7 Phase offset in the Hogge PD.

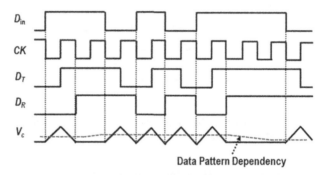

Figure 11.8 Data pattern dependency in the Hogge PD (V_c waveform exaggerated).

causes a phase offset, resulting in an extra pulse width at U output. On the other hand, the pulse with at D output is not affected by the DFF delay since both inputs of the XOR gate are delayed by the same amount. To mitigate this effect, an additional delay circuit needs to be added at the XOR gate to compensate for the DFF delay.

Another problem of the Hogge PD is data-pattern dependency. Even though the Hogge PD has a neutral state for the long absence of data transition, it still exhibits a pattern dependency when the phase lock occurs. Figure 11.8 illustrates how the pattern-dependency problem occurs during the locked state. When the clock and data transition edges are aligned, the U and D outputs are not enabled simultaneously but are generated sequentially. Accordingly, the charge pump generates a positive voltage and then a negative voltage, forming a triangular waveform with a positive charge whenever the data transition occurs. This positive charge generated by the Hogge PD must be compensated by the PLL to maintain the zero

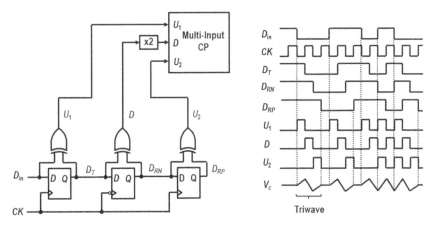

Figure 11.9 Triwave Hogge PD.

frequency error, resulting in the pattern-dependent jitter. The inherent offset problem of the Hogge PD can be mitigated by having additional DFF and XOR gate as illustrated in Fig. 11.9. The output of the XOR gate in the middle is weighted by two to provide twice the negative charge, while the XOR gate on the right generates a positive charge to compensate for the extra negative charge. In this way, a triwave waveform is formed with balanced positive and negative charges for every data transition, thus reducing the pattern-dependent jitter significantly. However, extra circuits with unequal weighting require a careful matching and could be a big burden for low-power design.

11.2.2 CDR with Binary Phase Detection

11.2.2.1 Binary Phase Detection Using a DFF

Consisting of at least two DFFs and two XOR gates, the Hogge PD has difficulty in achieving low-power high-speed operation since the received NRZ data has a low swing in most high-speed interfaces. In addition, the inherent path mismatch in the DFF and the pattern dependency problem in the conventional Hogge PD become critical for high-speed operation. To alleviate the speed and path mismatch problems, the Alexander PD that performs binary phase detection is dominantly used in high-speed CDR circuits.

To begin with, let us consider a simple bang-bang phase detector (BBPD) consisting of a single DFF. The DFF with normal clock and data configuration can work as a PD for periodic signals but fails to operate properly for non-periodic signals such as the NRZ data. If the clock and data inputs are swapped, that is, the NRZ data is fed to the clock path of the DFF as shown in Fig. 11.10(a), then the DFF provides the early-late information of the data in comparison with the clock

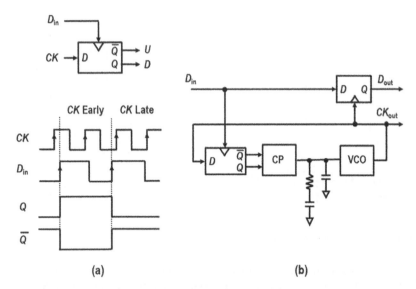

Figure 11.10 Binary phase detection: (a) DFF for binary phase detection and (b) CDR PLL example using two DFFs.

phase. Accordingly, the DFF can work as a PD for the CDR PLL. Figure 11.10(b) shows the simplest CDR PLL that use an input-swapped DFF for phase detection and another DFF for data retiming. However, this simple BBPD causes two serious problems for the CDR circuit. Firstly, the BBPD output stays high or low for a long period in case of the long sequence of identical bits in the NRZ data. Then, the control voltage of a VCO drifts for a long time and causes a serious pattern-dependent jitter. Secondly, the DFF for phase detection takes the NRZ data to the clock path of the DFF, while the DFF for data retiming takes it to the data path of the DFF. Such an asymmetric configuration becomes the source of a phase offset since the phase of a recovered clock is not well aligned with the data retiming in the DFF. Therefore, it is important to have the data retiming function embedded within the PD like the Hogge PD.

11.2.2.2 Alexander PD

Figure 11.11 shows the Alexander PD that outputs early-late information by having three data samples. For the k-th bit period, two data samples, D_k and S_k, are taken by the rising and falling edges of the clock. Hence, the Alexander PD is a BBPD with 2× oversampling. If there is no data transition in the next bit, that is $D_k = D_{k+1}$, both U and D outputs will be high or low, thus not producing a net charge to a loop filter. If there is data transition, the Alexander PD generates either U or D output by checking whether the falling edge of the clock comes earlier or later than the data transition between two rising edges of the clock. The early-late

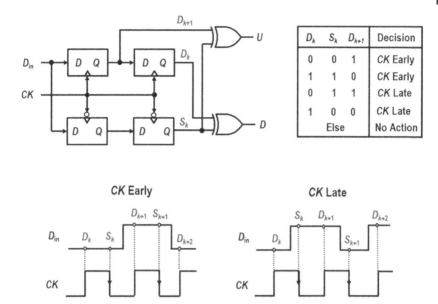

Figure 11.11 Alexander PD.

information can be obtained based on the falling-edge data sample S_k and two rising-edge data samples D_k and D_{k+1}. For example, if $D_k = 0$ and $D_{k+1} = 1$, it indicates that the data transition from low to high occurs between two rising edges of the clock. Accordingly, the falling-edge data sample S_k conveys the early-late information. If $S_k = 0$, it indicates that the falling edge of the clock is ahead of the data transition. If $S_k = 1$, the falling edge of the clock is behind the data transition. If $D_k = 1$ and $D_{k+1} = 0$, the data transition from high to low occurs between the rising edges of the clock, and the S_k value for the early-late decision is opposite in this case.

Compared with the simple BBPD in Fig. 11.10, the Alexander PD has a neutral state where no output is fed to a charge pump or a loop filter. Therefore, the CDR PLL with the Alexander PD has less pattern dependency than that with the simple BBPD. In addition, since both the data retiming and the phase detection are done by the same type of the DFF within the PD, the Alexander PD does not have an inherent clock skew due to the asymmetric DFF configuration like the CDR PLL having the simple BBPD or the Hogge PD. The BBPD is not a linear PD, so the loop gain of a CDR PLL depends on input jitter as discussed in Chapter 10. Assuming 50% transition density and Gaussian distribution, the BBPD gain K_{BBPD} with a supply voltage of V_{DD} can be directly obtained from (10.39). Considering the use of a CP-PLL structure instead of the DPLL, we use the mean value of the PD voltage V_p with the output range of $\pm V_{DD}/2$ instead of ± 1. Then, the BBPD

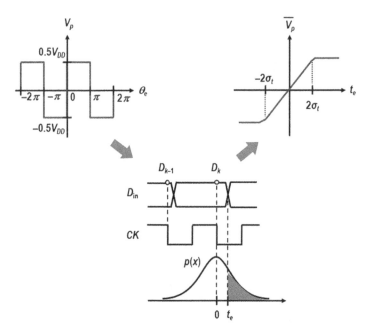

Figure 11.12 Binary PD gain.

gain K_{BBPD} is approximated as

$$K_{\mathrm{BBPD}} = \frac{V_{DD}}{\sigma_t \sqrt{2\pi}} \tag{11.10}$$

Hence, the BBPD can be linearized once noise variance σ_t^2 is well defined. Figure 11.12 illustrates how the BBPD is linearized due to the input jitter. As the BBPD gain depends on the input jitter, the loop gain and the JTRAN also depend on the input jitter. To warrant loop stability or low jitter peaking, it is important to design a heavily overdamped loop for BBPD-based CDR PLLs.

In Chapter 10, we also learned that the peak-to-peak phase error of the BB-DPLL increases with loop latency when the BB-DPLL operates in the limit-cycle regime. Similarly, the loop dynamics of the CDR PLL with the BBPD has a high dependency on the loop latency. In fact, the natural frequency of the PLL is governed by a limit-cycle oscillation whose amplitude is proportional to the loop latency. Figure 11.13 illustrates how the loop latency increases the amplitude of the limit-cycle oscillation. The bigger the loop latency τ_d, the larger is the peak phase deviation $\Delta\theta_{\mathrm{pk}}$. If the loop latency is substantial, a large peaking occurs in the jitter transfer function, resulting in high jitter peaking. Therefore, it is important to minimize the loop latency in the design of the CDR PLL with the BBPD.

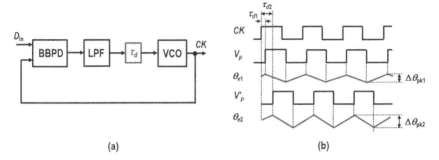

(a)   (b)

Figure 11.13 Limit cycle jitter: (a) BBPLL with loop latency and (b) increased peak-to-peak jitter due to the loop latency.

Some architectures to relax the loop latency by having a feedforward path will be discussed later.

11.2.2.3 Half-Rate Alexander PD

The bandwidth of the NRZ data is half the bandwidth of the clock. If quadrature clock outputs are available, the same number of data samples per UI can be obtained with half the VCO frequency. Therefore, the bandwidths of the data and clock are reduced by half, which relaxes the speed requirement of a PLL as well as the bandwidth of clock-distribution circuits. Figure 11.14 shows a half-rate Alexander PD where the NRZ data is sampled by the quadrature outputs of a VCO. If both the rising and falling edges of two quadrature clocks CK_I and CK_Q are used, data sampling can still be done with one UI of a full-rate clock. A control logic circuit provides up and down signals to control the VCO based on collected four logic outputs, U_1, D_1, U_2, and D_2. The retimed data in the quadrature

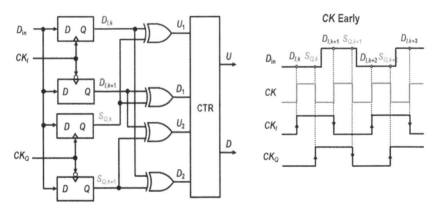

Figure 11.14 Half-rate Alexander PD.

paths are demultiplexed data at half rate, which can reduce the demultiplexing procedure if a CDR PLL is used for serializer-and-deserializer (SerDes) systems. If the quadrature clocks are distributed to multiple CDR circuits, the increased number of clock distribution circuits and matching requirement make it difficult for the half-rate CDR system to achieve significant power reduction except relaxing the maximum frequency of the PLL. The Hogge PD can also be designed as a half-rate PD. Compared with the Alexander PD that takes two data samples per bit, the half-rate Hogge PD can be implemented with complementary clock outputs only, thus not requiring a quadrature VCO. However, the motivation of the half-rate Hogge PD is not as strong as the half-rate Alexander PD when the critical performance of the CDR PLL is governed by the speed and power of the PD itself.

The drawback of the half-rate Alexander PD is the requirement of quadrature clock outputs. If an LC VCO is needed for high-frequency operation, the half-rate Alexander PD significantly increases the chip area. Another problem of the Alexander PD is increased loop latency due to reduced clock frequency. That is, an update rate is halved in the BBPLL, which will increase the amplitude of the limit-cycle oscillation, resulting in large peak-to-peak jitter. Note that the jitter requirement of the half-rate PD is not relaxed with reduced clock frequency since the required phase resolution for data sampling is still normalized to the period of a full-rate clock.

11.2.3 CDR with Baud-Rate Phase Detection

The BBPD achieves fast operation with binary decision but requires two data samples per bit or UI. A baud-rate PD takes one sample per UI as shown in Fig. 11.15(a) and has become a viable PD structure recently for high-speed CDR systems even though the basic concept was published in the 1970s. One of the most popular baud-rate PDs is a Mueller–Muller phase detector (MMPD). The basic operation principle of the MMPD is to estimate the impulse response $h(\tau)$ of a channel from baud-rate data samples and obtain timing information by equalizing the pre-cursor and post-cursor of the pulse response as illustrated in Fig. 11.15(b). For a given bit period T_b, we define a timing function $f(\tau)$ by

$$f(\tau) = (h_1 - h_{-1})/2 \tag{11.11}$$

where h_1 and h_{-1} are $h(\tau + T_b)$ and $h(\tau - T_b)$, respectively. From the Mueller's paper, it is shown that the timing function information can be obtained from binary-format signals by defining a timing estimate $\Delta\tau_k$ and taking the expectation value $E[\Delta\tau_k]$. That is,

$$\Delta\tau_k = (x_k d_{k-1} - x_{k-1} d_k)/2 \tag{11.12}$$

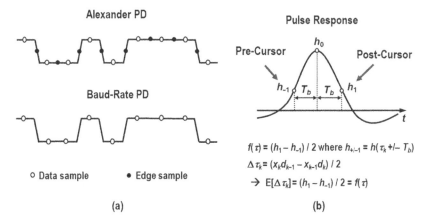

Figure 11.15 Baud-rate PD: (a) comparison with the Alexander PD and (b) concept of the Mueller and Muller PD.

and

$$E[\Delta\tau_k] = (h_1 - h_{-1})/2 = f(\tau) \tag{11.13}$$

where x_k is the kth data sample and d_k is the kth bit decision value of ± 1. If $E[\Delta\tau_k]$ is positive, the post-cursor h_1 is more dominant than the pre-cursor h_{-1}, indicating that the clock sampling is done earlier than the balanced position. Also, the negative $E[\Delta\tau_k]$ indicates that the clock sampling arrives later than the balanced position. Based on the early-late information of the clock sampling, the baud-rate detection can work as a binary baud-rate PD for the CDR PLL. As the post- and pre- cursors of the channel response are highly sensitive to inter-symbol interference (ISI), the MMPD is not as robust as the BBPD that performs binary phase detection based on two data samples per bit.

In practice, the analog data sample x_k is converted into a digital sample y_k by an analog-to-digital converter (ADC) in the MMPD. To further simplify hardware complexity, two-bit data conversion can be used with the following relation:

$$y_k = d_k(e_k + 1) \tag{11.14}$$

where $e_k = 1$ if $x_k > V_{\text{ref}}$ or $x_k < -V_{\text{ref}}$, and $e_k = -1$ if $-V_{\text{ref}} < x_k < V_{\text{ref}}$. Now, the timing estimate $\Delta\tau_k$ is approximated as

$$\Delta\tau_k = \frac{(x_k d_{k-1} - x_{k-1}d_k)}{2} \approx \frac{(y_k d_{k-1} - y_{k-1}d_k)}{2} = \frac{d_k d_{k-1}(e_k - e_{k-1})}{4} \tag{11.15}$$

Therefore, the baud-rate phase detection can be done by obtaining two-consecutive data sign bits and two error bits as illustrated in Fig. 11.16(a), which is also called a sign-sign MMPD. Figure 11.16(b) shows the block diagram of the sign-sign MMPD that consists of comparators, DFFs, and an XOR gate only.

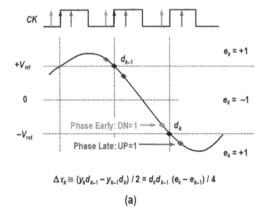

$$\Delta\tau_k \cong (y_k d_{k-1} - y_{k-1} d_k) / 2 = d_k d_{k-1} (e_k - e_{k-1}) / 4$$

(a)

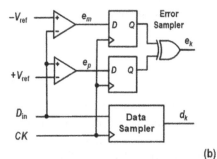

d_k	D_{k-1}	e_k	e_{k-1}	$\theta_{e,k}$
+1	−1	+1	−1	Late
−1	+1	+1	−1	Late
+1	−1	−1	+1	Early
−1	+1	−1	+1	Early
Else				No Action

(b)

Figure 11.16 Simplified MMPD with sign bits: (a) timing diagram and (b) circuit implementation.

If two consecutive sample errors, e_k and e_{k+1}, are different, then the early/late information is simply determined by e_k as shown in the logic table. Like the BBPD, the gain of the MMPD is inversely proportional to input jitter. Unlike the BBPD, the PD gain also depends on the bandwidth and response of the channel. In the sign-sign MMPD, finding an optimum reference voltage V_{ref} based on the pulse response is important for CDR performance.

11.3 Frequency Acquisition

Having the memory of a previous state, the phase-frequency detector (PFD) cannot be used for non-periodic signals. With other memory-less PDs, a type 2 PLL can still achieve frequency acquisition eventually by the pull-in behavior discussed in Chapter 3 unless the pull-in range (capture range) is limited by a circuit or the harmonic locking problem of a PD. However, the pull-in behavior is very slow and not useful in practice. Moreover, the CDR PLL working with an NRZ signal has more difficulty in frequency acquisition than that with periodic input signals. Therefore, a frequency acquisition aid circuit needs to be implemented to help the CDR PLL enter a lock-in range smoothly.

11.3.1 Frequency Detector

To understand how a frequency detector (FD) circuit is designed, let us consider a sinusoidal input with unit amplitude, that is, $\cos(\omega_0 t + \phi)$ with an initial phase offset ϕ for the sake of simplicity. If we multiply it with quadrature signals, $\cos\omega_1 t$ and $\sin\omega_1 t$, as illustrated in Fig. 11.17, the outputs of two mixers after a low-pass filter (LPF) will be $0.5\cos[(\omega_1 - \omega_0)t + \phi]$ and $0.5\sin[(\omega_1 - \omega_0)t + \phi]$, respectively. If $\omega_1 > \omega_0$, then $\cos[(\omega_1 - \omega_0)t + \phi]$ leads $\sin[(\omega_1 - \omega_0)t + \phi]/2$ by 90°, or vice versa. To obtain frequency information in the voltage domain, we differentiate the signal and multiply it with other quadrature signal, which is called a *quadricorrelator*. Since the amplitude of the mixer output depends on the initial phase offset ϕ, we use a symmetric configuration as shown in Fig. 11.18. Using the same signals from Fig. 11.17, the outputs of the upper and bottom mixers become

$$x_1(t) = -\frac{\omega_1 - \omega_0}{4}\sin^2[(\omega_1 - \omega_0)t + \phi] \tag{11.16}$$

and

$$x_2(t) = \frac{\omega_1 - \omega_0}{4}\cos^2[(\omega_1 - \omega_0)t + \phi] \tag{11.17}$$

After subtracting $x_1(t)$ from $x_2(t)$, we get

$$x_{\text{out}}(t) = \frac{1}{4}(\omega_1 - \omega_0) \tag{11.18}$$

To have the quadricorrelator properly work with the NRZ data, an edge detector circuit needs to be implemented in front of the quadricorrelator to convert the NRZ data into an RZ-like signal. Even though the quadricorrelator circuit successfully detects frequency difference, the use of mixers and differentiators in addition to the edge detector is not attractive for low-power CDR systems.

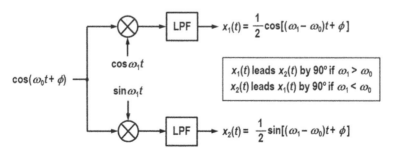

Figure 11.17 Frequency detection with quadrature mixers.

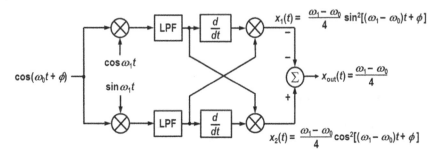

$$x_1(t) = \frac{\omega_1 - \omega_0}{4} \sin^2[(\omega_1 - \omega_0)t + \phi]$$

$$x_{out}(t) = \frac{\omega_1 - \omega_0}{4}$$

$$x_2(t) = \frac{\omega_1 - \omega_0}{4} \cos^2[(\omega_1 - \omega_0)t + \phi]$$

$\cos \omega_1 t$

$\cos(\omega_0 t + \phi)$

$\sin \omega_1 t$

Figure 11.18 Quadricorrelator with symmetric configuration.

Now, we consider the digital implementation of frequency detection. The operation of the simple BBPD shown in Fig. 11.10 is to sample a clock with the transition edge of the data, which is analogous to sampling the clock with the impulse pulse that is obtained by differentiating the data. In other words, the DFF having a data for the clock path and a clock for the data path behaves like a switching mixer. Note that the edge detection function is already included in the DFF. Then, the quadrature mixers in Fig. 11.17 can be converted to quadrature DFFs as shown in Fig. 11.19. To utilize both rising and falling transitions of the NRZ data, a double-edge-triggered (DET) flip flop can be used as depicted in Fig. 11.19. Since the frequency information is obtained by comparing the phases of the upper and bottom mixers in Fig. 11.17, we can also obtain the frequency information by comparing the phases of two DFFs. By having a DFF PD, a bang-bang FD to compare the data rate R_b with the clock frequency f_{CK} can be formed as shown in Fig. 11.20. The bang-bang FD helps the PLL acquire the frequency quickly, but the operating frequency range is limited especially with the NRZ data as the fast cycle slipping is not tolerable for consecutive identical bits.

Double-Edge-Triggered (DET)

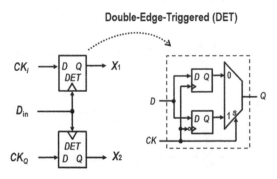

Figure 11.19 Frequency detection with double-edge DFFs.

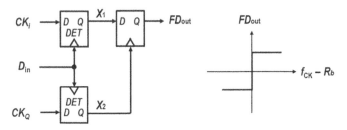

Figure 11.20   Bang-bang frequency detector.

11.3.2   CDR PLL with Frequency Acquisition Aid Circuits

When a CDR PLL is designed with overdamped loop dynamics to minimize the jitter peaking, a large capacitor should be used in the integral path of a type 2 PLL. Since the integral path of the type 2 PLL mainly performs frequency acquisition, the bang-bang FD in Fig. 11.20 can be added with a separate charge pump in the integral path to help frequency acquisition. Figure 11.21 shows the example of the CDR PLL that employs both the PD and the FD. The FD followed by another charge pump is directly connected to the integration capacitor of a loop filter. For the overdamped PLL, it helps reduce the charging and discharging times of a large capacitor during frequency acquisition. During the normal operation of the PLL, the FD circuits need to be disconnected by a control signal *EN* not to disturb the control voltage of a VCO. To further enhance the frequency acquisition, a dynamic bandwidth control could be considered. That is, we set a wider bandwidth during frequency acquisition and use a normal bandwidth optimized for the CDR performance after the frequency acquisition is done.

If a system clock is available for a CDR PLL, a PFD instead of the FD can be used in a frequency-acquisition loop since the FD with the NRZ data suffers from a limited frequency range. Figure 11.22 shows the CDR PLL architecture that employs a conventional CP-PLL for frequency acquisition. Unlike the FD-based acquisition loop, a frequency divider can be used in a feedback path. The CDR PLL employs a

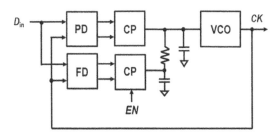

Figure 11.21   Frequency acquisition using the FD.

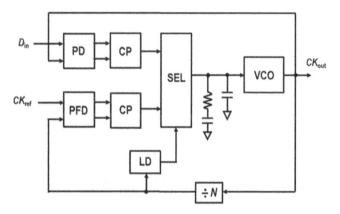

Figure 11.22 Frequency acquisition with an external reference clock.

lock detector to indicate whether the PLL operates within the lock-in range or not. If the PLL is out of the lock-in range, a system clock is selected as an input instead of the NRZ data, and the PLL topology is set for clock generation by the bottom loop that consists of the PFD and the frequency divider. To use the same loop filter, the current of a charge pump should be approximately N times higher than that of the upper charge pump to compensate for the bandwidth reduction due to the divide-by-N circuit. During frequency acquisition, the CDR PLL becomes the conventional CP-PLL that captures a wide frequency range with the PFD, having the VCO tuned close to a desired frequency. Once the PLL operates within the lock-in range, the lock detector selects the normal PD that takes the NRZ data as an input, and the PLL topology becomes a normal CDR PLL. The frequency acquisition loop is fully disconnected during normal operation so that the loop dynamics of the CDR PLL should not be affected by the frequency acquisition loop. We can also consider the dynamic bandwidth control to further reduce the frequency acquisition time with a wider loop bandwidth. However, a large bandwidth difference between two loops can cause overshooting during the transition from a wide-loop bandwidth to a normal loop bandwidth. Such a frequency peaking can make the CDR PLL suffer from additional settling time when it enters the normal operation region.

11.4 DLL-assisted CDR Architectures

The CDR PLL requires a narrow bandwidth to achieve good JTRAN performance with high-Q band-pass filtering, while a wide bandwidth is desired to have the agile phase tracking of an input signal for good JTOL performance. Even though the JTRAN and JTOL are the fundamental tradeoff in choosing a loop bandwidth, the CDR PLL assisted by a delay-locked loop (DLL) can overcome the trade-off. Also, if frequency acquisition can be provided, DLL-based CDR circuits can replace the PLL-based CDR circuits to avoid the crosstalk problem of multiple VCOs for multi-link applications.

11.4.1 Delay- and Phase-Locked Loop (D/PLL)

In the loop dynamics of a type 2 PLL, a zero should be added for loop stability. Because of the zero, the system transfer function always exhibits a jitter peaking problem near a loop bandwidth as illustrated in Fig. 11.2. For some applications like the SONET where a single clock handles data retiming across entire network with repeaters, the jitter peaking in the JTRAN could be problematic since it amplifies the jitter component near the loop bandwidth and deteriorate the overall jitter performance after several repeaters. For that reason, a heavily overdamped loop is designed to have jitter peaking less than 0.1 dB in the SONET system. To avoid the jitter peaking in the type 2 PLL, the zero of the loop dynamics should be eliminated. To realize a type 2 PLL without having the zero in the open-loop gain, a voltage-controlled delay line (VCDL) is employed in a feedforward path for stability. Figure 11.23(a) shows the block diagram of a CDR circuit based on a delay- and phase-locked loop (D/PLL). As depicted in Fig. 11.23(b), an effective zero $\omega_{z,\text{eff}}$ is formed by the DLL, while the PLL has a type 2 feedback system without the zero in the open-loop gain. As a result, the jitter peaking in the JTRAN is eliminated.

Even though the role of $\omega_{z,\text{eff}}$ in the D/PLL could be understood intuitively, let us check the jitter transfer function of the D/PLL based on a linear model shown in Fig. 11.24(a). The shared control voltage V_c of the VCDL and VCO is expressed as

$$V_c = \frac{K_d}{sC}(\theta_i - K_{vd}V_c - \theta_o) \tag{11.19}$$

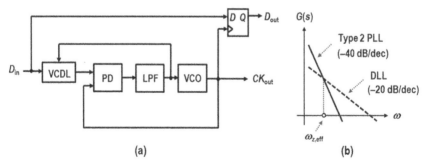

Figure 11.23 D/PLL-based CDR: (a) block diagram and (b) open-loop gain.

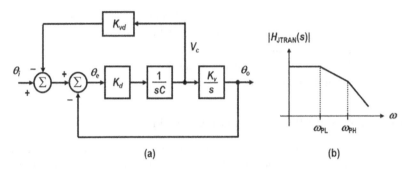

Figure 11.24 (a) Linear model of the D/PLL and (b) jitter transfer function.

where K_{vd} is the gain of the VCDL. Also, we have

$$\theta_o = V_c \frac{K_v}{s} \tag{11.20}$$

From (11.19) and (11.20), the jitter transfer function $H_{JTRAN}(s)$ of the D/PLL is given by

$$H_{JTRAN}(s) = \frac{\theta_o}{\theta_i} = \frac{1}{s^2 \frac{C}{K_d K_v} + s \frac{K_{vd}}{K_v} + 1} \tag{11.21}$$

As seen in (11.21), $H_{JTRAN}(s)$ does not contain any zero, while the denominator has the standard form of a second-order type 2 PLL. Therefore, the D/PLL does not exhibit jitter peaking as long as damping ratio is greater or equal to 0.707. By assuming an overdamped loop, that is large C, a simplified Bode plot shown in Fig. 11.24(b) can be obtained by

$$H_{JTRAN}(s) = \frac{1}{s^2 \frac{C}{K_d K_v} + s \frac{K_{vd}}{K_v} + 1} = \frac{1}{(1 + s/\omega_{PL})(1 + s/\omega_{PH})} \tag{11.22}$$

with approximated ω_{PL} and ω_{PH} values given by

$$\omega_{PL} \approx \frac{K_v}{K_{vd}}, \ \omega_{PH} \approx \frac{K_{vd} K_d}{C} \tag{11.23}$$

Then, the 3-dB bandwidth of $H_{JTRAN}(s)$ is approximated as

$$\omega_{3dB} = \omega_{PL} \approx \frac{K_v}{K_{vd}} \tag{11.24}$$

In the conventional PLL, the JTRAN bandwidth is determined by ω_{PH} from (11.4). As implied in (11.24), the D/PLL has a lower JTRAN bandwidth than the PLL, which could give better jitter filtering for a given CDR bandwidth.

Now let us consider the jitter tolerance performance of the D/PLL. From the linear model shown in Fig. 11.24(a), we obtain the jitter tracking function

$H_{\text{JTRACK}}(s)$ as

$$H_{\text{JTRACK}}(s) = \frac{\theta_e(s)}{\theta_i(s)} = \frac{s^2 \frac{C}{K_v K_d}}{s^2 \frac{C}{K_v K_d} + s\frac{K_{vd}}{K_v} + 1} = \frac{s^2/\omega_{\text{PL}}\omega_{\text{PH}}}{(1 + s/\omega_{\text{PL}})(1 + s/\omega_{\text{PH}})} \quad (11.25)$$

From (11.25), the 3-dB corner frequency ω_{JTRACK} of the high-pass filter $H_{\text{JTRACK}}(s)$ is given by

$$\omega_{\text{JTRACK}} \approx \omega_{\text{PH}} = \frac{K_{vd}K_d}{C} \quad (11.26)$$

Interestingly, the 3-dB corner frequencies of $H_{\text{JTRAN}}(s)$ and $H_{\text{JTRACK}}(s)$ are not the same in the D/PLL. In fact, this could be another feature of the D/PLL architecture. In the conventional PLL, the corner frequencies of both JTRAN and JTOR functions are set by a single frequency ω_{PH}. In the D/PLL, the corner frequency of the JTRAN function is narrower than that of the JTOR function. Therefore, the D/PLL can offer better jitter filtering for a given JTOR requirement or faster jitter tracking for the same JTRAN than the PLL.

Figure 11.25 shows the comparison of the JTRAN and JTOR performance in the type 2 PLL and the type 2 D/PLL. From (11.23), we can increase ω_{PH} by having a small C without affecting ω_{LH}. Therefore, in addition to the peaking-free JTRAN property, the D/PLL offers a degree of freedom in choosing an optimum bandwidth for the JTRAN and the JTOL. In the D/PLL, the main role of a PLL is to recover a clean clock from a noisy input data with good band-pass filtering in the frequency domain. On the other hand, the DLL can offer agile phase tracking in the phase domain. In the D/PLL, the open-loop gains of the PLL and DLL, $G_{\text{PLL}}(s)$ and $G_{\text{DLL}}(s)$, are expressed as

$$G_{\text{PLL}}(s) = \frac{K_d K_v}{s^2 C}, \quad G_{\text{DLL}}(s) = \frac{K_d K_{vd}}{sC} \quad (11.27)$$

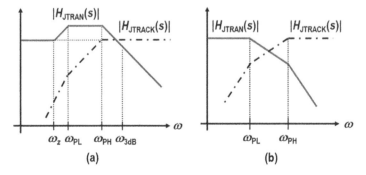

Figure 11.25 JTRAN and JTRACK comparison: (a) type 2 PLL and (b) type 2 D/PLL.

Therefore, the optimum bandwidths of the PLL and the DLL can be independently controlled for the JTRAN and the JTOL. In other words, a narrow bandwidth can be chosen for the PLL by considering the JTRAN and the JGEN, while a wide bandwidth is chosen for the DLL to have good JTOL. Nonetheless, a careful design of loop dynamics is required since the gain variations of a VCO and a VCDL could be substantial over process-voltage-temperature (PVT) variations. Another drawback of this architecture is significant power consumed by the VCDL that should cover a wide range of phase variation. To minimize the ISI problem caused by the VCDL, steep rising and falling edges in the NRZ data should be maintained for the differential delay circuits of the VCDL at the cost of increased power.

11.4.2 Phase- and Delay-Locked Loop (P/DLL)

To relax the stability problem and avoid the dual-loop control of the D/PLL, a cascaded CDR architecture based on a P/DLL could be considered. The operation principle is straightforward as the DFF used for data retiming is replaced with a DLL as shown in Fig. 11.26. Unlike the conventional CDR PLL, the clock recovery and the data retiming are not performed by a single PD like the Hogge PD or the Alexander PD. Compared with the D/PLL-based CDR architecture, the cascaded CDR topology does not have the dual-loop control, making it easier to design the loop dynamics of the PLL and DLL. In the P/DLL, the functions of clock recovery and data retiming are well divided by the PLL and the DLL. The first-stage PLL performs the clock recovery with an optimum bandwidth for JTRAN and JGEN, while the second-stage DLL has relatively a wide bandwidth to have agile phase tracking for JTOL. Since the data retiming is done with a separate loop after the clock recovery, it can also be called a clock-recovery and data-retiming (CRDR) circuit.

The main weakness of this architecture is to require multiple delay cells like the D/PLL-based CDR. For the design of a high-speed VCDL, a phase-interpolated

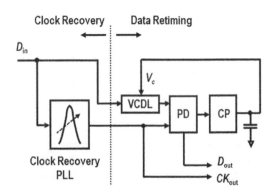

Figure 11.26 P/DLL-based CDR.

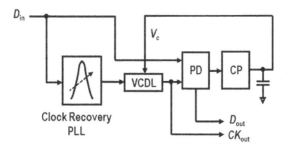

Figure 11.27   P/DLL CDR with a clock delay for JTOL and JGEN only.

delay circuit shown in Fig. 7.31 is mostly chosen as discussed. Since the tuning range of a single delay cell is limited, a large number of delay circuits are needed to achieve a wide phase-tuning range for the VCDL. Otherwise, there is a chance of false-locking even with the careful control of the initial voltage of the VCDL.

If a periodic clock is delayed instead of the NRZ data as shown in Fig. 11.27, the VCDL design could be relaxed since the ISI effect of the periodic clock signal is less than that of the NRZ data. However, if the clock delay is used instead of the data delay, we will lose the JTRAN property of the CDR PLL since the jitter of a retimed data is the same as that of an input data. Therefore, the P/DLL-based CDR with the clock delay should be considered only when an optimum bandwidth is considered for JTOL and JGEN only.

11.4.3   Digital DLL with Phase Rotation

Many electrical serial links for chip-to-chip communications do not require strict JTRAN. In addition, a frequency offset is not substantial since different chips still operate with a stable system clock. Nonetheless, CDR systems need to tolerate a frequency drift to a certain degree since a transceiver and a receiver use different crystal oscillators for the system clock. If a DLL can tolerate a certain amount of frequency variation and does not suffer from the out-of-lock problem, it would be highly attractive for CDR systems. The drawback of the PLL-based CDR system is the jitter generation and accumulation mainly caused by a VCO especially when the CDR circuit is embedded in a large digital system. For instance, phase variation due to a sudden supply jump by digital switching will be propagated to next clock periods in a type 2 PLL, resulting in jitter accumulation as shown in Fig. 5.29. When implementing multiple CDR systems is considered for multiple serial links, the use of multiple VCOs encounters a crosstalk problem among integrated VCOs. Unlike the VCO, the VCDL does not generate an integration factor in the phase domain, so it is more immune to the jitter accumulation and crosstalk problems.

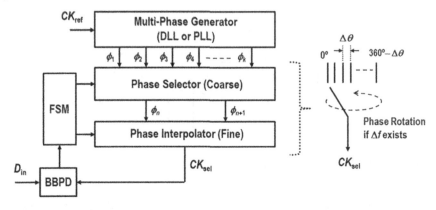

Figure 11.28 Frequency-tracking DLL.

The analog VCDL, however, suffers from a finite phase capture range, thus not being able to track the frequency change.

To make the DLL handle the frequency offset between a transmitter and a receiver, a CDR architecture based on a digital DLL[1] is developed. Figure 11.28 shows the functional block diagram of the DLL-based CDR system that is able to track a frequency change. A DLL or PLL generates multiple reference phases over one UI, and two of them are selected by a finite-state machine (FSM). The selected two phases are further interpolated to generate a phase between them with finer phase resolution $\Delta\theta$. That is, one UI is covered with digitized phases by having the coarse-resolution phase selector and the fine-resolution phase interpolator, and both of them are controlled by the FSM. As depicted in Fig. 11.28, by rotating the discrete phases after every 2π radian, unbounded phases can be realized, which makes it possible to track frequency variation under the assumption that the frequency difference between the data and clock is bounded and not substantial, that is plesiochronous. It is worth noting that the basic operation principle of the frequency tracking based on the multi-phase rotation is analogous to that of the multi-phase fractional-N divider shown in Fig. 9.6. A selected clock CK_{sel} is compared with an input data by a BBPD whose output controls the FSM. In that way, phase tracking is achieved by the digital DLL. The multi-phase generator provides reference phases and does not play a role in phase or frequency tracking. Therefore, the multi-phase generator can be shared by multiple digital DLLs in multi-link applications.

Figure 11.29 shows an example of how the PLL (or DPLL) and the digital DLL are combined to support multi-link interfaces. A multi-phase clock multiplier

---

1 Original paper used a term *semi-digital DLL* in early days. Based on the structure of the *digital PLL* these days, we also use the term *digital DLL* in this book.

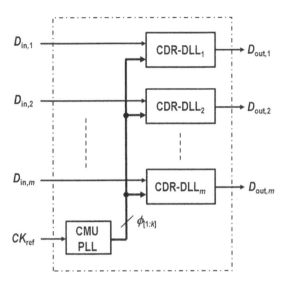

Figure 11.29 DLL-based CDR for multiple serial links.

unit (CMU) generates a high frequency from a reference frequency, which is typically designed based on a PLL with a ring VCO or a DPLL with a ring digitally-controlled oscillator (DCO). The multi-phase clock is distributed to digital DLLs that perform CDR function with frequency tracking capability. Since only a single PLL is used, there is no concern about the crosstalk among VCOs. The PLL-based multi-phase generator performs both frequency multiplication and multi-phase generation with a fixed output frequency, while the digital DLL tracks both phase and frequency variations with a wide bandwidth. Being a first-order type 1 feedback system, the DLL-based CDR circuit does not exhibit jitter peaking in the JTRAN. However, the JTRAN performance of the digital DLL is not as good as that of the PLL since the DLL is a first-order type 1 feedback system. Another drawback of the digital DLL is that it exhibits an algorithmic jitter due to quantized phases. It also causes more phase variation with a large frequency offset since the rotation rate of multiple phases needs to be increased with a large phase step as the frequency offset increases.

11.5 Open-Loop CDR Architectures

Some wireline transceivers require a very fast or immediate phase tracking for CDR systems. For example, if incoming data packet is sporadic, a fast data acquisition is desirable to reduce the length of preamble to maximize the data throughput. Another example is the passive optical network (PON) where a main optical cable is shared by multiple users through a passive optical splitter based on the time-domain multiple access (TDMA) scheme. A burst-mode data transmission is done by each user for a short period, and a slow lock time by a CDR circuit significantly increases a guard-band time, resulting in degraded data throughput. For those applications, a CDR system not based on a PLL but based on an open-loop structure needs to be considered. Two architectures are briefly introduced in the following sections.

11.5.1 Blind Oversampling CDR

If a large number of phases over a bit period is available, data transition could be detected by monitoring the data pattern of 1 and 0. To generate a large number of clock edges for data sampling, a multi-stage ring VCO is required. A PLL with the ring VCO is designed mainly to provide a high-frequency clock and the multi-phase sampling. If the number of phases is large enough, the operation frequency of the ring VCO can also be reduced by a factor of one-half or one-quarter with a half-rate or quad-rate CDR system. Even though the PLL is used for the multi-phase generation, the operation principle of the CDR system is based on an open-loop all-digital method and named as a blind oversampling CDR or an oversampling CDR circuit.

Figure 11.30 shows a functional block diagram. An input data is sampled by data samplers with high oversampling ratio and the set of sampled data are stored in a data register block. With a large number of the sampled data, a bit boundary can be detected by detecting a bit change. Once the bit boundary is determined, bit decision can be made by the data sample that is far from two boundary edges as depicted in Fig. 11.30. There could be ambiguity for the bit decision, but its

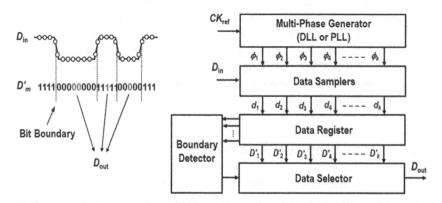

Figure 11.30 Blind oversampling CDR.

impact becomes negligible as the number of sampled data increases. Unlike the digital DLL that uses the phase interpolation and rotation, the phase selection of the blind oversample CDR circuit is not controlled by a feedback loop. Therefore, an immediate data retiming for optimum bit decision can be achieved.

The drawback of the blind oversampling CDR system is the limited data rate as it requires much higher sampling rate than data rate. In addition, it increases hardware complexity significantly for data storage and digital processing, consuming large power and area. Strictly speaking, the blind oversampling CDR system does not perform clock recovery but provides the retimed data only, but the term CDR is commonly used at the system level.

11.5.2 Burst-Mode CDR

Another open-loop CDR system is the burst-mode CDR that employs a gated oscillator whose phase is instantaneously locked to the first edge of an incoming data. Figure 11.31 shows the block diagram of a burst-mode CDR system for instantaneous data retiming. The NRZ data is used as a gating signal to enable oscillation. At the output of the gated oscillator, a continuous clock is recovered and used for the data retiming. The burst-mode CDR system features a simple architecture and lower power, while achieving instant data retiming. Unlike the blind oversampling CDR, the burst-mode CDR generates a recovered clock even though minimal jitter filtering is achievable in this kind of an open-loop method. Since the gated VCO is only activated during the first arrival of a data edge, the burst-mode CDR is immune to a long sequence of consecutive identical digital bits (CIDs). However, the frequency drift of the gated VCO over PVT variations causes a frequency offset between the data and the clock, which still limits the maximum number of the CIDs. When there is a frequency offset Δf between the data and the gated VCO,

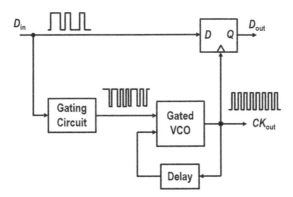

Figure 11.31 Burst-mode CDR.

the maximum number of CIDs denoted by N_{CID} is given by

$$N_{CID} = \frac{\varphi_m R_b}{\Delta f} \qquad (11.28)$$

where φ_m is the phase margin in UI and R_b is the data rate. Therefore, an on-chip tuning scheme must be provided to adaptively tune the gated VCO.

Figure 11.32 shows two burst-mode CDR architectures with an on-chip tuning block for the gated VCO. The first one shown in Fig. 11.32(a) employs a conventional CP-PLL to generate a PVT-insensitive control voltage for the gated VCO. The CP-PLL with a PFD and a frequency divider exactly tunes the frequency of the gated VCO that is the same as the gated VCO in the burst-mode CDR circuit. Since the control voltage V_{ctr} is shared by two gated VCOs, the output frequency of the gated VCO in the CDR circuit should be the same as that in the CP-PLL as long as they are perfectly matched. In practice, it is difficult to achieve perfect matching between two gated VCOs in different loops. For instance, the gated VCO in the burst-mode CDR circuit is more sensitive to supply noise coupling than that in the CP-PLL. It is because the gated VCO in the CP-PLL can suppress low-frequency noise within a loop bandwidth. In addition, a small amount of mismatch in two gated VCOs during gating still generates a substantial frequency offset due to high sensitivity of the ring VCO. To avoid using two gated VCOs, the on-chip tuning of the gated VCO is done by having a frequency-locked loop (FLL) instead of the PLL. As shown in Fig. 11.32(b), an FD generates binary outputs, and an FSM generates a control word to tune the gated VCO. To convert the digital control word to an analog control voltage, a $\Delta\Sigma$ DAC is implemented. The $\Delta\Sigma$ DAC provides fine

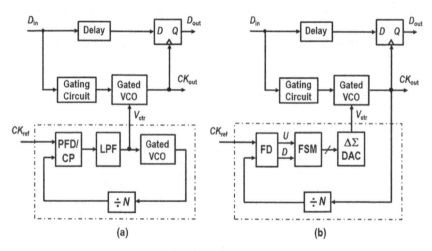

(a)   (b)

Figure 11.32 Burst-mode CDR circuits with improved control of the gated VCO: (a) using a PLL-based voltage biasing and (b) using an FLL-based voltage biasing.

resolution with noise-shaping, thus not affecting the gated VCO. However, data retiming performance is still affected by PVT variations due to high sensitivity of the gated VCO.

References

1 D. H. Wolaver, *Phase-Locked Loop Circuit Design*, Prentice Hall, Englewood Cliffs, NJ, 1991.
2 B. Razavi, *Monolithic Phase-Locked Loops and Clock Recovery Circuits*, IEEE Press, New York, 1996.
3 B. Razavi, *Design of Integrated Circuits for Optical Communications*, Wiley, New York, 2012.
4 B. Razavi, *Design of CMOS Phase-Locked Loops: From Circuit Level to Architecture Level*, Cambridge University Press, United Kingdom, 2020.
5 W. Rhee (ed.), *Phase-Locked Frequency Generation and Clocking: Architectures and Circuits for Modern Wireless and Wireline Systems*, The Institution of Engineering and Technology, United Kingdom, 2020.
6 L. De Vito, "A versatile clock recovery architecture and monolithic implementation," in Razavi, B. (ed.), *Monolithic Phase-Locked Loops and Clock Recovery Circuits*, IEEE Press, New York, 1996.
7 B. Casper and F. O'Mahony, "Clocking analysis, implementation and measurement techniques for high-speed data links—A tutorial," *IEEE Transactions on Circuits and Systems I*, vol. 56, no. 1, pp. 17–39, 2009.
8 P. Hanumolu, "Clock and data recovery architectures and circuits," *IEEE International Solid-State Circuits Conference*, Tutorial, Feb. 2015.
9 A. Amirkhany, "Basics of clock and data recovery circuits," in *IEEE Solid-State Circuits Magazine*, pp. 25–38, Winter Issue, 2020.
10 C. Hogge, "A self correcting clock recovery circuit," *Journal of Lightwave Technology*, vol. 3, no. 6, pp. 1312–1314, 1985.
11 L. DeVito, J. Newton, R. Croughwell *et al.*, "A 52 MHz and 155 MHz clock recovery PLL," in *Proc. IEEE International Solid-State Circuits Conference*, Feb. 1991, pp. 142–143.
12 R.C. Walker, T. Hornak, and C.-S. Yen, "A chipset for gigabit rate data communication," in *Proc. IEEE Bipolar/BiCMOS Circuits and Technology Meeting*, 1989, pp. 288–290
13 R. C. Walker, "Designing bang–bang PLLs for clock and data recovery in serial data transmission systems," in Razavi, B. (Ed.), *Phase-Locking in High-Performance Systems*, IEEE Press, Piscataway, NJ, 2003, pp. 34–45.
14 J. D. H. Alexander, "Clock recovery from random binary signals," *Electronics Letters*, vol. 11, no. 22, pp. 541–542, 1975.

15 J. Lee, K. S. Kundert and B. Razavi, "Analysis and modeling of bang–bang clock and data recovery circuits," *IEEE Journal of Solid-State Circuits*, vol. 39, no. 9, pp. 1571–1580, 2004.

16 N. Da Dalt, "A design-oriented study of the nonlinear dynamics of digital bang-bang PLLs," *IEEE Transactions on Circuits and Systems I*, vol. 52, no. 1, pp. 21–31, 2005.

17 N. Da Dalt, "Linearized analysis of a digital bang-bang PLL and its validity limits applied to jitter transfer and jitter generation," *IEEE Transactions on Circuits and Systems I*, vol. 55, no. 11, pp. 3663–3675, 2008.

18 M. Zanuso, D. Tasca, S. Levantino *et al.*, "Noise analysis and minimization in bang-bang digital PLLs," *IEEE Transactions on Circuits and Systems II*, vol. 56, no. 11, pp. 835–839, 2009.

19 S. Levantino, "Bang-bang digital PLLs," in *Proc. European Solid-State Circuits Conference*, Sept. 2016, pp. 329–334.

20 J. Savoj and B. Razavi, "A 10-Gb/s CMOS clock and data recovery circuit with a half-rate linear phase detector," *IEEE Journal of Solid-State Circuits*, vol. 36, pp. 761–768, 2001.

21 M. Rau, T. Oberst, R. Lares *et al.*, "Clock/data recovery PLL using half-frequency clock," *IEEE Journal of Solid-State Circuits*, pp. 1156–1159, 1997.

22 K. Mueller and M. Muller, "Timing recovery in digital synchronous data receivers," *IEEE Transactions on Communications*, vol. 24, no. 5, pp. 516–531, 1976.

23 F. Spagna, L. Chen, M. Deshpande *et al.*, "A 78 mW 11.8 Gb/s serial link transceiver with adaptive RX equalization and baud-rate CDR in 32 nm CMOS," in *Proc. IEEE International Solid-State Circuits Conference*, Feb. 2010, pp. 366–367.

24 D. Messerschmitt, "Frequency detectors for PLL acquisition in timing and carrier recovery," *IEEE Transactions on Communications*, vol. CMO-27, no. 9, pp. 1288–1295, 1979.

25 A. Pottbacker, U. Langmann and H. Schreiber, "A Si bipolar phase and frequency detector IC for clock extraction up to 8 Gb/s," *IEEE Journal of Solid-State Circuits*, vol. 27, no. 12, pp. 1747–1751, 1992.

26 R. Inti, W. Yin, A. Elshazly *et al.*, "A 0.5-to-2.5Gb/s reference-less half-rate digital CDR with unlimited frequency acquisition range and improved input duty-cycle error tolerance," in *Proc. IEEE International Solid-State Circuits Conference*, Feb. 2011, pp. 438–450.

27 J. Savoj and B. Razavi, "A 10-Gb/s CMOS clock and data recovery circuit with frequency detection," in *Proc. IEEE International Solid-State Circuits Conference*, Feb. 2001, pp. 78-79.

28 S.B. Anand and B. Razavi, "A 2.75 Gb/s CMOS clock recovery circuit with broad capture range," in *Proc. IEEE International Solid-State Circuits Conference*, Feb. 2001, pp. 214–215.

29 T. Lee and J. Bulzacchelli, "A 155-MHz clock recovery delay- and phase-locked loop," *IEEE Journal of Solid-State Circuits*, vol. 27, no. 12, pp. 1736–1746, 1992.

30 D. Dalton, K. Chai, E. Evans *et al.*, "A 12.5-Mb/s to 2.7-Gb/s continuous-rate CDR with automatic frequency acquisition and data-rate readback," *IEEE Journal of Solid-State Circuits*, vol. 40, pp. 2713–2725, 2005.

31 W. Rhee, H. Ainspan, S. Rylov *et al.*, "A 10-Gb/s CMOS clock and data recovery circuits using a secondary delay-locked loop," in *Proc. IEEE Custom Integrated Circuits Conference*, Sept. 2003, pp. 81–84.

32 S. Sidiropoulos and M. Horowitz, "A semidigital dual delay-locked loop," *IEEE Journal of Solid-State Circuits*, vol. 32, pp. 1683–1692, 1997.

33 R. Farjad-Rad, C.-K. K. Yang, M. A. Horowitz *et al.*, "A 0.3-µ CMOS 8-Gb/s 4-PAM serial link transceiver," *IEEE Journal of Solid-State Circuits*, vol. 35, no. 5, pp. 757–764, 2000.

34 J. L. Zerbe, C. W. Werner, V. Stojanovic *et al.*, "Equalization and clock recovery for a 2.5–10-Gb/s 2-PAM/4-PAM backplane transceiver cell," *IEEE Journal of Solid-State Circuits*, vol. 38, no. 12, pp. 2121–2130, 2003.

35 D. Zheng, X. Jin, E. Cheung *et al.*, "A quad 3.125 Gb/s/channel transceiver with analog phase rotators," in *Proc. IEEE International Solid-State Circuits Conference*, Feb. 2002, pp. 70–71.

36 T. Beukema, M. Sorna, K. Selander *et al.*, "A 6.4-Gb/s CMOS SerDes core with feed-forward and decision-feedback equalization," *IEEE Journal of Solid-State Circuits*, vol. 40, no. 12, pp. 2633–2645, 2005.

37 R. Payne, P. Landman, B. Bhakta *et al.*, "A 6.25-Gb/s binary transceiver in 0.13-µ CMOS for serial data transmission across high loss legacy backplane channels," *IEEE Journal of Solid-State Circuits*, vol. 40, no. 12, pp. 2646–2657, 2005.

38 K. Krishna, D. A. Yokoyama-Martin, A. Caffee *et al.*, "A multigigabit backplane transceiver core in 0.13- µ CMOS with a power-efficient equalization architecture," *IEEE Journal of Solid-State Circuits*, vol. 40, no. 12, pp. 2658–2666, 2005.

39 B. Kim, D. N. Helman and P. R. Gray, "A 30-MHz hybrid analog/digital clock recovery circuit in 2- µm CMOS," *IEEE Journal of Solid-State Circuits*, vol. 25, pp. 1385–1394, 1990.

40 C.-K. K. Yang, R. Farjad-Rad and M. A. Horowitz, "A 0.5-µm CMOS 4.0-Gb/s serial link transceiver with data recovery using oversampling," *IEEE Journal of Solid-State Circuits*, vol. 33, pp. 713–722, 1998.

41 J. Kim and D. K. Jeong, "Multi-gigabit-rate clock and data recovery based on blind oversampling," *IEEE Communications Magazine*, pp. 68–74, 2003.

42 M. van Ierssel, A. Sheikholeslami, H. Tamura *et al.*, "A 3.2 Gb/s CDR using semi-blind oversampling to achieve high jitter tolerance," *IEEE Journal of Solid-State Circuits*, vol. 42, pp. 2224–2234, 2007.

43 M. Banu and A. E. Dunlop, "Clock recovery circuits with instantaneous locking," *Electronics Letters*, vol. 28, no. 23, pp. 2127–2130, 1992.

44 M. Banu and A.E. Dunlop, "A 660 Mb/s CMOS Clock Recovery Circuit with Instantaneous Locking for NRZ data and Burst-Mode Transmission," in *Proc. IEEE International Solid-State Circuits Conference*, Feb. 1992, pp. 102–103.

45 Q. Le, S.-G. Lee, Y.-H. Oh *et al.*, "A burst-mode receiver for 1.25-Gb/s ethernet PON with AGC and internally created reset signal," *IEEE Journal of Solid-State Circuits*, vol. 39, pp. 2379–2388, 2004.

46 K. Nishimura, H. Kimura, M. Watanabe *et al.*, "A 1.25-Gb/s CMOS burst-mode optical transceiver for ethernet PON system," *IEEE Journal of Solid-State Circuits*, vol. 40, pp. 1027–1034, 2005.

47 M. Nakamura, Y. Imai, Y. Umeda *et al.*, "1.25-Gb/s burst-mode receiver ICs with quick response for PON systems," *IEEE Journal of Solid-State Circuits*, vol. 40, pp. 2680–2688, 2005.

48 M. Nogawa, K. Nishimura, S. Kimura *et al.*, "A 10 Gb/s burst-mode CDR IC in 0.13 μm CMOS," in *Proc. IEEE International Solid-State Circuits Conference*, Feb. 2005, pp. 228–229.

49 J. Terada, K. Nishimura, S. Kimura *et al.*, "A 10.3 Gb/s burst-mode CDR using a ΔΣ DAC," *IEEE Journal of Solid-State Circuits*, vol. 43, pp. 2921–2928, 2008.

50 T. Gabara, "A 3.25 Gb/s injection locked CMOS clock recovery cell," in *Proc. IEEE Custom Integrated Circuits Conference*, Sept. 1999, pp. 521–524.

51 J. Lee and M. Liu, "A 20-Gb/s burst-mode clock and data recovery circuit using injection-locking technique," *IEEE Journal of Solid-State Circuits*, vol. 43, pp. 619–630, 2008.

Index

Phase-Locked Loops: System Perspectives and Circuit Design Aspects, First Edition.
Woogeun Rhee and Zhiping Yu.
© 2024 The Institute of Electrical and Electronics Engineers, Inc. Published 2024 by John Wiley & Sons, Inc.

Printed and bound by CPI Group (UK) Ltd, Croydon, CR0 4YY

16/04/2025

14658419-0003